2013 IEEE International Conference on Microelectronic Test Structures

(ICMTS 2013)

Osaka, Japan
25 – 28 March 2013

IEEE Catalog Number: CFP13MTS-PRT
ISBN: 978-1-4673-4845-4

Copyright © 2013 by the Institute of Electrical and Electronic Engineers, Inc
All Rights Reserved

Copyright and Reprint Permissions: Abstracting is permitted with credit to the source. Libraries are permitted to photocopy beyond the limit of U.S. copyright law for private use of patrons those articles in this volume that carry a code at the bottom of the first page, provided the per-copy fee indicated in the code is paid through Copyright Clearance Center, 222 Rosewood Drive, Danvers, MA 01923.

For other copying, reprint or republication permission, write to IEEE Copyrights Manager, IEEE Service Center, 445 Hoes Lane, Piscataway, NJ 08854. All rights reserved.

***This publication is a representation of what appears in the IEEE Digital Libraries. Some format issues inherent in the e-media version may also appear in this print version.**

IEEE Catalog Number:	CFP13MTS-PRT
ISBN 13:	978-1-4673-4845-4
ISSN:	1071-9032

Additional Copies of This Publication Are Available From:

Curran Associates, Inc
57 Morehouse Lane
Red Hook, NY 12571 USA
Phone: (845) 758-0400
Fax: (845) 758-2633
E-mail: curran@proceedings.com
Web: www.proceedings.com

2013 IEEE International Conference on Microelectronic Test Structures

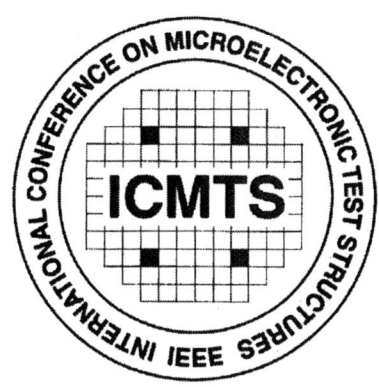

Conference Proceedings

March 25-28, 2013
Nakanoshima Center
Osaka University, Osaka, Japan

2013 IEEE International Conference on Microelectronic Test Structures (ICMTS)

Conference Proceedings

Sponsored by
IEEE Electron Devices Society (EDS)

In cooperation with
The Institute of Electronics, Information and Communication Engineers (IEICE)
The Japan Society of Applied Physics (JSAP)

Welcome Letter

Dear Colleagues,

The 2013 International Conference on Microelectronic Test Structures (ICMTS 2013) will be held at Osaka University Nakanoshima Center, Osaka in Japan. This is the 26th anniversary of the ICMTS and the 8th conference to be held in Japan.

Test structure is key characterization vehicle for process development, device development, circuit design and its production and its importance is increasing. ICMTS has brought engineers and researchers to discuss recent development and future directions of test structure over the past two decades. Topics of ICMTS have been expanded as semiconductor technologies and application advance. They include material and process characterization, replicated feature metrology, manufacturing of integrated circuits, reliability and product failure analysis, characterization of Memory, MEMS, sensors, RF or Power IC, Si or Organic devices, Capacitance, 3D structure with TSV & circuit modeling, parameter extraction, variability, yield enhancement, production process control, etc.

The ICMTS 2013 consists of 40 papers in 10 oral sessions. Details of the sessions are described in this program. I believe all papers turn on your interest and help your work. The sessions will be preceded by a one-day Tutorial Short Course on microelectronic test structure and there will be an equipment exhibition in parallel with the sessions.

Osaka is a historical city in Japan and includes Osaka Castle with beautiful Golden Tea Room and a lot of nice foods. You will be able to enjoy some Osaka foods and many Cherry blossoms on March. Osaka University Nakanoshima Center is very easy access from hotels and you can easily visit most of historical place, shopping and food area. I'm sure that everyone will get benefit by attending the various sessions and also enjoy Osaka. For further information, please visit the ICMTS web site at http://www.if.t.u-tokyo.ac.jp/ICMTS13/.

I look forward seeing you at the ICMTS 2013 in Osaka.

Sincerely,

Tatsuya Ohguro
General Chairman

Table of Contents

SESSION 1: MEMS

Tuesday, March 26, 9:10-10:50
Co-Chairs: Kjell Jeppson, *Chalmers University of Technology, Sweden*
Hiroaki Matsui, *University of Tokyo, Japan*

1.1 An Integrated CMOS-MEMS Probe having Two-Tips per Cantilever for Individual Contact Sensing and Kelvin Measurement with Two Cantilevers
Kota Hosaka, Satoshi Morishita, Isao Mori, Masanori Kubota, and Yoshio Mita
University of Tokyo, Japan — 3

1.2 Characterization and Integration of Parylene as an Insulating Structural Layer for High Aspect Ratio Electroplated Copper Coils
R. Walker, E. Sirotkin, I. Schmueser, J.G. Terry, S. Smith, J.T.M. Stevenson, and A.J. Walton
University of Edinburgh, UK — 7

1.3 Micromechanical Test Structures for the Characterization of Electroplated NiFe Cantilevers and their Viability for use in MEMS Switching Devices
G. Schiavone[1], S. Smith[1], J. Murray[1], J.G. Terry[1], M.P.Y. Desmulliez[2], and A.J. Walton[1]
[1]*University of Edinburgh, UK,* [2]*Heriot-Watt University, UK* — 13

1.4 Investigation of Devices for In-Vivo Energy Harvesting through Blood-Flow-Like Excitation
Rosemary O'Keeffe, Nathan Jackson, Alan Mathewson, Kevin G. McCarthy
University College Cork, Ireland — 19

1.5 A New Measurement Set-up to Investigate the Charge Trapping Phenomena in RF MEMS Packaged Switches
Marco Barbato, Valentina Giliberto, and Gaudenzio Meneghesso
University of Padova, Italy — 25

SESSION 2: TSV and 3D

Tuesday, March 26, 11:20-12:20
Co-Chairs: Yosiho Mita, *University of Tokyo, Japan*
Kevin McCarthy, *University College Cork, Ireland*

2.1 Test Structure and Analysis for Accurate RF-Characterization of Tungsten Through Silicon Via (TSV) Grounding Devices
Volker Blaschke and Hadi Jebory
TowerJazz, USA — 33

2.2 Test Structures for Electrical Evaluation of High Aspect Ratio TSV Arrays Fabricated Using Planarised Sacrificial Photoresist
R. Zhang[1], Y. Li[1], J. Murray[1], A. S. Bunting[1], S. Smith[1], C. C. Dunare[1], J. T. M. Stevenson[1], M.P. Desmulliez[2], and A. J. Walton[1]
[1]*University of Edinburgh, UK,* [2]*Heriot-Watt University, UK* — 37

2.3 A Novel Silicon Interposer for Measuring Devices Requiring Complex Two-sided Contacting
Jaber Derakhshandeh, Negin Golshani, Loek A. Steenweg, Wim van der Vlist, and Lis K. Nanver
Delft University of Technology, The Netherlands — 43

SESSION 3: Capacitance

Tuesday, March 26, 13:50-15:10
Co-Chairs: Larg H. Weiland, *PDF Solutions, USA*
Alain Toffoli, *CEA-LETI, France*

3.1 Characterization of Capacitance Mismatch Using Simple Difference Charge-Based Capacitance Measurement (DCBCM) Test Structure
Ken Sawada[1], Geert Van der Plas[2], Yuichi Miyamori[3], Tetsuya Oishi[4], Cherman Vladimir[2], Abdelkarim Mercha[2], Verkest Diederik[2], and Hiroaki Ammo[4]
[1]Sony Corporation to IMEC, [2]IMEC, Belgium
[3]Sony Semiconductor Corporation, Japan, [4]Sony Corporation, Japan   49

3.2 Comparison of C-V Measurement Methods for RF-MEMS Capacitive Switches
Jiahui Wang, Cora Salm, and Jurriaan Schmitz
University of Twente, The Netherlands   53

3.3 Effective Channel Length Estimation Using Charge-Based Capacitance Measurement
Katsuhiro Tsuji and Kazuo Terada
Hiroshima City University, Japan   59

3.4 A New Ultra-Fast Single Pulse Technique (UFSP) for Channel Effective Mobility Evaluation in MOSFETs
Z. Ji[1], J. Gillbert[2], J. F. Zhang[1], and W. Zhang[1]
[1]Liverpool John Moores University, UK, [2]Keithley Instruments, UK   64

SESSION 4: Noise and RF

Tuesday, March 26, 15:40-17:00
Co-Chairs: Hi-Deok Lee, *Chungnam National University, Korea*
Tatsuya Ohguro, *Toshiba Corporation*

4.1 Optical High Frequency Test structure and Test Bench Definition for on Wafer Silicon Integrated Noise Source Characterization up to 110 GHz based on Germanium-on-Silicon Photodiode
S. Oeuvrard[1,2], J.-F. Lampin[2], G. Ducournau[2], L. Virot[1,3], J.M. Fedeli[3], J.M. Hartmann[3], F. Danneville[2], Y. Morandini[4], and D. Gloria[1]
[1]STMicroelectronics, France, [2]IEMN, France
[3]CEA LETI, France, [4]DOLPHIN INTEGRATION, France   73

4.2 Measurements of SRAM Sensitivity against AC Power Noise with Effects of Device Variation
Takuya Sawada[1], Kumpei Yoshikawa[1], Hidehiro Takata[2], Koji Nii[2], and Makoto Nagata[1,3]
[1]Kobe University, Japan
[2]Renesas Electronics Corporation, Japan
[3]CREST, JST, Japan   77

4.3 On the Length of THRU Standard for TRL De-embedding on Si Substrate above 110 GHz
A. Orii, M. Suizu, S. Amakawa, K. Katayama, K. Takano, M. Motoyoshi, T. Yoshida, and M. Fujishima
Hiroshima University, Japan   81

4.4 Evaluation of 1/f Noise Variability in the Subthreshold Region of MOSFETs
Hans Tuinhout and Adrie Zegers-van Duijnhoven
NXP Semiconductors, The Netherlands   87

SESSION 5: Variability and Yield

Wednesday, March 27, 9:00-10:20
Co-Chairs: Yoichi Tamaki, *CASMAT, Japan*
Hans Tuinhout, *NXP Semiconductors, The Netherlands*

5.1 Newly Developed Test-Element-Group for Detecting Soft Failures of the Low-Resistance-Element using Doubly Nesting Array
Shingo Sato, Hiroki Shinkawata, Atsushi Tsuda, Tomoaki.Yoshizawa, and Takio Ohno
Renesas Electronics Corporation, Japan **95**

5.2 New Methodology for Drain Current Local Variability Characterization using Y Function Method
L. Rahhal[1,2], A. Bajolet[1], C. Diouf[1,2], A. Cros[1], J. Rosa[1], N. Planes[1], and G. Ghibaudo[2]
[1]*STMicroelectronics, France*
[2]*IMEP-LAHC, France* **99**

5.3 A Novel BJT Structure for High-Performance Analog Circuit Applications
Seon-Man Hwang[1], Hyuk-Min Kwon[1], Jae-Hyung Jang[1], Ho-Young Kwak[1], Sung-Kyu Kwon[1], Seung-Yong Sung[1], Jong-Kwan Shin[1], Jae-Nam Yu[1], In-Shik Han[2], Yi-Sun Chung[2], Jung-Hwan Lee[2], Ga-Won Lee[1], and Hi-Deok Lee[1]
[1]*Chungnam National University, Korea*
[2]*MagnaChip Semiconductor, Korea* **104**

5.4 Reconsideration of the Threshold Voltage Variability Estimated with Pair Transistor Cell Array
Kazuo Terada, Naoya Higuchi, and Katsuhiro Tsuji
Hiroshima City University, Japan **108**

SESSION 6: Thermal and Power

Wednesday, March 27, 10:50-12:10
Co-Chairs: Satoshi Habu, *Agilent Technologies Japan, Japan*
Stewart Smith, *University of Edinburgh, UK*

6.1 Comparison of Electrical Techniques for Temperature Evaluation in Power MOS Transistors
A. Ferrara[1], P.G. Steeneken[2], K. Reimann[2], A. Heringa[3], L. Yan[3], B.K. Boksteen[1], M. Swanenberg[2], G.E.J. Koops[3], A.J. Scholten[2], R. Surdeanu[3], J. Schmitz[1], and R.J.E. Hueting[1]
[1]*University of Twente, The Netherlands*
[2]*NXP Semiconductors, The Netherlands,* [3]*NXP Semiconductors, Belgium* **115**

6.2 Measurement and Investigation of Thermal Properties of the On-Chip Metallization for Integrated Power Technologies
Martin Pfost[1], Cristian Boianceanu[2], Dan-Ionuţ Simon[2], and Sebastian Sosin[2]
[1]*Reutlingen University, Germany*
[2]*Infineon Technologies, Romania* **121**

6.3 Investigation on Safe Operating Area and ESD Robustness in a 60-V BCD Process with Different Deep P-Well Test Structures
Chia-Tsen Dai and Ming-Dou Ker
National Chiao-Tung University, Taiwan **127**

6.4 A Test Structure for Analysis of Temperature Distribution in CMOS LSI with Sensing Device Array
T. Matsuda[1], H. Hanai[1], H. Iwata[1], D. Kondo[1], T. Hatakeyama[1], M. Ishizuka[1], and T. Ohzone[2]
[1]*Toyama Prefectural University, Japan*
[2]*Dawn Enterprise, Japan* **131**

SESSION 7: Parameter Extraction

Wednesday, March 27, 13:50-15:10
Co-Chairs: Luca Selmi, *University of Udine, Italy*
Colin McAndrew, *Freescale Semiconductor, USA*

7.1 **Analysis of Narrow Gate to Gate Space Dependence of MOS Gate-Source/Drain Capacitance by Using Contact-less and Drawn-out Source/Drain Test Structure**
Yasuhisa Naruta and Shigetaka Kumashiro
Renesas Electronics Corporation, Japan 137

7.2 **Three- and Four-Point Hamer-type MOSFET Parameter Extraction Methods Revisited**
Kjell O. Jeppson
Chalmers University of Technology, Sweden 141

7.3 **Die-to-Die and Within-Die Variation Extraction for Circuit Simulation with Surface-Potential Compact Model**
Y. Ohnari, A.A. Khan, A. Dutta, M. Miura-Mattausch, and H. J. Mattausch
Hiroshima University, Japan 146

7.4 **BSIM4 Parameter Extraction for Tri-gate Si Nanowire Transistors**
Chika Tanaka, Masumi Saitoh, Kensuke Ota, and Toshinori Numata
Toshiba Corporation, Japan 151

SESSION 8: Emerging Technologies

Wednesday, March 27, 15:40-17:00
Co-Chairs: Anthony J. Walton, *University of Edinburgh, UK*
Bill Verzi, *Agilent Technologies, USA*

8.1 **Benchmarking of a Surface Potential Based Organic Thin-Film Transistor Model against C_{10}-DNTT High Performance Test Devices**
T. K. Maiti, T. Hayashi, H. Mori, M. J. Kang, K. Takimiya, M. Miura-Mattausch, and H. J. Mattausch
Hiroshima University, Japan 157

8.2 **Electrical and Mechanical Characterization of a Large-area Printed Organic Transistor Active Matrix with Floating-gate-based Non-uniformity Compensator**
Tsuyoshi Sekitani, Tomoyuki Yokota, Takeyoshi Tokuhara, Naoya Take, and Takao Someya
University of Tokyo, Japan
ERATO, JST, Japan 162

8.3 **Greek Cross Test Structures for Ink Jet Printed Thin Films**
Elkin Díaz, Eloi Ramon, and Jordi Carrabina
Universitat Autònoma de Barcelona, Spain 167

8.4 **Process Control Monitors for Individual Single-walled Carbon Nanotube Transistor Fabrication Processes**
Kiran Chikkadi, Miroslav Haluska, Christofer Hierold, and Cosmin Roman
ETH Zurich, Switzerland 173

SESSION 9: Memory

Thursday, March 28, 9:00-10:20
Co-Chairs: Kazuo Terada, *Hiroshima City University, Japan*
Alexey Kovalgin, *University of Twente, The Netherlands*

9.1 Automatic Test Methodology to Optimize Operating Conditions and Reliability of Conductive Bridge RAM
A. Toffoli, E. Vianello, G. Molas, L. Perniola, B. De Salvo, and G. Reimbold
CEA- LETI, France .... N/A

9.2 A Proper Approach to Characterize Retention-after-Cycling in 3D-Flash Devices
Fengying Qiao[1], Antonio Arreghini[2], Pieter Blomme[2], Geert Van den bosch[2], Liyang Pan[1], Jun Xu[1], and Jan Van Houdt[2]
[1]*Tsinghua University, China*
[2]*IMEC, Belgium* 187

9.3 A Novel Test Structure to Implement a Programmable Logic Array Using Split-Gate Flash Memory Cells
Henry Om'mani, Mandana Tadayoni, Nitya Thota, Ian Yue, and Nhan Do
Silicon Storage Technology, USA 192

9.4 On-wafer Integrated System for Fast Characterization and Parametric Test of New-Generation Non Volatile Memories
Erika Covi[1], Alessandro Cabrini[1], Loris Vendrame[2], Luca Bortesi[2], Roberto Gastaldi[2], and Guido Torelli[1]
[1]*University of Pavia, Italy*
[2]*Micron Semiconductor Italia, Italy* 195

SESSION 10: Arrays and Ring Oscillators

Thursday, March 28, 10:40-12:00
Co-Chairs: Tsuyoshi Sekitani, *University of Tokyo, Japan*
Christopher Hess, *PDF Solutions, USA*

10.1 Tr Variance Evaluation induced by Probing Pressure and its Stress Extraction Methodology in 28nm High-K and Metal Gate process
T. Okagaki, T. Hasegawa, H. Takashino, M. Fujii, A. Tsuda, K. Shibutani, Y. Deguchi, M. Yokota, and K. Onozawa
Renesas Electronics Corporation, Japan 203

10.2 Efficient Technique for Si Validation of Level Shifters
Puneet Sharma[1], Brad Smith[2], Donald Hall[2], Mike Nelson[2], and Umesh Lohani[1]
[1]*Freescale Semiconductor, India,* [2]*Freescale Semiconductor, USA* 207

10.3 Mosaic SRAM Cell TEGs with Intentionally-added Device Variability for Confirming the Ratio-less SRAM Operation
Hitoshi Okamura, Takahiko Saito, Hiroaki Goto, Masahiro Yamamoto, and Kazuyuki Nakamura
Kyushu Institute of Technology, Japan 212

10.4 Characterization and Simulation of NMOS Pass Transistor Reliability for FPGA Routing Circuits
Christopher S. Chen and Jeffrey T. Watt
Altera Corporation, USA 216

Conference Officials

ICMTS 2013 Conference Committee

General Chairman:

OHGURO, Tatsuya
Semiconductor & Storage Products Company
Toshiba Corporation
8, Shinsugita-Cho, Isogo-ku,
Yokohama, 235-8522, JAPAN
Phone: +81-45-776-5697
FAX : +81-45-776-4104
e-mail: tatsuya.ooguro@toshiba.co.jp

Technical Chairman:

TAKEUCHI, Kiyoshi
LSI Research Laboratory
Renesas Electronics Corporation
1120 Shimokuzawa, Chuou-ku,
Sagamihara, 252-5298, JAPAN
Phone : +81-42-771-0690
FAX: +81-42-771-0692
e-mail : kiyoshi.takeuchi.zn@renesas.com

Tutorial Chairman:

MITA, Yoshio
Dept. Electrical Engineering
University of Tokyo
2-11-16, Yayoi, Bunkyo-ku,
Tokyo, 113-8656, JAPAN
Phone: +81-3-5841-6730
FAX: +81-3-5841-6730
e-mail: mita@ee.t.u-tokyo.ac.jp

Local Arrangements:

SEKITANI, Tsuyoshi
Dept. Electrical & Electronic Engineering
University of Tokyo
7-3-1 Hongo, Bunkyo-ku,
Tokyo, 113-8656, JAPAN
Phone: +81-3-5841-0413
FAX: +81-3-5841-6709
e-mail: sekitani@ee.t.u-tokyo.ac.jp

Equipment Exhibition:

TANIGUCHI, Jun
Agilent Technologies International Japan
9-1, Takakura-cho, Hachioji,
Tokyo, 192-0033, JAPAN
Phone: +81-42-660-3215
FAX: +81-42-660-8430
e-mail: jun_taniguchi@agilent.com

Asian Representative:

ASADA, Kunihiro
VLSI Design and Education Center
University of Tokyo
2-11-16, Yayoi, Bunkyo-ku,
Tokyo, 113-8656 JAPAN
Phone: +81-3-5841-6671
FAX: +81-3-5841-8911
e-mail: asada@silicon.u-tokyo.ac.jp

European Representative:

WALTON, Anthony J.
Scottish Microelectronic Centre
School of Engineering
Kings Bldg., University of Edinburgh
Edinburgh, EH9 3JF, UK
Phone: +44-131-650-5620
FAX: +44-131-650-6554
e-mail: Anthony.Walton@ee.ed.ac.uk

USA Representative:

LINHOLM, Loren W.
e-mail: linhlw@comcast.net

Steering Committee

Kunihiro Asada	*University of Tokyo, Japan*
Loren Linholm	*USA*
Luca Selmi	*University of Udine, Italy*
Yoichi Tamaki	*Japan*
Bill Verzi	*Agilent Technologies, USA*
Anthony J. Walton	*University of Edinburgh, UK*

Technical Committee

Richard Allen	*NIST, USA*	Yoshio Mita	*University of Tokyo, Japan*
Cros Antoine	*STMicroelectronics, France*	Jerome Mitrad	*IMEC, Belgium*
Satoshi Habu	*Agilent Technologies, Japan*	Tatsuya Ohguro	*Toshiba Corporation, Japan*
Christopher Hess	*PDF Solutions, USA*	Luigi Pantisano	*IMEC, Belgium*
Kjell Jeppson	*Chalmers University of Technology, Sweden*	Mark Poulter	*Texas Instruments, USA*
		Dieter Schroder	*Arizona State University, USA*
Won-Young Jung	*Dongbu Hitek, Korea*	Tsuyoshi Sekitani	*University of Tokyo, Japan*
Chang-Yong Kang	*SEMATECH, USA*	Luca Selmi	*University of Udine, Italy*
Mark Ketchen	*IBM, USA*	Brad Smith	*Freescale Semiconductor, USA*
Johan Klootwijk	*Phillips Research Europe, The Netherlands*		
		Stewart Smith	*University of Edinburgh, UK*
Alexey Kovalgin	*University of Twente, The Netherlands*	Lee Stauffer	*Keithley Instruments, USA*
		Kiyoshi Takeuchi	*Renesas Electronics Corporation, Japan*
Choongho Lee	*Samsung Electronics, Korea*		
Hi-Deok Lee	*Chungnam National University, Korea*	Yoichi Tamaki	*Japan*
		Kazuo Terada	*Hiroshima City University, Japan*
Loren Linholm	*USA*		
Emilio Lora-Tamayo	*Autonomous University of Barcelona, Spain*	Alain Toffoli	*CEA-LETI, France*
		Bing-Yue Tsui	*National Chiao Tung University, Taiwan*
Alan Mathewson	*University College Cork, Ireland*	Hans P. Tuinhout	*NXP Semiconductors, The Netherlands*
Hiroaki Matsui	*University of Tokyo, Japan*		
Colin McAndrew	*Freescale Semiconductor, USA*	Bill Verzi	*Agilent Technologies, USA*
		Anthony J. Walton	*University of Edinburgh, UK*
Kevin McCarthy	*University College Cork, Ireland*	Larg Weiland	*PDF Solutions, USA*
		Greg Yeric	*ARM, USA*

SESSION 1: MEMS

978-1-4673-4845-4/13 $31.00 © 2013 IEEE

An Integrated CMOS-MEMS Probe having Two-Tips per Cantilever for Individual Contact Sensing and Kelvin Measurement with Two Cantilevers

Kota Hosaka, Satoshi Morishita, Isao Mori, Masanori Kubota, and Yoshio Mita

School of Engineering,
The University of Tokyo
Tokyo, JAPAN
mems@if.t.u-tokyo.ac.jp

Abstract—the MEMS-made probe cards can drastically improve semiconductor wafer test quality as compared to traditional tungsten probe. To further take advantage of MEMS technology, the authors propose a CMOS-MEMS integrated probe card, to solve the tradeoff problem of measurement precision and excess pad damage by skating, by 4-terminal (Kelvin) measurement with two-tracks-per-cantilever needle. Putting two tips on each cantilever enables us to detect electrical contact and to decrease skating. And by this structure, electrical properties of a device under test are measured precisely with 4-terminal measurement which can eliminate track resistance and contact resistance. We measured the resistance of a gold thin film. With 2-terminal method, the resistance was measured to be about 74 ohms. However with Kelvin measurement, the resistance was 0.012–0.022 ohms. This result shows the successful implementation of 4-terminal measurement probe with MEMS technology.

Keywords—*probe card; MEMS probe; cantilever; Kelvin measurement; wafer test*

I. INTRODUCTION

In the semiconductor industry, not only in Large Scale Integration (LSI) but also in discrete components (such as LED chips) fabrication, each device manufactured on a wafer is tested before shipping. Probe cards are used for acceptance tests: the electrical characteristics of a device under test (DUT) are measured by contacting the probe needles to the electrical pads on the device. The probe cards were traditionally made with tungsten probe needles (figure 1 left). According to the miniaturization of semiconductors in recent years, probe pad size and pitch are under continuous shrinking. In addition, more and more probe needles should be integrated on a probe card to simultaneously measure several chips on the wafer for high throughput [1,2]. To satisfy these demands, recently, MEMS probe cards are replacing traditional ones and are being optimized towards low contact force and low contact resistance [3–6]. The nature of MEMS batch process helps obtaining many needles on one probe [7].

II. THE PROBLEM OF TRADITIONAL TUNGSTEN AND MEMS PROBE CARDS

The fundamental problem of the conventional probe cards is that it is unclear whether the probe needle touching the device is in a good electrical contact or not [2,8,9]. This

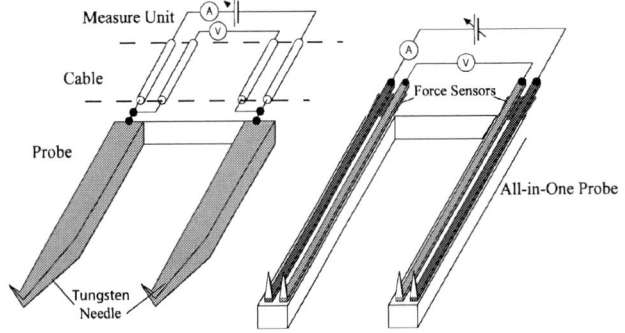

Figure 1. Conventional Tungsten Probe (left) and Proposed CMOS-MEMS Probe (right). Four-terminal measurement is possible only with two tips needles. In addition to intrinsic Kelvin Measurement by two needles, higher functionality such as Force Sensing and Contact Detection will become possible due to integrated circuit.

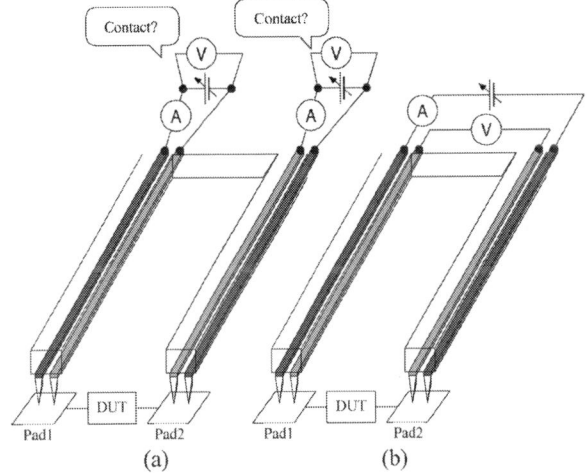

Figure 2. Conctact detection and Kelvin measuremen system
(a) Contact detection between two tips on a cantlever
(b) Kelvin measurement with two cantilevers.

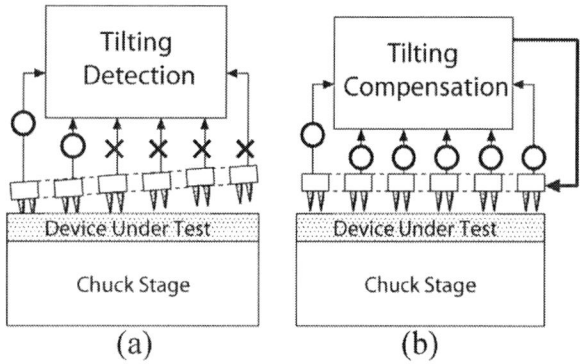

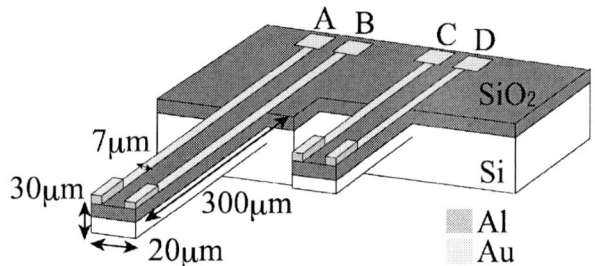

Figure 4. Schematic of the demonstration device

Figure 3. Tilting compensation system
(a)reads signals of contact sensing
(b)Tilting compensation with contact signal.

problem is much more remarkable on a warped substrate with high internal stress (such as LED wafer) because the contact force differs drastically according to pad locations. Therefore higher force is applied to the probe to cause so-called "skating" [2,8], in order to scratch the pad for a good contact. As a result, there are several inconveniences such as pad breakage and excess of total force per probe card due to pin count increase. The conventional MEMS probes cannot solve the problem either, because they are miniaturized tungsten probe in principle.

III. PROPOSED CMOS-MEMS PROBE CARD WITH TWO NEEDLES PER CANTILEVER FOR KELVIN MEASUREMENT

To solve the problems of electrical contact, the authors suggest CMOS-MEMS integrated probe card as shown in figure 1 (right). The probe needle is in a form of a silicon cantilever and accommodates two electrical tracks, made with CMOS-wiring metal layer. At the end of the cantilever, there are two tracks and two tips on a cantilever. At the clamping of the cantilever, a piezo-resistive force sensing circuit may be integrated. Wiring and detection circuits are fabricated by CMOS technology and probe needles and cantilevers are fabricated by MEMS post-processing. Putting two tips on each cantilever has two critical advantages:

• Four-terminal (a.k.a. Kelvin) measurement is possible.

• Probe contact is electrically detectable.

A deployment scenario is as follows (figure 2)

• Using two tips on a cantilever touching the same pad, two-terminal measurement is performed to electrically detect the contact.

• Once the contact is sensed for all pads, the circuit is then switched to Kelvin measurement mode.

• On the same time, by combining stress information from the integrated contact force sensor, tilting between DUT and probes is identified and compensated (figure 3).

Due to CMOS-MEMS integrated transistors, contact sensing mode to resistance measurement mode switches as well as contact sensing circuits can easily be integrated on same chip.

IV. CONTACT-SENSE-THEN-KELVIN-MEASUREMENT SCHEME DEMONSTRATION

In order to confirm the effect of electrical contact sensing and Kelvin measurement, the authors manufactured a simple demonstration device of two tips per cantilever. Cantilevers having gold wires on dielectric layer (SiO$_2$) were fabricated. Aluminum tips were embedded at the extremity of the gold wire in order to touch down the pad easily. A schematic of the device is shown in figure 4. The touch force to overdrive distance is defined by the stiffness k of the cantilever, which is defined by the following equation;

$$k = Ebh^3 / 4l^3.$$

E, b, h and l are Young's module, width, thick and length. Thus, this demonstration probe's k is 925 N/m. The k characteristic can be optimized by changing by h (thickness) or l (length), in keeping b (width) according to the pad size. The value can be simply reduced down to around 1N/m by making the thickness 10 times less important for example.

A. Process flow (figure 5)

The demonstration device was made on a standard 4-inch bulk silicon wafer with 420 nm-thick thermal SiO$_2$ layer (1080 °C, 0.5 L/min and 50 min). A 500 nm-thick Al layer was deposited, then positive E-Beam resist OEBR-CAP112PM (TOKYO OHKA KOGYO Co., Ltd) was spin-coated and exposed with a high-throughput electron beam lithography system (F5112+VD01 from Advantest Corporation) (dose 6 μC/cm^2). The resist was developed in 2.38% Tetramethylammonium hydroxide. Then the Aluminum was wet etched (phosphoric acid, nitric acid and acetic acid) and the resist was stripped with a standard resist stripper.

A 5 nm-thick Cr adhesion layer and a 150 nm-thick Au layer were deposited by EB evaporation. A resist layer formed and patterned by the same EB lithography. The gold layer was etched by Argon milling (400 W, 3 min) with a high-density plasma etching system (CE-300I from ULVAC Corporation).

978-1-4673-4845-4/13 $31.00 © 2013 IEEE

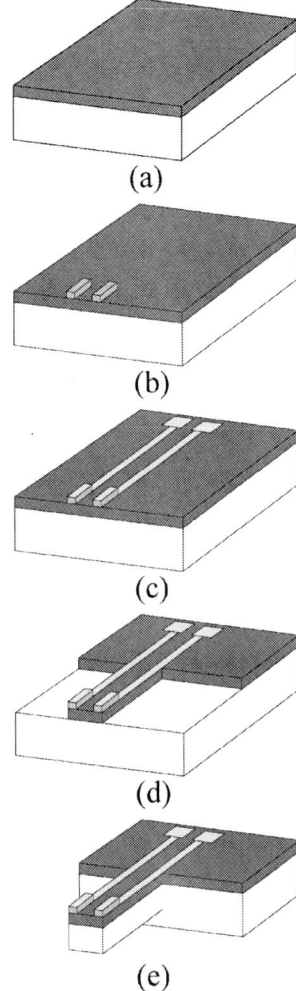

(a)

(b)

(c)

(d)

(e)

Figure 5. Process flow: (a) thermal oxidation of a silicon wafer, (b) Al tips patterning, (c) patterning of Au tracks, (d) cantilever patterning on the SiO2 layer, (e) cantilever etching and release

Cantilever pattern mask was formed with the same method described above. The oxide was etched by CHF_3 (150 W, 20 sccm and 5 min) with CE-300I machine. Then the cantilever was fabricated by etching the Si with a deep reactive ion etching system (MS-100, Alcatel): a 30 μm-depth vertical etching using Bosch process followed by isotropic etching was applied to form and release the cantilevers. Finally CF passivation layer and the resist were removed by O_2 plasma ashing.

B. Measurement

While measuring the resistance between terminal A and B (adjacent tracks on the same cantilever, cf. figure 4), the probe was brought to contact with the 1 μm-thick gold-coated glass plate. The resistance got much lower when the two tips got in contact with the plate. Next, the resistance between another pair of terminal (such as C and D) was measured with the 2-terminal mode. The DUT was tilted until the tips got in contact

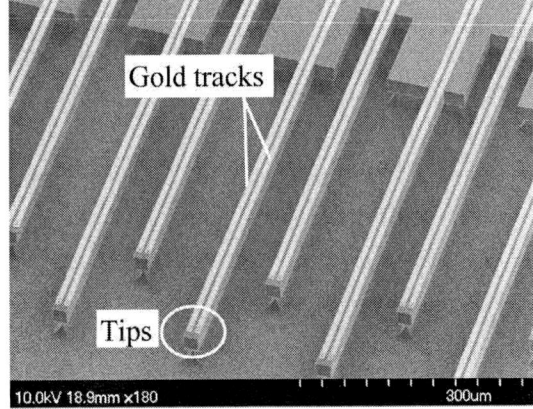

Figure 6. SEM photography of the probes

with the plate. Once the four tips were in contact, the resistance between terminals belonging to different cantilevers (for instance the resistance between B and C) was measured with the 4-terminal mode.

C. Measurement results and discussion

The probe device was fixed on tilting stage. The open resistance between adjacent terminal A and B was measured for reference and calibration (figure 7 (a)). The value was a few Giga-ohms. Then the probe was slowly moved toward the gold- coated glass, and the A–B resistance dropped to 74 ohms, indicating that the two tips were in contact with the pad. Then, the measurement target was switched to the C–D pair of terminals on the other cantilever. This resulted in a value of 74 ohms, indicating that the two tips on this cantilever where also in contact with the gold layer (figure 7 (b)).

The measurement mode was shifted from contact detection to two-terminal mode (figure 7 (c)). The resistance between B and C measured in this mode was about 74 ohms (figure 8). The gold-coated glass sample was then pulled up in order to perform calibration for Kelvin measurement. Then the probe was brought in contact again with the sample (confirmed by contact detection) and Kelvin measurement was performed. The values were in a range of 0.012–0.022 ohms (figure 8).

The calculated resistance per gold track was 28.3 ohms. Then, the wiring resistance for 2-terminal measurement was at least 56.6 ohms, so that 2-terminal measurement showed not useable for low-resistance DUT. Resistance was measured several times with both modes, but the result of 2-terminal mode was always unstable while Kelvin measurement showed higher stability (figure 9). The cause is suspected to be the contact condition and the contact resistance change.

This result shows that the proposed method enables precise measurement of resistance by eliminating the contribution from high wiring resistance and unsuitable contact resistance.

V. CONCLUSIONS

A method was proposed for contact sensing and precise measurement of resistance (current–voltage characteristics). The contact sensing solves the problem of skating found with

978-1-4673-4845-4/13 $31.00 © 2013 IEEE

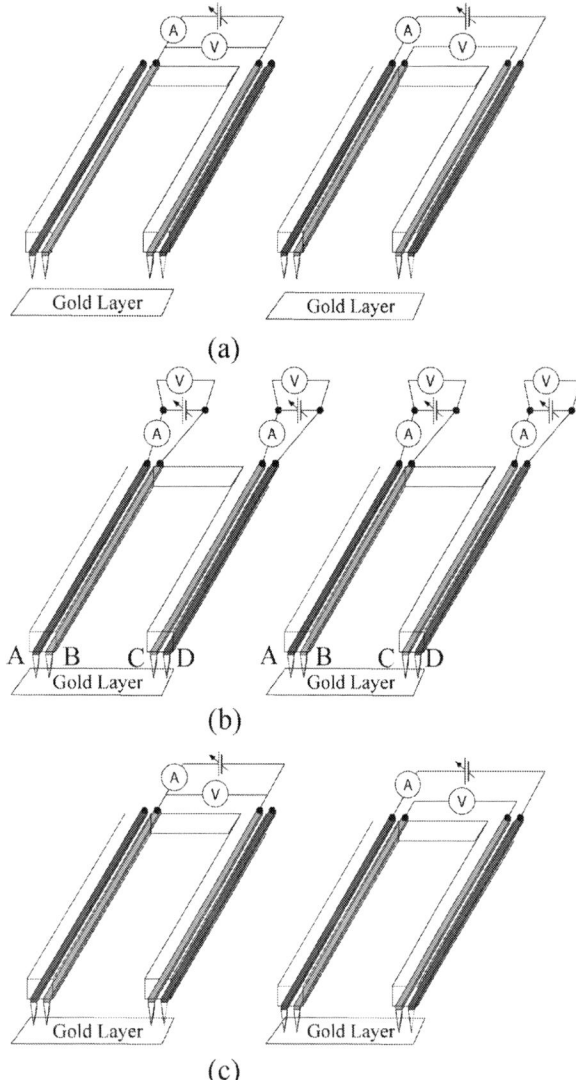

(a)

(b)

(c)

Figure 7. Process of the demonstration measurement
(a) system calibration on open state
(b) contact sensing for each cantilever
(c) measurement of resistance between two cantilevers

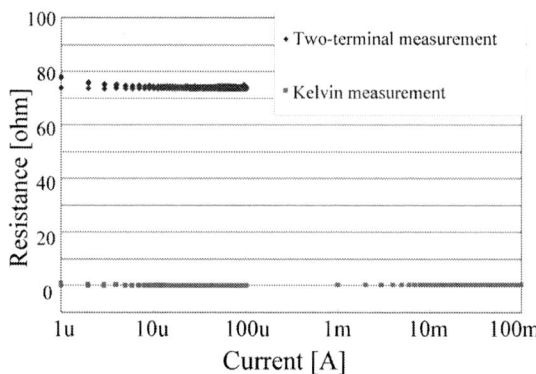

Figure 8. Result of resistance-current characteristics with two-terminal measurement or Kelvin measurement

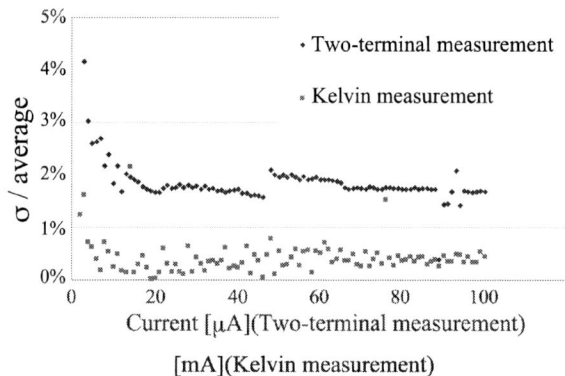

Figure 9. Two-terminal measurement showed larger instability of resistance.

traditional probe card and the 4-terminal method enables to get rid of the contact and wiring resistance in resistance measurement. This method makes the wafer test more precise.

REFERENCE

[1] John Heck, Denald Adams, Nickolai Belov, Tsung-Kuan A. Chou, Byong Kin, Kevin Kornelsen, Qing Ma, Valluri Rao, Simone Severi, Dean Spicer, Ghassan Tchelepi and Ann Witvrouw, "Ultra-high density MEMS probe memory device", Microelectronic Engineering 87, pp.1198-1203, 2010

[2] Chau-Shing Wang, Wen-Ren Yang, Cheng-Ten Chung and Wen-Liang Chang, "Application of Image Processing to Wafer Probe Mark Area Calculation", IEEE Conference on Industrial Electronics and Applications, 2010

[3] Bong-Hwan Kim and Jong-Bok Kim, "Design and fabrication of a high manufacturable MEMS probe card for high speed testing", Journal of Micromechanics and Microengineering, 2008

[4] T.Lai and C.Tsou, "A CLAW TYPE OF MEMS PROBE CARD FOR THE ELECTRICAL TESTING OF MICRO-SOLDER BALL", MEMS 2012, pp.345-348

[5] K.Kataoka, S.Kawamura, T.Itoh and T.Suga, "LOW CONTACT-FORCE AND COMPLIANT MEMS PROBE CARD UTILIZING FRITTING CONTACT",MEMS 2002, pp.364-367

[6] K.Soejima, M.Kimura, Y.Shimada and S.Aoyama, "New Probe Microstructure for Full-Wafer, Contact-Probe Cards",Electronic Components and Technology Conference, pp.1175-1180,1999

[7] Tsung-Hung Lin, Hasharng Yang, Ching-Kong Cho and Mau-Shiun Yeh, "New fabrication method for micro-pyramidal vertical probe array for probe card.", Microsystem Technologies, v 16, n7, p 125-1220, 2010

[8] S.Maekawa, M.Takemoto, Y.Kashiba Y.Deguchi, K.Miki and T.Nagata,"Highly Reliable Probe Card for Wafer Testing", Electronic Components and Technology Conference, 2000

[9] Bong-Hwan Kim and Jong-Bok Kim, "Design and fabrication of a high manufacturable MEMS probe card for high speedtesting", Journal of Micromechanics and Microengineering, 2008

978-1-4673-4845-4/13 $31.00 © 2013 IEEE

Characterisation and Integration of Parylene as an Insulating Structural Layer for High Aspect Ratio Electroplated Copper Coils

R. Walker, E. Sirotkin, I. Schmueser, J.G. Terry, S. Smith, J.T.M. Stevenson and A.J. Walton

Institute of Integrated Micro and Nano Systems, School of Engineering,
Scottish Microelectronics Centre, The University of Edinburgh, EH9 3JF, UK,   Email: Ross.Walker@ed.ac.uk

Abstract – **This paper reports the development of processing methods and test structures for the characterisation and evaluation of Parylene-C as an insulating structural layer material for integration with planar micro-inductors. The process involves the filling of high aspect ratio gaps between copper structures with Parylene and subsequent chemical mechanical planarisation. A test chip has been designed to characterise this process and the results presented. Subsequently complete micro-inductors, with magnetic cores, have been fabricated to demonstrate the capability of the process.**

I. Introduction

High aspect ratio metal structures are important for the development of high quality integrated passive devices, such as micro-inductors. Electroplating processes are used in combination with thick photoresist to create copper coils. SU-8 is commonly used for both the electroplating mould and the inter-coil insulation [1,2,3]. However there are intrinsic problems with SU-8 due to adhesion and mechanical stress [4]. While these can often be mitigated, there is clearly an interest in identifying and characterising new processes and materials, both for filling of high aspect ratio gaps between metal coils and features, as well as minimising any mechanical stress resulting from the required thick layers. This is particularly the case when integrating thick copper coils with thick NiFe films.

One possible candidate to resolve these issues is the polymer, Parylene. This material is widely used as a protective coating on printed circuit boards (PCB) [5] as it can be conformally deposited from a vapour phase at room temperature. The properties of Parylene as an interlayer dielectric, along with its barrier properties, and inherent low stress deposition [6], make it an ideal candidate to replace thick SU-8 films as the insulator between the electroplated copper coils in micro-inductors. An example of this is the fabrication of Parylene-coated 2-D coils that were then folded on top of each other to produce a 14.6-mm-diameter 96-turn three-layer copper 3-D coil [7].

This paper reports the development of test structures and processes to characterise the suitability of Parylene for integration into a planar inductor fabrication process that is compatible with post processing on a CMOS or bipolar integrated circuit.

Test structures consisting of trenches, tracks and planar coils have been used in this work to assess the gap-filling and insulation properties of vapour deposited Parylene and adhesion tests performed to augment previously reported work [8]. Micro-mechanical test structures have also been fabricated to evaluate the type and level of stress in Parylene [9]. Finally, a complete micro-inductor, incorporating NiFe layers and Parylene, is fabricated and presented here.

II. Characterisation of Parylene

One method of making copper coils with SU-8 is to define the mould using photolithography and then fill the trenches in the resulting mould with electroplated copper. The attraction with this approach is that the SU-8 becomes part of the structure and provides the inter-coil insulator. However, SU-8 suffers from adhesion and stress problems, which make integrating it into a process a significant challenge.

In the architectures proposed in this paper, where Parylene is used as an inter-coil dielectric, photoresist that can be removed after electroplating must be used as a mould for the electroplating of the thick copper tracks. Once the photoresist is removed, the Parylene will be conformally deposited to insulate adjacent coils. Fortunately, because the deposition process is conformal, the thickness of Parylene that is deposited is only required to be half the separation of the coils. This reduces the amount of polymer that needs to be removed in the chemical mechanical planarisation (CMP) step that follows the Parylene deposition. Once the surface is planarised to the top of the copper coils, another layer of Parylene is deposited before the final step of adding a layer of magnetic material, such as NiFe alloy, to act as a core.

A. Adhesion Testing

Standard "Scotch-tape" tests and optical microscopy inspections have been carried out on various patterned and unpatterned layers, covered with Parylene films, in order to assess its adhesion to materials commonly used in MEMS processing.

978-1-4673-4845-4/13 $31.00 © 2013 IEEE

Table 1 presents these results, which indicate that, with the appropriate surface treatment, good adhesion can be achieved between all layers tested

TABLE 1. SCOTCH TAPE TESTS, TESTING THE ADHESION OF PARYLENE TO VARIOUS MATERIALS. A TICK REPORTS A PASS.

Material	Patterned	Unpatterned	Successful Completion of Scotch Tape Test
Silicon		✓	✓
Silicon Dioxide (Thermal Oxide)		✓	✓
Silicon Nitride (PECVD)	✓	✓	✓
Parylene-C	✓	✓	✓
Copper (Sputtered)	✓	✓	✓
Copper (Electroplated)	✓	✓	✓
Titanium (Sputtered)	✓	✓	✓
NiFe (Electroplated)	✓	✓	✓
SU-8 3005	✓	✓	✓

B. Design of test structures for Parylene fill efficiency and Chemical Mechanical Planarisation

A test chip design, consisting of arrays of stripes with a range of widths and separations, has been produced to evaluate the fill performance of Parylene, and is presented in Figure 1. The resulting mask was used to produce high aspect ratio trenches in a thick negative photoresist (NR2-20000P) to provide a mould for subsequent copper electroplating. After plating, this photoresist was then removed to leave a pattern of thick copper tracks with trenches of various widths, and these have been used to evaluate: (1) filling efficiency of Parylene and (2) dishing between copper tracks resulting from CMP processing.

The copper tracks forming the trenches have widths varying between 10µm and 50µm, while the width of the trenches vary from 2µm to 50µm. A SEM image of a cross section through copper tracks with a 3:1 aspect ratio trench is presented in figure 2. It is clear that the gap between tracks has been conformally filled with Parylene.

The CMP of Parylene is a process which is not detailed in literature and hence the choice of slurry was not obvious. Klebosol 30HB50 with custom additives was selected as it had previously been successfully used for CMP of SU-8. Parameters used for the successful CMP of Parylene were as follows: back pressure – 0.4 Bar; down force – 0.17 Bar; pressure on wafer – 0.25 Bar; platen speed – 30 RPM; wafer speed – 35 RPM. This resulted in an average removal rate of 833 ± 33 nm/min of Parylene on top of the copper stripe structures. The average surface roughness (R_a) of Parylene after CMP measured using atomic force microscopy (AFM) was 7.9 nm.

An evaluation of Parylene dishing between 50 µm tracks, with varying separation, was carried out. The dishing profile was measured using a Dektak surface profiling tool and confirmed using AFM. A maximum dishing profile of 619 ± 70nm was measured.

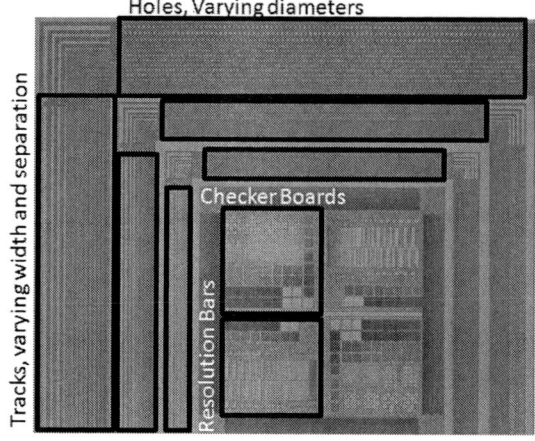

Figure 1; Layout of the test chip (10x10mm) with arrays of stripes with various widths and separation between stripes.

Figure 2; SEM cross section of copper tracks coated with Parylene.

D. Parylene Absorption

During fabrication, the Parylene layers are exposed to a number of solvent and acids and it is important to know how the material will react to these different solutions.

A silicon wafer was cut into squares with 25 mm side length. These were coated with approximately 1µm of Parylene on both sides to prevent liquid access to the silicon-Parylene interface. After soaking in the liquid being tested, the thickness of the Parylene layer was monitored over time using a Nanospec reflectometer. Table 2 lists liquids used and the soaking and measurement times for sample M0-M13. The results for Parylene thickness changes over time are presented in figure 3 and a summary of the changes in thickness presented in table 3.

TABLE 2. CHARACTERISATION CONDITIONS FOR EACH PARYLENE SAMPLE (NOMINAL THICKNESS 1µm)

Measurment	1st Soaking Liquid	2nd Soaking Liquid	1st Soaking Time	2nd Soaking Time	Measurment time
M01	Acetone	-	30 min	-	45 min
M02	Acetone	-	60 min	-	45 min
M03	IPA	-	15 min	-	36 min
M04	Methanol	-	15 min	-	10 min
M05	Ethanol	-	15 min	-	10 min
M07	PGMEA	-	15 min	-	10 min
M06	DI Water	-	15 min	-	10 min
M08	KOH	-	15 min	-	10 min
M09	H_2SO_4	-	15 min	-	5 min
M10	Acetone	IPA	15 min	5 min	45 min
M11	Acetone	Methanol	15 min	5 min	45 min
M12	Acetone	Ethanol	15 min	5 min	45 min
M13	Acetone	DI Water	15 min	5 min	10 min

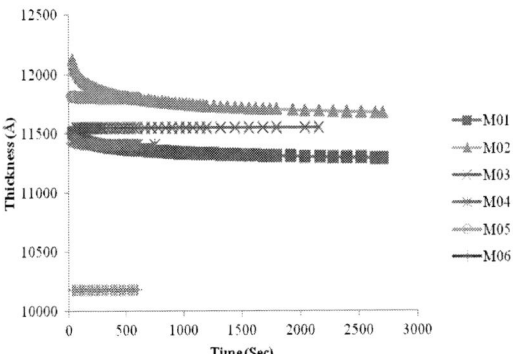

(a)

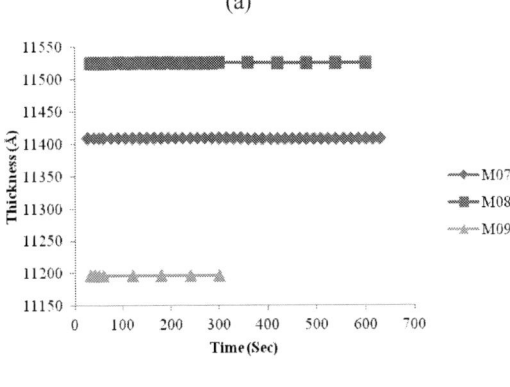

(b)

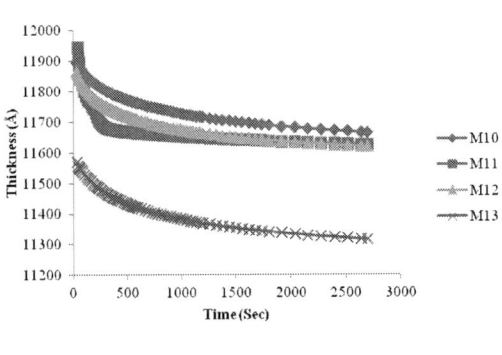

(c)

Figure 3. Thickness change in parylene over time, as a result of soaking in different liquids (see Table 2). (a) solvents; (b) DI water, acids and bases; (c) multiple soakings.

TABLE 3. SUMMARY OF THICKNESS MEASURMENTS FOR THE CONDITIONS PRESENTED IN TABLE 2. MEASUREMENTS M03, M04, M07, M08 AND M09 HAVE NOT BEEN PRESENTED DUE TO NEGLIGIBLE EFFECT FROM THE SOAKING LIQUID.

Measurement	Initial Thickness (Å)	Maximum Thickness (Å)	Increase in Thickness (Å)	Increase in Thickness (%)
M01	11250	11950	700	6.2
M02	11600	12830	1230	10.6
M05	11780	11826	46	0.4
M06	11400	11510	110	1
M10	11550	11950	400	3.5
M11	11600	12300	700	6
M12	11550	11900	350	3
M13	11150	11700	550	4.9

It can be observed that soaking in acetone resulted in the greatest thickness change (10.2% in the case of M02). However, this sample returned to approximate

its original thickness within 45 min of being removed from the liquid. In every case, the Parylene films remained intact; no delaminating or cracking was observed.

E. *Strain Test Structures*

The Young's modulus of Parylene is reported to be in the range of 2 to 5.29 GPa [10,11] and it is important to confirm its value for the thin films deposited in this study. A nano-indentor was used to measure the data presented in figure 4, which shows the relation between Young's modulus and the temperature to which the film has been heated during processing. This heating represents the process used to anneal the NiFe alloy used as a magnetic core for a microscale inductor. The results show that the value of Young's modulus increases with annealing temperature; indicating the importance of monitoring the thermal budget experienced by any devices that incorporate Parylene [9].

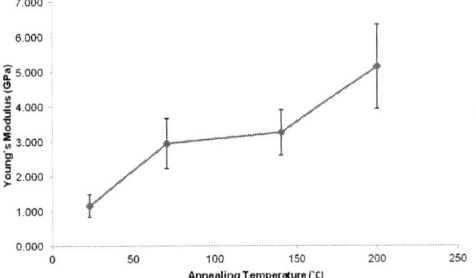

Figure 4, Young's modulus of Parylene after annealing for 1 hour at the given temperature, error bars indicate standard deviation.

For these initial evaluations, to wafer map the strain in Parylene, strain indicator test structures were fabricated using masks designed for wafer mapping the strain in NiFe films [12].

The process involved using a Ti(30nm)-Cu(700nm)-Ti(30nm) sacrificial layer, which also acted as a seed layer for electroplating. NR2-20000P was used as a negative mould for electroplating copper to a thickness of 5µm. Following this, the photoresist was stripped and 5µm of Parylene deposited to conformally coat the copper pattern. CMP was then used to polish the Parylene back to the level of copper mould. Finally, the copper mould and the seed layers were etched using 15% nitric acid for copper and 1% HF for Ti in order to release the structures. It should be noted that this wet etch release of the very fragile Parylene structures would result in many damaged structures, Figure 5 shows a diagrammatic cross-section through the pointer arm of the completed device. However, it should be noted that the wet release would almost certainly result in stiction, and if this is the case the assumption is the pointer arm would be fixed in the position defined by the strain.

Figure 6 shows a successfully fabricated strain indicator structures with 5 µm thick Parylene and it can be seen to exhibit tensile stress. It can be noted that the angle of rotation is much greater than that observed with NiFe films [12]. This is clearly a result of the different material properties of Parylene, rather than higher stress level as there was no observed bow on the wafer resulting from the deposited film.

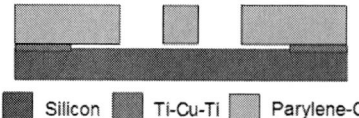

Figure 5; Cross-section diagram of stress indicator structure.

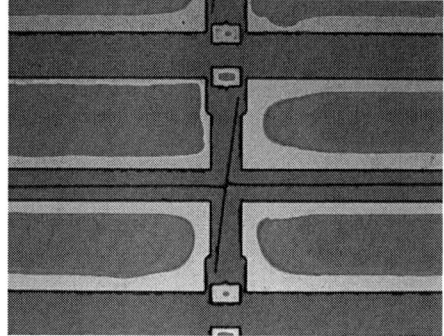

Figure 6; Parylene stress indicator structure showing tensile stress.

In order to evaluate spatial variation, 1,407 of the structures shown in figure 6 were used to generate strain wafer maps for 3 inch silicon wafers. As was the case for the nano-indentation measurements, strain measurements were performed on unannealed wafers and wafers annealed at 70, 140 and 200°C. The resulting wafer maps, which were automatically measured [12] are presented in Figure 7, apart from the 200°C wafer for reasons described later.

The unannealed test structures show an average deflection of 8.15° which compares with 8.49° for wafers annealed at 70°C. However, considering the 1.78GPa increase in Young's modulus between non-annealed and 70°C annealed Parylene, a higher level of stress is assumed in the annealed Parylene.

It can be observed from figure 7 that wafers annealed at 140°C exhibit a significantly lower yield (44%) than non-annealed (64%) and wafers annealed 70°C (76%). The reason for this is the larger rotation at 140°C, which results in the pointer arms being constrained as can be seen in figure 8. This is the result in many of the indicator structures breaking due to strain in the film forcing the deflection past the 14° maximum deflection. Those that survived, such as the device shown in figure 8, typically had their pointer arms bent, and would have rotated through a larger angle if they had not been constrained. Hence, these structures are not able to provide quantitative value of deflection for these films.

When annealed at 200°C, none of the indicator structures survived the release processing step. It is postulated that their rotation angle would be significantly larger than the 140°C annealed structures; as a result they would experience a high degree of strain. A 398% increase in Young's modulus between non-annealed Parylene and material annealed at 200°C suggests a significantly higher stress will result in these films. This stress is assumed to exceed the yield strength for Parylene, 59MPa [13].

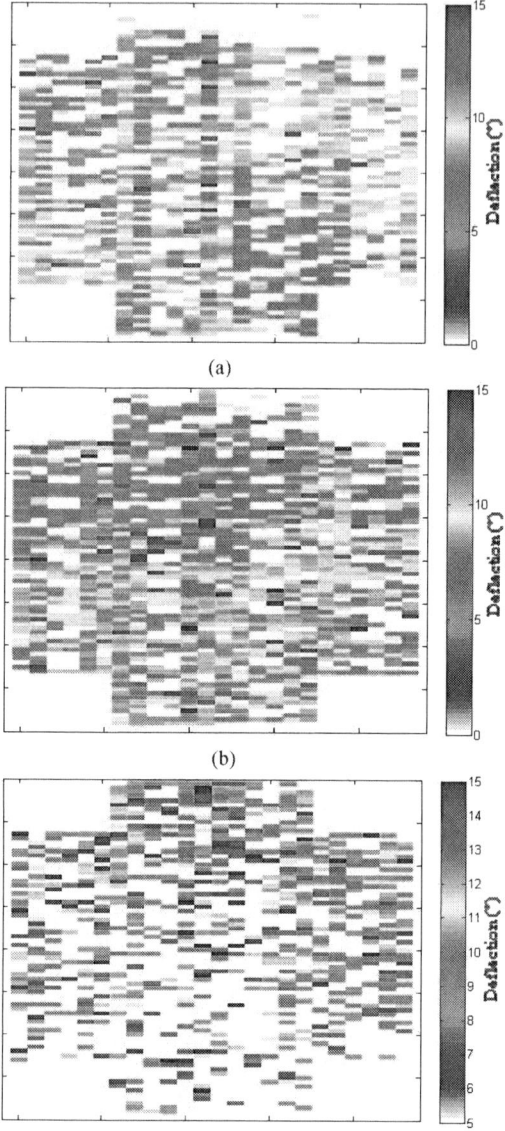

Figure 7; Strain wafer maps; (a) no annealing; (b) annealed at 70°C; (c) annealed at 140°C (note the white areas indicate damaged structures and sites with stress stuctures

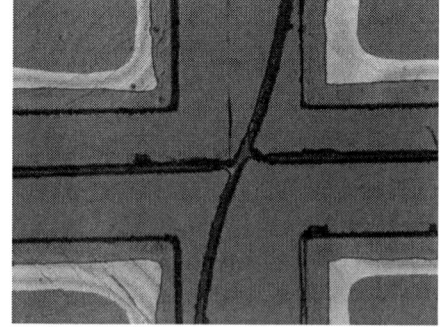

Figure 8; Constrained Strain indicator structure annealed at 140°C.

III: DESIGN OF MICRO-INDUCTOR

A. Micro-inductor fabrication process

In order to confirm the compatibility of Parylene with the whole sequence of the required process architecture, a complete micro-inductor has been fabricated. Firstly, a Parylene layer was used as a structural material and mould for electroplating NiFe, which will act as a magnetic core below the inductor coils. The NiFe and Parylene were then planarised using CMP before a further 5µm of Parylene was deposited as an insulating layer. A 30µm thick layer of NR2-20000P was patterned with standard lithography to form a negative mould for electroplating the 30µm thick copper coils. Once the photoresist was stripped, following electroplating, the underlying seed layer was etched and Parylene was deposited over the coils. 35µm of NR2-20000P was then used as a mask to etch holes in the Parylene layer over contact pads and areas required to produce magnetic vias to the bottom layer of NiFe. Following this step, the NR2-20000P mask was stripped off and the Parylene was planarised to the top of the copper coils. An additional passivation layer of Parylene was deposited over the copper before the Parylene etching process was repeated. Then, a 35µm thick NR2-20000P layer was again used as a mould for electroplating of a top NiFe layer before it was stripped off to uncover the seed layer which is removed to complete the structure. A schematic cross-section and optical microscopy image of the completed micro-inductor are presented in Figure 9 (a) and (b).

B. DC measurements on coil structures

Inductors with 4 and 10 turn coils (Figure 10) have been fabricated. The track widths are 50µm and 20 µm for 4 and 10 turn coils respectively. Separation between windings, for both coils is 10µm.

DC electrical measurements on these coils when coated with 5µm of Parylene are presented in Figure 8. Measurements were carried out using a HP4156B Precision Semiconductor Parametric Analyzer. The measured values of resistance are in good agreement with calculated values of 1.60Ω and 3.14Ω for the 4 turn and 10 turn coils respectively. These results confirm that the seed layer between windings has been completely removed and Parylene acts as a good insulator.

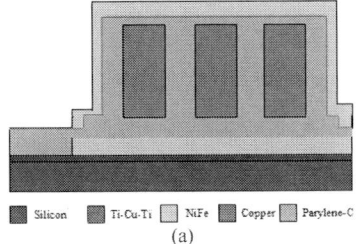

Silicon ▪ Ti-Cu-Ti ▪ NiFe ▪ Copper ▪ Parylene-C

(a)

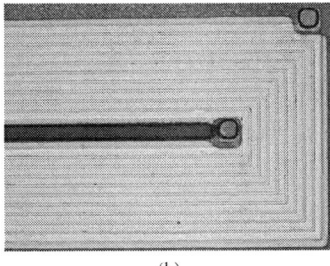

(b)

Figure 9: (a) cross-section of the completed micro-inductor and (b) optical microscopy image of the top view.

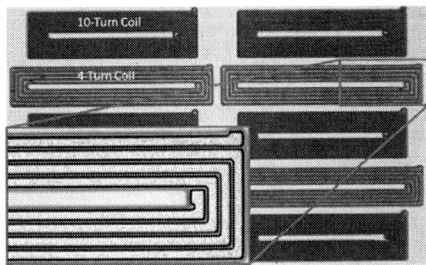

Figure 10; Photo of 10-turn and 4-turn copper coil structures.

IV. CHARACTERISATION OF THE MICRO-INDUCTOR

Fully operational inductors, as shown in figure 9(b) have been measured using an HP4275A LCR over the 100 kHz to 10 MHz range. The results are presented in Figure 12 as impedance spectra. The iron content measured here across the wafer using X-ray fluorescence (XRF) spectrometry varies between 10 and 15%. Therefore, the values for inductance presented in Figure 12 are an average measured from different micro-inductors on the wafer. This show that the average variation of the for 4 and 10 turn inductors is 61.2 nH and 67.6 nH respectively, at 10 MHz. It is believed the major source of the variation is the NiFe cores whose iron concentration varies between 10% and 15%.

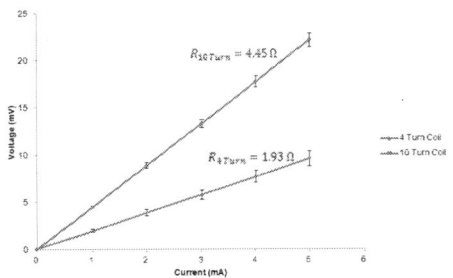

Figure 11; DC electrical measurements of coil structures coated with 5µm Parylene.

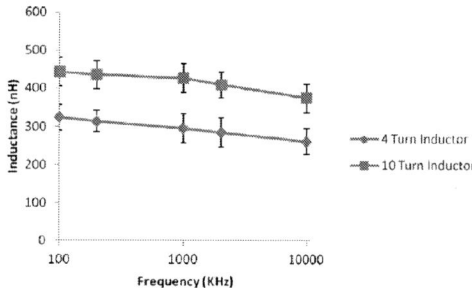

Figure 12; Inductance spectra for 4-turn and 10-turn micro-inductors loaded with NiFe core, error bars indicate standard deviation.

V. CONCLUSIONS AND FURTHER WORK

This work has characterised Parylene and demonstrated that, this material together with subsequent CMP processing can be successfully used to fabricate planar micro-inductors with thick electroplated copper coils and NiFe alloy cores.

Young's modulus and strain in these layers has also been examined and the need to carefully consider thermal budget, when using Parylene, has been highlighted. In addition to their change in thickness due to exposure to different liquids, commonly used in MEMS processing, has also been evaluated.

The evidence of excellent compatibility of Parylene with other intrinsic materials and processes has been confirmed by the successful fabrication of a fully operational micro-inductor. This micro-inductor incorporates Parylene as an insulating and structural layer, and to date has not exhibited any delamination that is often experienced with SU-8.

Electrical results show that fabricated micro-inductors produce and inductance of 298.4 nH and 428.3nH at 1MHz for 4 an 10 turn micro-inductors respectively.

Future work will involve the incorporation of Parylene into laminated micro-inductors. In addition to this further work will be carried out using the strain indicator structures and Young's modulus measurements to compare stress in Parylene to that of SU-8.

REFERENCES

[1] S.C.O. Mathuna, T. O'Donnell, N. Wang, K. Rinne, "Magnetics on silicon: An enabling technology for power supply on chip," *IEEE Transactions on Power Electronics.*, vol. 20, no. 3, pp. 585-592, 2005.

[2] N. Wang, T. O'Donnell, S. Roy, P. McCloskey, C. O'Mathuna, "Micro-Inductors Integrated on Silicon for Power Supply on Chip,"*Journal of Magnetism and Magnetic Materials.*, vol. 316, no. 2, pp. 233-237, 2007.

[3] P. Artillan, M. Brunet, D. Bourrier, J. P. Laur, N. Mauran, L. Bary, M. Dilhan, B. Estibals, C. Alonso, J. L. Sanchez, "Integrated *LC* Filter on Silicon for DC–DC Converter Applications," *IEEE Trans. Power Electronics.*, vol. 26, no. 8, pp. 2319-2325, 2011.

[4] S. Smith, et. al., "Fabrication and Measurement of Test Structures to Monitor Stress in SU-8 Films," *IEEE Trans. Semicond. Manuf.*, vol. 25, no. 3, pp. 346–354, 2012.

[5] *M.E.* Cosens. "Parylene conformal coatings [PCB protection]," *Electronic Production*, vol. 12, no. 3, pp. 12-13, 1983.

[6] M. Brunet, T. O'Donnell, J. O'Brien, P. McCloskey, S. Cian Ó Mathuna[2]., "Thick photoresist development for the fabrication of high aspect ratio magnetic coils," *J. Micromech. Microeng.*, vol. 12, no. 4, pp. 444-449, 2002.

[7] F. Herrault, S. Yorish, T.M. Crittenden, C-H. Ji, Mark G. Allen,, "Parylene-Insulated Ultradense Micro-fabricated Coils", IEEE Journal of Microelectro-mechanical Systems, Vol. 19, No. 6, December 2010 pp 1277-1283

[8] M. Liger, D. Rodger and Y.C. Tai, "Robust Parylene-to-Silicon Mechanical Anchoring," Proceedings, The Sixteenth IEEE International Conference on Micro Electro Mechanical Systems (MEMS '03), Kyoto, Japan, Jan. 19-23, pp. 602-605, 2003.

[9] T.A. Harder, T-J. Yao, Q. He, C-Y. Shih, Y-C. Tai, "Residual Stress in Parylene C", Proceedings of the 15th IEEE International Conference on Micro Electro Mechanical Systems (MEMS '02) pp. 435-438.

[10] J.G. Cham, J. G. Cham, Z. Nenadic, S. Musallam, Y. C. Tai, J. W. Burdick, R. A. Andersen, "A New Multi-Site Probe Array with Monolithically Integrated Parylene Flexiable Cable for Neural Prosthesis," *Engineering in Medicine and Biology Society*, pp. 7114-7117, 2005.

[11] D. Wright, D. Wright, B. Rajalingam, S. Selvarasah, Y. Ling, J. Yeh, R. Langer, Mehmet R. Dokmeci, A. Khademhosseini, "Reusable, Reversable, Sealable Parylene Membranes for Cell and Protein Patterning," *J. Biomedical Materials Research*, vol. 85A, no. 2, pp. 530- 538, 2007.

[12] J. Murray, S. Smith, G. Schiavone, J. G. Terry, A. R. Mount, A. J. Walton, "Characterisation of Electroplated NiFe Films using Test Structures and Wafer Mapping Measurments," *IEEE Conference on Microelectronic Test Structures*, pp. 63–68, 2011.

[13] V. C. Y. Shih, T.A. Harder, Y. C Tai, "Yield Strength of Thin-Film Parylene-C," *Symposium on Design, Test, Integration and Packaging of MEMS/MOEMS*, pp. 394-398, 2003.

Micromechanical test structures for the characterisation of electroplated NiFe cantilevers and their viability for use in MEMS switching devices

G. Schiavone[†], S. Smith[†], J. Murray[†*], J.G. Terry[†], M.P.Y. Desmulliez[§] and A.J. Walton[†]

[†]Institute for Integrated Micro and Nano Systems, Joint Research Institute for Integrated Systems, School of Engineering, Scottish Microelectronics Centre, The University of Edinburgh, Edinburgh, EH9 3JF, UK
[*]School of Chemistry, Joseph Black Building, The University of Edinburgh, EH9 3JJ, UK
[§]MIcroSystems Engineering Centre, Joint Research Institute for Integrated Systems, School of Engineering & Physical Sciences, Heriot-Watt University, Edinburgh EH14 4AS, UK

ABSTRACT

This paper presents the fabrication of a series of test devices designed to prove the viability of electroplated NiFe freestanding structures for use in magnetically actuated MEMS switches. Preliminary results show promising actuation responses and further release optimisation and testing will enable the quantitative measurement of the desired characteristics. In addition, this will potentially enable the mechanical characterisation of freestanding structures in other materials by means of magnetic actuation, simply by depositing small quantities of NiFe or other magnetic materials in convenient areas of existing devices.

INTRODUCTION

Alloys of nickel and iron, and in particular Permalloy (Ni:Fe, 80%:20%), are important materials for the microfabrication of integrated passive magnetic devices such as inductors, due to their high magnetic permeability and low coercivity [1,2]. NiFe alloys can also be used to build micro-electromechanical switches with magnetic actuation [3,4]. This particular architecture offers many advantages over other actuation schemes for MEMS switches, particularly electrostatic actuation. These include large actuation forces, allowing for higher contact gaps and thus providing better isolation in the off-state, and greater robustness to stiction and other wear and failure mechanisms. The high voltages required to achieve electrostatic actuation are also removed, creating the opportunity to employ low voltage electrostatic latching mechanisms, subsequent to magnetic actuation, for the retention of the switch state with near zero power consumption (except for the charging energy) [3].

The overall purpose of this work is to investigate the release of these structures and the testing of electroplated NiFe structures which can be employed as the driven section of magnetically actuated MEMS switches. The operation of these devices requires the integration of reliable and manufacturable micromechanical magnetic structures with integrated inductor coils. This combination provides opportunities to develop novel devices which can offer greater performances in terms of off-state isolation, lower voltage actuation for compatibility with standard silicon integrated circuits, and lower power consumption, even with respect to semiconductor switches if employed with high frequency signals [5]. This paper reports the development of test structures and processes employed to characterise the release process step and the mechanical and magnetic performances of electroplated NiFe structures that can be employed in a MEMS switch device framework.

MAGNETIC TEST STRUCTURES AND TEST METHODOLOGY

The test structures designed to characterise the behaviour and performance of electroplated NiFe structures are based on cantilevers anchored to the substrate at one end. The cantilever is deposited through a photoresist mould onto a patterned sacrificial mesa, which is subsequently removed to release the freestanding structure. Fig. 1 presents a schematic of the process flow for the fabrication of the test structures.

The chosen substrate is a 3" (75 mm) Si wafer coated with a ~20 μm thick SU-8 layer, in order to monitor any process incompatibilities in terms of temperature, stress and etching selectivity. For a complete device the thick SU-8 mould would contain the embedded actuation part consisting of copper coils and NiFe cores, see fig. 2, bottom.

978-1-4673-4845-4/13 $31.00 © 2013 IEEE

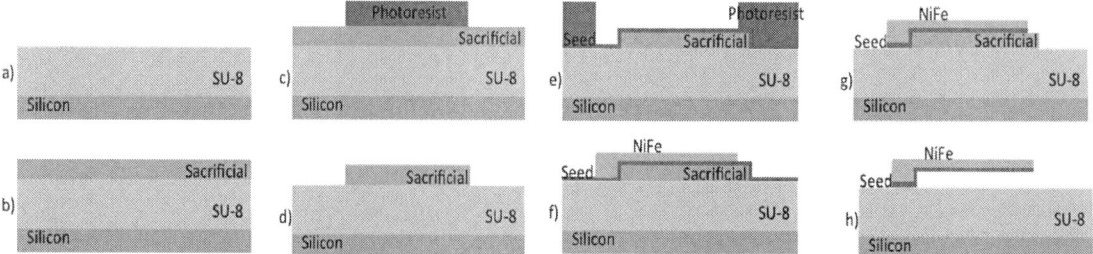

Figure 1 – Process flow for the fabrication of the test cantilevers. a) Preparation of the substrate with 20 μm thick SU-8. b) Blanket deposition of 2 μm thick sacrificial material. c) Patterning of photoresist mask to define the sacrificial structures that will support the cantilever. d) RIE of the sacrificial material and removal of the photoresist mask. e) Blanket sputtering deposition of a 30 nm Ti and 300 nm Cu seed layer and patterning of photoresist mask to define the electroplated cantilever. f) ~2 μm thick NiFe electroplating and removal of the photoresist mould. g) Wet etch of the seed layer over the areas previously covered by the photoresist mask. h) Removal of the sacrificial material and release of the NiFe cantilever.

Electroplating of the NiFe alloy is performed in a solution prepared using the components and the quantities reported in Table 1, for a total volume of 35 litres.

Component	Concentration
NiCl$_2$.6H$_2$O	110 g/l
FeCl$_2$.4H$_2$O	8.0 g/l
Citric acid	5.25 g/l
Tri-sodium citrate	27 g/l
Saccharin	2 g/l
SDS	0.1 g/l

TABLE 1 – ELECTROPLATING BATH COMPOSITION.

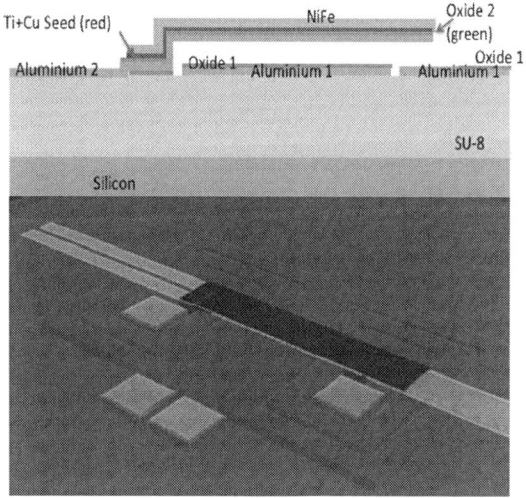

Figure 2 – Scheme of a test device for the actuated section of the switching device (top) and 3D view of the device with the embedded actuation section (bottom).

Following the results obtained from previously reported studies [6], a combination of mild agitation and low current density is employed for the electroplating in order to achieve a reliable balance between thickness uniformity across the wafer surface, desirable Ni:Fe ratio ranges and low stress.

The fabricated cantilevers can be used to study both the mechanical and magnetic properties of the structures in response to an externally applied magnetic field. A magnetic measurement rig is currently being developed, which uses a current controlled electromagnet incorporated into a Zygo white light phase shifting optical interferometer. Copper wires are wound around a magnetic core which is placed underneath the wafer under test, while deflection profiles can be acquired from the top side via the interferometer.

Deflection curves of the cantilevers as a function of the different DC current regimes in the excitation section can be built for cantilevers of different dimensions (310 μm length, 65 to 95 μm width and 1.5 to 2.5 μm thickness). These can be useful for the case of the response of a cantilever employed in a switching device if the displacement values are correlated to actual force values, by either analytically computing the distributed force exerted by the excitation section or by comparing deflections obtained by testing cantilevers in a force controlled measurement as previously reported [7]. The required force to obtain the observed deflection can thus be used as a reference goal for the design of the actuation section of a complete MEMS switching device.

RELEASE OF THE CANTILEVERS

The initial design consisted of solid NiFe cantilevers, in order to maximise the electrical conductivity and the volume of magnetically

978-1-4673-4845-4/13 $31.00 © 2013 IEEE

responsive material of the structure, and consequently reduce the resistance of the electrical path for the switched signal. As part of the switching device design phase, NiFe cantilevers of different widths were contemplated. In principle, neglecting boundary effects, the linear increase in the cantilever stiffness k due to a greater width w is balanced by a simultaneous increase in the magnetic response of the structure due to a decrease in the reluctance R, inversely proportional to the cross-sectional area A. Expression (1) denotes the spring constant k of a cantilever of width w, length l and thickness t fabricated in a material with Young's modulus E and subject to an excitation in the direction parallel to its thickness [8,9], while equation (2) indicates the reluctance R of a magnetic structure of relative permeability μ_r, length l and cross section A.

$$k = E\frac{wt^3}{4l^3} \qquad (1)$$

$$R = \frac{l}{\mu_0\mu_r A} \qquad (2)$$

Although these two opposing effects balance each other, a wider structure used as an active conduction path ensures a greater electrical conductivity, providing benefit to the performances of the switching device. On the other hand, from a fabrication point of view, the release process for suspended structures of greater width is significantly more demanding. Specifically, in the case of dry etching, wider structures require the etching vapour to reach further distances underneath the magnetic material in order to provide a complete release.

A range of candidate materials has been considered for use as the sacrificial layer, with the key parameter being compatibility with materials used in the full switch structure, while preserving the NiFe integrity and functionality. The current focus is on the use of polymers, including Parylene-C and polyimide, which can be removed by means of an oxygen plasma etch process and amorphous silicon, attacked by a XeF$_2$ vapour etch process. Figure 3 shows a test structure indicating that a 70 μm wide cantilever can be released.

According to previous studies [6], NiFe shows a

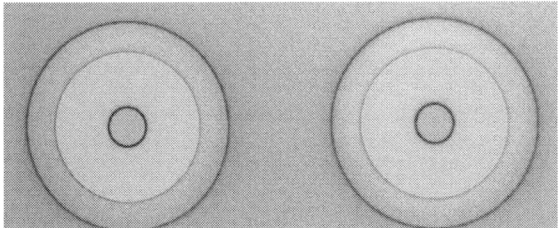

Figure 3 – Test structure for evaluating XeF$_2$ sacrificial etch for a membrane release (36 μm etch entrance hole, structure diameter = 190 μm). Courtesy of Memsstar.

considerable increase in stress when exposed to temperatures around 200 °C for several minutes. Analogous temperature constraints come in force when the underlying SU-8 mould containing the embedded coils is present to provide the actuation mechanism in a complete switching device.

When cured, polymers show a smearing effect on the edges of the mesa structures, providing a smoother and less sharp rest surface for the NiFe cantilever to lie on (see fig. 4). This allows a reduction of the mechanical stress that would otherwise be concentrated on sharp cornered features in the fabricated cantilevers, constituting a prominent failure mechanism. The disadvantage of using polymers is that the oxygen plasma etch process raises the sample temperatures to values around 200 °C, with higher temperature regimes yielding a higher etch rate. As observed in previous studies, these conditions increase the stress levels of NiFe films and catalyse the oxidation of the material.

Figure 4 – SEM image of a NiFe cantilever anchor section. Notice on the left the gentle slope that leads the suspended structure from a height of 2 μm to the substrate level.

On the other hand, using an XeF$_2$ etch on amorphous silicon enables a lower temperature etch process that better preserves the NiFe and other materials employed in the final switching device thanks to its great selectivity [10]. Note that this requires a slow etch recipe as the XeF$_2$ etch is exothermic. The sacrificial mesa structures in this case, will have a steeper profile and this can potentially cause unwanted fractures, with cantilevers breaking at the joint point.

A pair of test masks have been designed with the purpose of investigating the release process for cantilevers with different widths and employing etch release holes of various dimensions. These holes are distributed on the NiFe structures using different patterns (see fig. 5). The masks include cantilever designs of 300 μm lengths with widths ranging from 100 μm to 20 μm, in steps of 10 μm. Solid cantilevers are included, as well as cantilevers with release holes.

978-1-4673-4845-4/13 $31.00 © 2013 IEEE

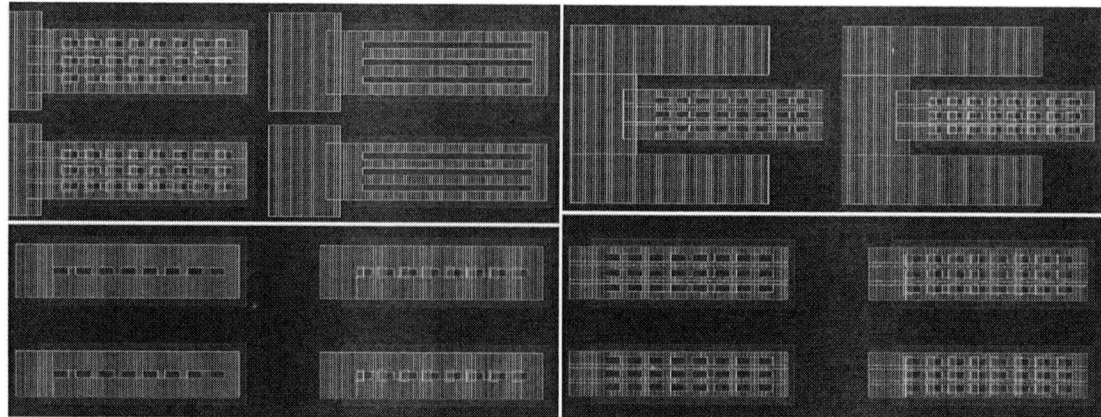

Figure 5 – Snippets of the test mask layout for the investigation of the release process for cantilevers of different widths, with different etch hole sizes and patterns and different anchor shapes.

The available hole sizes are (in μm) 10×20 with a 10 μm separation, 10×10 with a 20 μm separation, 10×230 as a continuous hole, 5×20 with a 10 μm separation, 5×10 with a 20 μm separation, 5×230 as a continuous hole, 2×20 with a 10 μm separation, 2×10 with a 20 μm separation and 2×230 as a continuous hole. The holes are arranged in a single row (fig. 5, bottom left) or in three rows spanning the whole cantilever width (fig. 5, bottom right). Four different types of anchor are included in the design, comprising simple anchors of the same width as the cantilevers (fig. 5, bottom left), slightly wider anchors (fig. 5, top left), a whole single anchor for each set of cantilever widths (not shown) and an anchor that expands on the side to provide better stability (fig. 5, top right).

This mask set provides the ability to undertake experiments to select the best compromise between ease of processing, NiFe preservation and device performances.

ANALYTICAL MODEL
FOR THE MAGNETIC FORCE

The 'current model' for the calculation of the forces on magnetised bodies uses a generalisation of the Lorentz force expression for the case of no currents. Starting from the standard expression:

$$\vec{F} = q\ \vec{u} \times \vec{B} \qquad (3)$$

for the force F exerted on a charge q moving with a velocity u in a space characterised by a magnetic flux density B, and substituting the equivalent volume and surface current densities definitions introduced in the 'current model' for magnetised bodies [11]:

$$\vec{J}_m = \nabla \times \vec{M}, \quad \vec{j}_m = \vec{M} \times \hat{n} \qquad (4)$$

where M is the magnetisation of the body with equivalent volume current density (J_m) and surface current density (j_m), one can derive the following expression for the for the integral (F) and distributed (f) magnetic force on the body:

$$\vec{F} = \int_V (\vec{M} \cdot \nabla)\ \vec{B}\ d\vec{v}$$
$$\vec{f} = (\vec{M} \cdot \nabla)\ \vec{B} \qquad (5)$$

where V is the volume of the body and v the set of volume integration variables. Since the magnetisation M of the Permalloy structures occurs as a response to the external excitation B, it is possible to write the former as a function of the latter. Considering a simplified geometry for the cantilevers as parallelepipeds with:

- Length = 310 μm
- Width = 65 to 95 μm
- Thickness = 1 to 3 μm

and the corresponding demagnetisation field arising due to shape anisotropy,

$$\vec{M} = \chi_m \left(\vec{H}_{ext} - N_{demag} \vec{M} \right)$$
$$\vec{M} = \frac{\chi_m}{\chi_m N_{demag} + 1} \vec{H}_{ext} \qquad (6)$$

where N_{demag} is the demagnetising factor corresponding to a specific alignment between the NiFe cantilever and the external field, χ_m is the magnetic susceptibility of NiFe and H_{ext} is the external applied magnetic field. The relevant value for N_{demag} in the case of an excitation field parallel to the thickness of the cantilevers, as it occurs with the electromagnet placed underneath the wafer in the measurement setup, can be obtained from the tables reported in [12]. For the dimensions employed in this work, the obtained values range from 0.995 to 0.93. This geometrical configuration does not therefore favour magnetisation of the material along the thickness axis. In this case the magnetisation M can be assumed to have the same value of the external field H_{ext}. It is hence possible to obtain the final expression for the force distribution over the volume of the cantilever:

$$f_z = M_z \frac{\partial}{\partial z} B_z = H_{ext,z} \frac{\partial}{\partial z} B_z \qquad (7)$$

which only depends on the external excitation, fully known and custom-designed to have high field intensity and field gradient.

The analysis of the deflection of cantilevers of different dimensions resulting from different excitation currents will reveal information on the magnetic responsiveness of the fabricated test structures.

MICRO-SWITCH TEST STRUCTURES

In order to confirm the viability of freestanding electroplated NiFe test structures for use in MEMS switches, the actuated portion of a magnetic switch has been fabricated to test the on-state effectiveness, quality and reliability of the contact between two parts of a switched line, one lying on the substrate and the other running along the bottom surface of NiFe cantilevers. Additional capacitor structures (metal plus insulator stack) are fabricated underneath the freestanding cantilever in order to introduce an electrostatic latching mechanism able to preserve the on-state once the magnetic actuation has already driven the cantilevers down to the substrate. Figure 2 shows a 2D and 3D diagram of the test devices.

The fabrication of such test devices follows the process described and illustrated in figure 1, with the addition of patterning and etching steps required to define the metal and insulator structures for the contacts and the latches. Aluminium has been chosen as the metal for lines and capacitor plates as it is less prone to oxidation when compared to copper, while being easier to etch and pattern. A combination of overlapped structures on the masks and controlled etch allows for a precise patterning of the metal and insulator structures, avoiding etch selectivity problems. Figures 6 and 7 show SEM images of fabricated switching test structures.

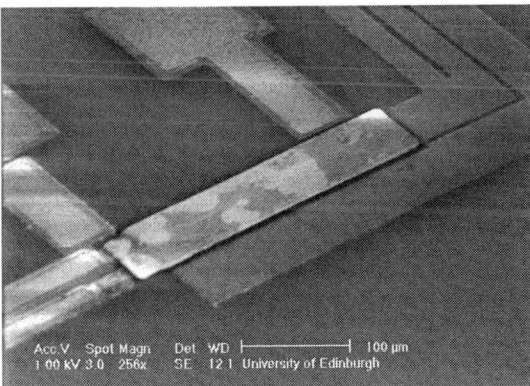

Figure 6 – SEM image of a switching test device. The NiFe cantilever is anchored to the substrate on top of the latching and signal aluminium lines (left). The image also shows the aluminium output signal line exiting below the edge of the cantilever and the Al bottom latching plate coming out along the cantilever side. The sacrificial layer is still present under the cantilever, as visible on the darker shading along its edges.

Figure 7 – SEM image of a NiFe cantilever after removal of the Parylene-C sacrificial structure by means of an Oxygen plasma process. The apparent bigger gap underneath the portion of the cantilever closer to its anchor (left image) might be due to stress relief in the cantilever stack. The image on the right shows a detail of the tip of the cantilever. The top surface shows the contours of the structures built on the bottom surface, i.e. the contact line and the latching line (insulated from the rest by a SiO_2 layer).

PRELIMINARY RESULTS

Observations and recordings have been made of cantilever movement when actuated via an applied magnetic field generated by a permanent magnet. However, the release process has not been robust and hence the mask design shown in figure 5 is in the process of being used to characterise the design/process. Once completed, the next step is to measure the movement and quantify the response, however the visual inspection shown in figure 8 is promising.

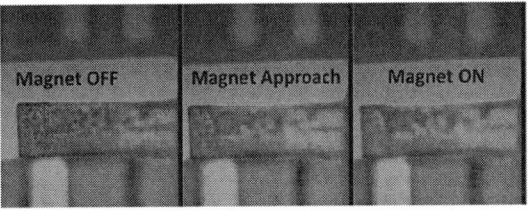

Figure 8 – The cantilever tip is pulled to the substrate as the magnet approaches the bottom of the wafer, as indicated by the focus difference between the right and left frames.

CONCLUSIONS AND FURTHER WORK

This paper presents the fabrication of a series of test devices designed to prove the viability of electroplated NiFe freestanding test structures for use in MEMS switching applications. Preliminary results show promising actuation response and further release optimisation and testing will enable the quantitative measurement these characteristics. Successful implementation of the described test routine will prove the viability of the proposed structures for use as switching devices. Furthermore it may enable the mechanical characterisation of other freestanding structures by means of magnetic actuation by simply depositing small quantities of

978-1-4673-4845-4/13 $31.00 © 2013 IEEE

NiFe or other magnetic materials in convenient areas of existing devices.

ACKNOWLEDGEMENTS

The authors would like to acknowledge the financial support of the Edinburgh Research Partnership in Engineering and Mathematics, EPSRC/IeMRC through DTA funding DTA/15/2007, IeMRC/EPSRC (FS/01/02/10) through the Smart Microsystems grant, Dr Nathan Brockie of Texas Instruments for his help with electroplating, Mr Daniel Drysdale and Dr Tony O'Hara of Memsstar for their help with the vapour etching processes, and Dr Enrico Mastropaolo of SMC, University of Edinburgh for his help with the polyimide preparation.

REFERENCES

[1] J. K. Luo, M. Pritschow, A. J. Flewitt, S. M. Spearing, N. A. Fleck, and W. I. Milne, "Effects of Process Conditions on Properties of Electroplated Ni Thin Films for Microsystem Applications", *Journal of The Electrochemical Society*, 153, pp D155-D161, 2006.

[2] K.I. Arai and T. Honda, "Micromagnetic actuators", *Robotica*, Vol. 14, pp 477-481. Cambridge University Press, 1996

[3] M. Ruan, J. Shen, and C.B. Wheeler, "Latching microelectromagnetic relays," *Sensors and Actuators*, A(91), pp 346 – 350, 2001.

[4] Il-Joo Cho, Taeksang Song, Sang-Hyun Baek, and Euisik Yoon, "A low-voltage and low-power rf mems series and shunt switches actuated by combination of electromagnetic and electrostatic forces", *IEEE Transactions on Microwave theory and techniques*, 53(7), pp 2450 – 2457 , July 2005.

[5] Gabriel M. Rebeiz, "RF MEMS Switches, status of the technology", *The 12th International Conference on Solid State Sensors, Actuators and Microsystems*, Boston, pp 1726 – 1729, June 8-12, 2003

[6] J. Murray, G. Schiavone, S. Smith, J. Terry, A.R. Mount, A.J. Walton, "Characterisation of Electroplated NiFe Films using Test Structures and Wafer Mapped Measurements", *24th International Conference on Microelectronic Test Structures*, Amsterdam, pp 63 – 68, 4th – 7th April 2011.

[7] G. Schiavone, M.P.Y. Desmulliez, S. Smith, J. Murray, E. Sirotkin, J.G. Terry, A.R. Mount, A.J. Walton, "Quantitative wafer mapping of residual stress in electroplated NiFe films using independent strain and Young's modulus measurements", *24th International Conference on Microelectronic Test Structures*, San Diego, pp 105 – 110, 19th – 22nd March 2012.

[8] Beer F P and Johnston E R 1981 *Mechanics of Materials* (New York: McGraw-Hill).

[9] Cottrell A H 1964 *The Mechanical Properties of Matter* (New York: Wiley).

[10] Brazzle J.D., Dokmeci M.R., Mastrangelo C.H., "Modeling and characterization of sacrificial polysilicon etching using vapor-phase xenon difluoride", *17th IEEE International Conference on Micro Electro Mechanical Systems (MEMS)*, pp 737 – 740, 2004.

[11] E. P. Furlani. "Permanent magnet and electromechanical devices – materials, analysis and applications." Academic Press, 2001.

[12] Chen Du-Xing, Enric Pardo, and Alvaro Sanchez, "Demagnetizing factors of rectangular prisms and ellipsoids", *IEEE Transactions on Magnetics*, Vol. 38, N. 4, pp 1742 – 1752, 2002.

Investigation of Devices for In-Vivo Energy Harvesting through Blood-Flow-Like Excitation

Rosemary O'Keeffe[1], Nathan Jackson[1], Alan Mathewson[1], Kevin G. McCarthy[2]

1. Tyndall National Institute,
University College Cork,
Lee Maltings, Dyke Parade, Cork, Ireland
2. Department of Electrical and Electronic Engineering,
University College Cork,
College Rd., Cork, Ireland

Abstract – **Test structures for energy harvesting through piezoelectric materials are designed using FEM. The results of the modelling are compared to those for fabricated devices which were tested using a perfusion machine to simulate blood flow. The results show considerable accuracy between measured characteristics and simulated ones which only differed by approximately 4%. The modelling also established the best design for energy harvesting using aluminium nitride (AlN) in blood flow.**

Keywords – AlN, piezoelectric, energy harvesting, FEM

I. INTRODUCTION

In previous work [1], a first generation of cantilever test structures, designed to test energy harvesting in air, was used to develop accurate finite element modelling (FEM) of energy harvesting structures using COMSOL Multiphysics[R] [2]. The results from physical tests on the devices were used to tune the models until the results which were achieved accurately predicted the behaviour of the devices. These structures were used to determine the effect of various geometries of AlN piezoelectric structures when vibrated at resonant frequency in air. Using the results from these test structures to update and improve the accuracy of the FEM enabled the creation of well characterised models which could be used to design subsequent generations of energy harvesting devices. However, the new structures that relate to this work were targeted at the harvesting of energy from blood flow where the stress on the piezoelectric material is generated by the oscillating pressure of the blood on the device and not by vibrating the devices at resonant frequency but instead are moved by the ebb and flow of the fluid.

Various groups have performed research into human powered energy harvesting using piezoelectric devices, these range from smart clothing such as piezoelectric energy harvesting in shoes discussed by Priya et al. in [3] to knee piezelements in knee replacement joints by Aton et al. in [4]. Neither of which use the piezoelectric element for energy harvesting in a fluid. However, other research has been conducted in a] air – a piezoelectric device in a Helmholtz resonator by Matova et al. [5], and b] water, an energy harvesting eel by Taylor et al. [6].

There are considerable regulations and constraints associated with testing devices in blood so resonant frequency was used as a parameter to measure to evaluate the models and compare them with measured results. After confidence in the numerical simulations of the devices was achieved by obtaining a good fit between measured and simulated performance, in air, these models were used to design prototype devices for performing energy harvesting through blood flow. Since a small disruption in blood flow of even ~5% can be fatal, it was very important that the models for these devices and their deployment environment, describe the full system during operation.

To study this fully, the use of FEM models of the devices as well as testing fabricated devices using a perfusion machine were undertaken. This allowed the extraction of information on the voltage output of the devices as well as the effect of placing the device into the blood flow to be established. A specially designed holder, shown in figure 8, was used to place the fabricated devices in the pulsating flow. All the devices were fabricated on a single die and this had to be placed in the holder in its entirety. This meant that the effect on the flow for individual devices could not be established using this test fixture. However, previous models of devices in an artery showed that an accurate measure of the effect on the flow by a device can be achieved through this simulation approach [7].

Tests were carried out on aluminium nitride (AlN) devices which were grown on a silicon (Si) wafer with titanium (Ti) under layer and an aluminium (Al) top layer. Tests on a sheet of polyvinylidene fluoride (PVDF) were also performed for confirmation of the accuracy of the models as well as information on a less stiff piezoelectric material. These tests were able to determine the pressure applied to the devices in the system also. The thickness of the mass on the cantilever consists of Si of 425μm.

978-1-4673-4845-4/13 $31.00 © 2013 IEEE

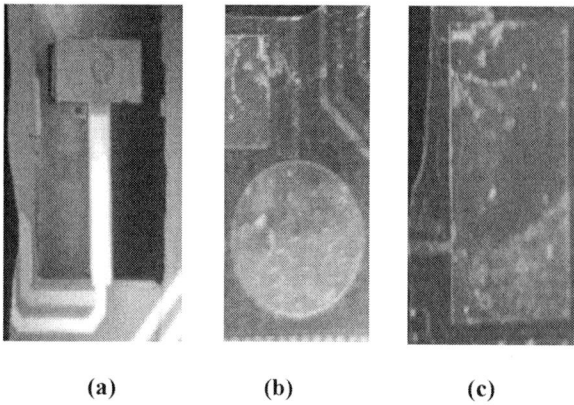

| (a) | (b) | (c) |

Figure 1: Structures for Energy Harvesting from Blood Flow (a) Cantilever (b) Diaphragm (c) Fixed Beam

The addition of Si layers made the AlN devices stiff compared to the PVDF sheet which is a polymer and therefore flexible. AlN materials and devices are an excellent choice for implantable structures because they are biocompatible as well as CMOS compatible and have a good conversion efficiency when grown properly. Previous work by Jackson et al. has shown that devices fabricated in a similar method to those in this paper can be used in x-ray diffraction measurements to determine the Full Width Half Maximum (FWHM) of the rocking curve in [7], and therefore the piezoelectricity, generation capability of the devices. Tests on this type of material have shown good FWHM values (< 2). The cantilever device fabrication was similar to that shown in the work by Jackson et al [8]. The AlN structures discussed in this paper are shown in figure 1. Figure 1((a) to (c)) show the fabricated devices, these devices were all drawn to be of a similar surface area which is not evident in the images above. The die is connected to a PCB and then wire is soldered to the pads to measure the voltage output of the devices.

I. THEORY

Piezoelectric devices respond to an applied force by generating an electric potential. This is an AC signal which can be used to power devices. The designs discussed here are for use with in-vivo energy harvesting and so the material choice was very important because the material needed to be biocompatible as well as piezoelectric. A further concern was that the energy harvesting device needed to be compatible with CMOS processing for placement into a full energy harvesting system and needed to provide high energy conversion efficiency because the geometry is limited significantly when dealing with in-vivo

systems. Most piezoelectric devices are designed for use in a vacuum, for these devices the excitation comes from vibrations at the resonant frequency of the devices. However, for devices which are designed for use in fluid, such as water or blood, the devices are stressed due to fluctuations in the pressure of the fluid on the device and so the resonant frequency of the devices is not such a critical factor. For this family of devices the critical factors are the area of the piezoelectric material and the stress on the material imposed by the fluid flow (these two factors determine the amount of energy which can be extracted).

A sinusoidal waveform was used to model the flow of blood using equation 1, although the flow is not really sinusoidal in shape. This is an approximation of the pulsating flow of the blood in an artery and this equation has been used in previous work by Coppola et al. in [9].

$$P_{in} = P_{min} + (P_{max} - P_{min})\sin(\omega t) \qquad (1)$$

where P_{in} is the pressure in the artery, P_{max} is the systolic pressure (or maximum pressure) and P_{min} is the diastolic pressure (or minimum pressure) of the fluid in the artery, t is time, ω is the frequency in rad/s and since the fluid is flowing at 60bpm the frequency is 1Hz and therefore $\omega = 2\pi$. In this case an artery of 1cm diameter was chosen for the models. This is an accurate representation of the typical artery (Aorta) in which this system would be deployed.

In this study, the impacts of diaphragm and fixed beam energy harvesting structures have been compared to results for a cantilever design placed perpendicular to the flow. The diaphragm and fixed beam were placed in line with the flow and on the lower surface of the artery and the results of this investigation provided information on the best geometries and configuration for in-vivo energy harvesting so that it can achieve maximum power and minimum disruption to flow. Research has been conducted by Khaligh et al. into the use of square and circular structures for energy harvesting in fluid but this did not take into account the impact of putting the device into the flow [10].

II. MODELLING RESULTS

The simulations were designed using COMSOL Multiphysics[R]. They consist of two models; the first determined the effect of the device on the change in fluid as well as the stress on the device due to the pressure of the fluid. For these models the artery is modelled as an empty tube (i.e. a tube with fluid but without a piezoelectric device) of 1cm diameter and the fluid chosen was water order to mirror the experimental set-up where water is also used.

978-1-4673-4845-4/13 $31.00 © 2013 IEEE

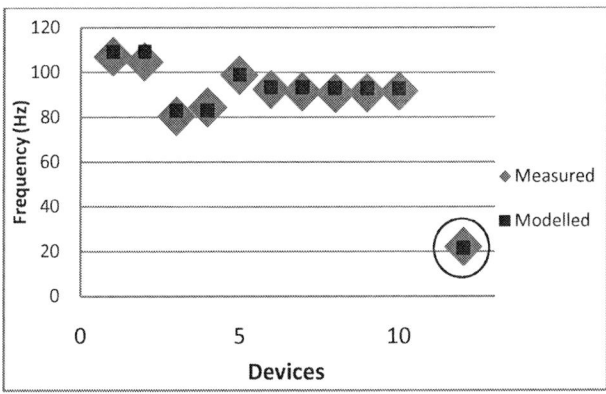

Figure 2: Comparison of Modelled and Measured Resonant Frequencies for Devices of 500nm AlN thickness

The devices were considered to be purely Si structures in the first model to reduce the complexity and the solution time. A time dependent study was completed over 1second to determine the results of a single heartbeat at 60bpm (typical resting heart rate). This information was then used to determine the voltage output by applying the same stress, as a boundary load, to models of the full piezoelectric structure with all the layers. For these devices the Si thickness was 50μm, Ti thickness was 100nm and Al thickness of 1μm. The AlN thickness was 500nm for devices fabricated on wafer 5 and 1μm for wafer 6.

Figure 2 shows the resonant frequencies of the modelled devices as well as the measured resonant frequency results obtained in air from the fabricated devices. The devices consist of cantilevers and triangular beam structures as well as more unusual devices. This is a good indication of the accuracies of the models when compared to actual fabricated devices (each point represents the resonant frequency of a fabricated device and the calculated value from the modelled results). As can be seen here, the results are very close with a maximum error of 4%. The point circled shows a very interesting result. During fabrication one of the devices broke (a doubly tethered device broke at one tether and became a cantilever) but the resulting device could still be measured. This new device was also modelled to determine whether the model could predict the behaviour and as is seen here the measured and modelled results are in very good agreement.

Figure 3 shows the results of modelling a tube (i.e. an artery) with fluid flowing through it which has a pulsating excitation but without a piezoelectric device. This model was designed to simulate an artery without an energy harvesting device present for comparison purposes. The results of this model are then compared to the results obtained when various piezoelectric devices are introduced into the flow. These then give approximations on the effect which can be expected on the flow in terms of the change in velocity and the increase in vorticity which will be discussed here.

These two are critical features for the fluid in an artery as well as a good factor for establishing the lifetime of the test structures, i.e. the more effect on the flow the more likely that particles will be deposited on the device and the structure will fail earlier than when there is less effect on the flow. Blood consists of many small particles and when the blood flow is slowed down these particles can be deposited which could cause a build-up on the devices and cause them to fail over time. This is also a danger factor for causing a deposition of plaque in an artery which can lead to major health problems and require surgery to reopen the artery. For these reasons the change in the flow of the fluid needs to be investigated and using accurate numerical models are a good way to achieve this.

Figure 4 shows the velocity profiles (i.e. the areas of various different velocities of the fluid) for the fluid in the artery when each of the piezoelectric devices is placed into it. The tubes are the same length and diameter in all cases. The aim of this figure is to give a view of the full length of the fixed beam so the structures look as if they are all of different lengths. The velocity profile is changed by the fluid impacting with the device and flowing around it. This also reduces the velocity as shown in table 1, which shows the maximum velocities achieved in the system. The profile shows the least amount of change for the fixed beam device, however, table 1 shows that this device causes the greatest decrease in the maximum velocity of the fluid. This apparent decrease in velocity is worrying because, as previously stated, a slow-down in the velocity can ultimately cause particle build up which could lead to health problems in the patient. The diaphragm also causes minimum change to the velocity profile and the change in velocity is the lowest seen for all the devices. This suggests that a device of this design would be the best solution to the

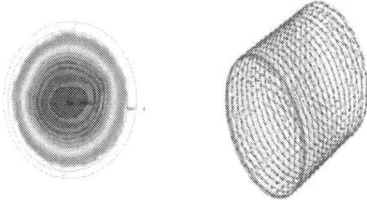

Figure 3: Velocity and Vorticity Profile for an Empty Artery

978-1-4673-4845-4/13 $31.00 © 2013 IEEE

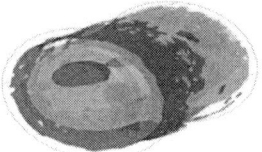 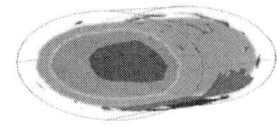 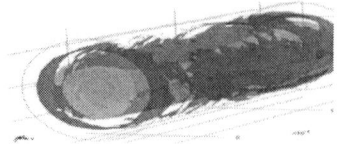

Figure 4: Velocity profile of fluid for cantilever structure, diaphragm and fixed beam obstruction in flow at t=0.6s

problem. The cantilever causes a large change in the velocity profile as well as decreasing the maximum velocity significantly.

Device (AlN thickness = 500nm)	Maximum Velocity at t=0.6s (m/s)	Voltage at t = 0.6s (V)
No Device (Empty Artery)	0.4602	-
Cantilever	0.39018	5.5
Diaphragm	0.4326	1.8
Fixed Beam	0.234323	0.9

Table 1: Maximum Velocity of Fluid and Voltage of Device at t=0.6s

Figure 5 shows the vorticity of the fluid when obstructed by the piezoelectric structures. It is important to examine this because the vorticity provides a good description of the effect that the device has on the flow. The greater the difference in vorticity, the more eddies are created in the fluid. These eddies interact with each other and can cause a major turbulence in the flow of the fluid. This is a factor which affects the fluid velocity as well as inducing the possible build-up of particles both from particles released from the blood and particles which have

been removed from the artery wall due to the impact of the eddies. What is clear from these images is that the diaphragm causes the least change in vorticity with the fixed beam also remaining pretty stable. However, the cantilever causes a big change in the vorticity and so is therefore not an ideal structure for energy harvesting from the fluid flow.

Figure 6 shows the stress on the devices due to the fluid flow. The maximum point of stress is at the anchor points, the tethered end of the beam for the cantilever, the edges of the diaphragm and the long edges of the fixed beam. This is excellent for energy harvesting because this is the point when the energy will be extracted. The geometry of the structure allows for more or less area of piezoelectric material. For the cantilever structure the AlN lies along the beam only and is not present on the mass. This means that only a small area is available for energy harvesting. However, the stress on the structure is very high due to the large mass. The diaphragm structure has an AlN layer over the entire structure and so there is a much greater area for energy harvesting although the stress is reduced. The fixed beam structure has the largest area of AlN as well as high stress due to the effect of the pulsating flow over such a large area.

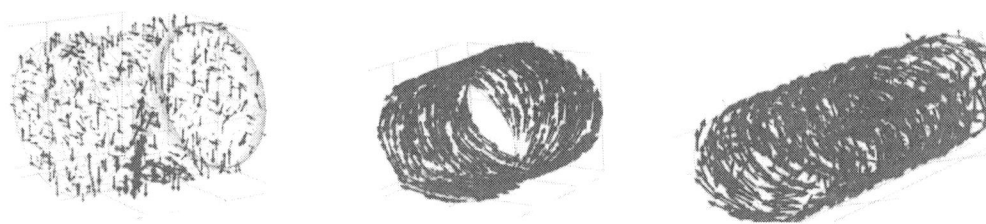

Figure 5: Vorticity of flow in artery for a cantilever, diaphragm and fixed beam

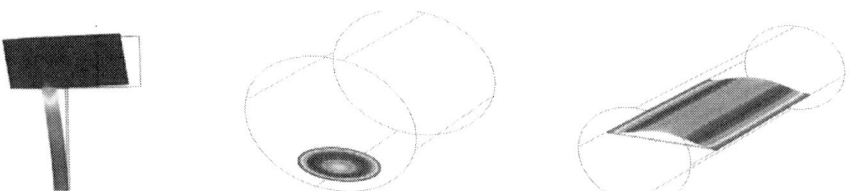

Figure 6: Stress on cantilever, diaphragm and fixed beam from fluid flow

978-1-4673-4845-4/13 $31.00 © 2013 IEEE

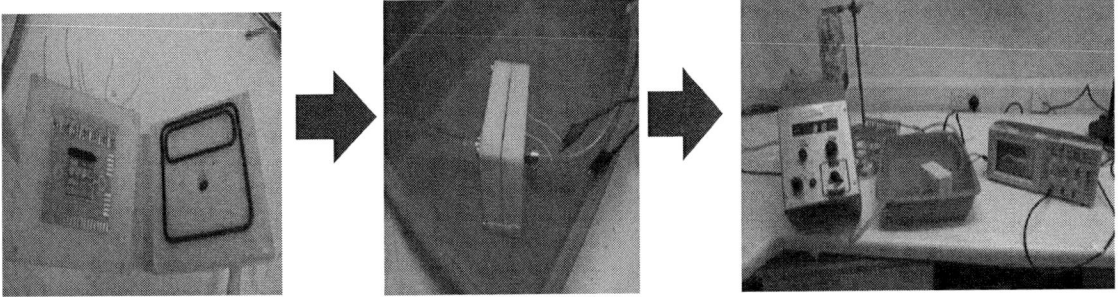

Figure 7: Experimental Set-Up

III. EXPERIMENTAL RESULTS

Figure 7 shows the experimental set-up. The devices (full die on PCB) are placed in the holder. The two halves are connected and the holder is placed in shallow water. The fluid flows into and out of the holder through the plastic tubing shown. This allows the fluid to flow over and through the various devices. The tubes are connected to the perfusion machine which provides the excitation of the devices by creating a pulsating flow. The velocity can be adjusted as well as, to a lesser extent, the pressure. The voltage out is read by connecting the wires (soldered to the PCB) to an oscilloscope.

A range of test wafers were fabricated and this paper presents results from two sample wafers, referred to as wafer 5 (AlN thickness 500nm) and wafer 6 (AlN thickness 1µm). Wafer 5 gave an output of 900mVpp when the perfusion machine was operating at the highest pressure achievable (30mmHg systolic and 25mmHg diastolic). The noise in this case was 500mVpp. This result was the same for all devices which was not what was expected from the results of the models. Figure 8 shows a sketch of the output waveform obtained. Over a 1s period the output increases to a maximum and then returns to a minimum in the next second. This shows a periodic output of 1s which is to be expected. Also there is a significant amount of noise in the system and this is also to be expected due to the environment in which the device is placed. For wafer 6 the output was 1.5Vpp with a noise level of 500mVpp. The noise level is the same in both cases which may be due to the relatively small difference in the thickness of the AlN compared to the full structure, even though the increase in AlN thickness is significant from the energy harvesting point of view. This increase in output is due to the increase in thickness of the AlN. The devices are stiff due to the thick Si which may be one reason for the low output voltage and future designs would benefit from a more flexible substrate such as a polymer. To investigate the use of a polymer and determine an accurate reading of the pressure a PVDF film with a 2mm hole, as shown in figure 8, was placed in the flow. The results from this were compared to those of a FEM and it was determined that the pressure in the system was $P_{max} = 30$mmHg and $P_{min} = 25$mmHg. The output from the PVDF was about 1.5Vpp and this corresponded to a pressure profile of 30mmHg to 25mmHg. This was also compared to the pressure sensor on the system and found to show the same value. This is less than that which is experienced by the device in the blood flow (typically 80 to 120mmHg).

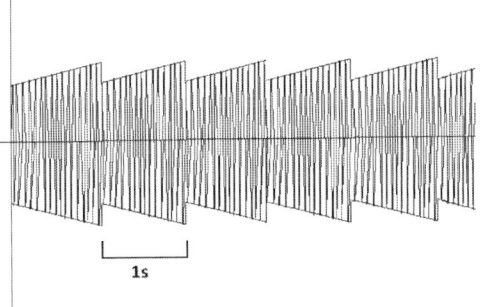

Figure 8: Typical Output Waveform

Figure 8: PVDF Test Structure

IV. CONCLUSION

The energy harvesting test structures performed as expected from the models when used in air and it was found that a good fit between measured and simulated characteristics was established. Due to the low pressure capabilities of the experimental set-up, the performance of the models was very important because this gave a view of what results could be achieved at higher pressure. This also showed the limitations of a model when compared to a fabricated device which can provide more insights into excitation problems as well as such issues as device breaking which are difficult to determine through modelling alone. The results obtained from the test structure did show that increasing the AlN surface area increases the voltage output. The PVDF device and model provided information on the pressure of the flow. Due to the limitations on the precision of the pressure sensor it was found that the PVDF could be used to give a clearer reading of the exact values of the pressure.

The modelled structures provided information on the expected performance of the devices at full blood pressure as well as information on the effect they had on flow. Future designs on the test structures will include devices using a flexible polymer instead of Si which is very stiff. This should increase the usefulness of the test set-up. Further developments will be to consider how to increase the pressure available. Currently the diaphragm structure looks to be the most viable for energy harvesting with minimum effect on the flow and future generations will focus on this structure. Future tests and modelling will be conducted using a fluid closer to the viscosity of blood for more accurate results, however, it is not expected that this will have a large effect on the results as the pressure is the major factor affecting the stress on the devices.

V. ACKNOWLEDGEMENT

This work was funded by the Collaborative Centre for Applied Nanotechnology (CCAN) as part of the Government's Strategy for Science Technology and Innovation through Enterprise Ireland and the International Centre for Graduate Education in Micro- and Nano-Engineering (ICGEE).

VI. REFERENCES

[1] A. Rivadeneyra, R. O'Keeffe, N. Jackson, J. Lopez-Villanueva, M. O'Neill and A. Mathewson, "Frequency response variants of a cantilever beam," in *The International Conference on Synthesis, Modeling, Analysis and Simulation Methods and Applications to Circuit Design*, Seville, Spain, 2012.

[2] COMSOL, "COMSOL," COMSOL Multiphysics R, [Online]. Available: www.comsol.com. [Accessed 11 January 2013].

[3] S. Priya, "Advances in Energy Harvesting using Low Profile Piezoelectric Transducers," *Journal of Electroceramics*, vol. 19, no. 1, pp. 165-182, 2007.

[4] S. R. Anton and H. A. Sodano, "A Review of Power Harvesting using Piezoelectric Materials (2003-2006)," *Smart Materials and Structures*, vol. 16, no. 3, pp. 1-21, 2007.

[5] S. Matova, R. Elfrink, R. Vullers and R. van Schaijk, "Harvesting energy from airflow with a michromachined piezoelectric harvester inside a Helmholtz resonator," *Journal of Micromechanics and Microengineering*, vol. 21, no. 10, pp. 1-6, 19 July 2011.

[6] G. Taylor, J. Burns, S. Kammann, W. Powers and T. Welsh, "The Energy Harvesting Eel: A Small Subsurface Ocean/River Power Generator," *IEEE Journal of Oceanic Engineering*, vol. 26, no. 4, pp. 539-547, October 2001.

[7] R. O'Keeffe, *Investigation into the use of Accelerometers for the Detection of Restenosis*, Masters Thesis: University College Cork, 2010.

[8] N. Jackson, R. O'Keeffe, R. O'Leary, F. Waldron and A. Mathewson, "A Diaphragm based Piezoelectric AlN Film Quality Test Structure," in *International Conference on Microelectronic Test Structures*, San Diego, CA, USA, 2012.

[9] N. Jackson, R. O'Keeffe, F. Waldron, M. O'Neill and A. Mathewson, "CMOS compatible Low-Frequency Aluminium Nitride MEMS Piezoelectric Energy Harvesting Device," in *SPIE Microtechnologies*, Grenoble, France, 2013.

[10] G. Coppola and K. Liu, "Study of Compliance Mismatch within a Stented Artery," in *COMSOL Conference*, Boston, 2008.

[11] A. Khaligh, P. Zeng and C. Zheng, "Kinetic Energy Harvesting Using Piezoelectric and Electromagnetic Technologies - State of the Art," *IEEE Transactions on Industrial Electronics*, vol. 57, no. 3, pp. 850-860, 10 June 2009.

A new measurement set-up to investigate the charge trapping phenomena in RF MEMS packaged switches

Marco Barbato, Valentina Giliberto and Gaudenzio Meneghesso

Department of Information Engineering, University of Padova, Via Gradenigo 6/B, 35131, Padova, Italy
Italian Universities Nano-Electronics Team (IUNET), 40125 Bologna, Italy
gaudenzio.meneghesso@dei.unipd.it

Abstract—In this work we present a new measurement set up able to predict the lifetime of packaged ohmic RF MEMS submitted to long actuation periods. Experimental results were carried out for a relatively long time period in order to verify the degradation law relates to charge trapping and stiction on cantilever and clamped-clamped switches. Thanks to the use of a microcontroller we have been able to reach a complete control of the timing during the stress phase. Furthermore, the characterization phase has been remarkably reduced in order to minimally influence the charge trapping during the characterization of stressed device. Results are carried out on two different MEMS designs (clamped-clamped and cantilever configurations).

Keywords—ohmic RF MEMS; charge trapping; reliability; long term actuation.

I. INTRODUCTION

Micro-Electro-Mechanical-Systems (MEMS) for radio frequency (RF) applications are becoming interesting devices for their better performances than solid state transistor. In particular we mentioned low power consumption, high frequency signals, good linearity and low production cost. Despite the many advantages, RF-MEMS switches have not yet taken the market because of their problems related to reliability issues [1]. Moreover the lack of standardization in the reliability tests bring to the difficulty in predicting the lifetime and so in predicting the devices performances compared to standard silicon solid state devices.

Reliability issue in RF-MEMS is a big challenge due to the large variety of problems: (i) mechanical: shocks, vibrations, bending of suspended part [2]; (ii) electrical: stiction due to charge trapping, microwelding and power handling [3]; (iii) environmental: humidity, temperature and radiation [4]. One of the most critical problem related to the electrical functionality is the charge trapping. The latter is mainly due to the high electric field present in the MEMS structure during the actuation phase (> 1 MV/m). This filed is sufficiently high to induce charge trapping in the oxide layer of the device [5].

In the last years the development of RF-MEMS switches went toward dielectric-less devices with the aim of reducing the charges trapping phenomenon. However the charge trapping occurs also in the oxide grown over the substrate around the actuator and the charge trapping problem is also present in dielectric-less devices [6]. Charge trapping induces narrowing and shift of the actuation and release voltages of the switches. When the charge trapped is sufficiently high the release voltage can become negative leading to stiction (i.e. devices remain actuated when no voltage are applied) [7].

Standard wafer-based measurement processes take several seconds to characterize a device. In this work we develop an appropriate measurement set up using a microcontroller able to stress the devices and to characterize the devices in only 1.6 seconds, period that minimally influences the long stress test. The duration of the characterization period can be varied from 0.8 s to 28.8 s setting opportunely a prescaler in the microcontroller. This technique guarantee a good timing over all the measurement period and a good synchronization in the determination of the actuation and release voltages. We show some experimental results related to lifetime test in term of charge trapping phenomenon on dielectric-less RF-MEMS switches. Moreover the measurement set up can be used on packaged devices, which cannot be made with standard wafer-lever measurement set up. This makes possible a prolonged study (even months) without the use of expensive equipment.

II. DEVICES DESCRIPTION

This paper is based on the reliability analysis of two types of dielectric-less ohmic RF-MEMS. The first is clamped-clamped and the second is a cantilever. A photo of the devices is shown in Figure 1 (a) and Figure 1 (b). Devices stressed are manufactured by FBK IRST (Trento, Italy); the production process is composed of an eight masks surface micromachining process [8]. A schematic representation of the deposition process is presented in Figure 1 (c).

Cantilever switches are based on a suspended membrane (110 μm wide and 190 μm long) above an interrupted coplanar line. This membrane, reinforced in its central part with thicker

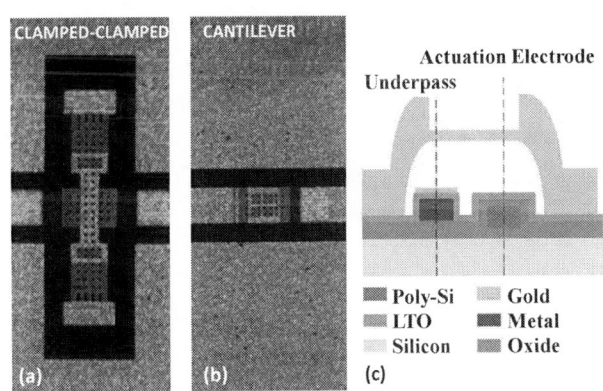

Figure. 1.   Photo of devices and technological process. (a) Clamped-Clamped configuration, (b) cantilever configuration and (c) construction process.

978-1-4673-4845-4/13 $31.00 © 2013 IEEE

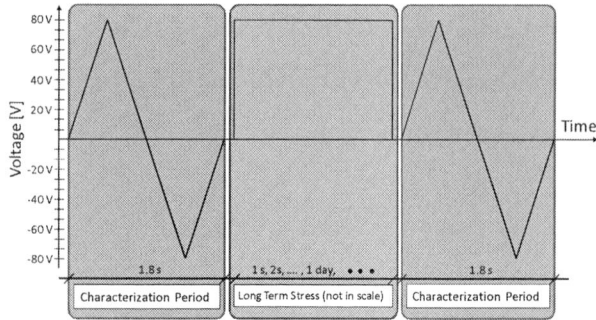

Figure. 2. Schematic representation of the stress procedure.

gold, is anchored at one end and closes the other side. Clamped-Clamped devices have a different structure: the gold membrane is 100 µm wide and 500 µm long, it is 90 degree rotated respect to the coplanar waveguide and the contact with the coplanar line occurs in both lateral points of the bridge. The two configurations are very different and manifest different performances to long actuation stress test.

Generally speaking one could think that dielectric less devices don't suffer from charge trapping. This is false because the charge can remain blocked in the oxide layer over the substrate all around the actuation pad. This is confirmed by previous analysis [9-10].

III. EXPERIMETAL SET UP

The measurement set up was appositely developed with the aim of optimizing the stress procedure. The stress procedure timeline is presented in Figure 2: the devices are submitted to a time increasing fixed voltage stress. After each time step a "characterization period" was performed in order to investigate any deviation of the main electrical parameters (V_{ACT+}, V_{REL+}, V_{ACT-}, V_{REL-}) from the initial ones. It is very important to maintain the time between "characterization period" and "long term stress period" as short as possible in order to have less impact on the charge detrapping phenomenon. With the aim of

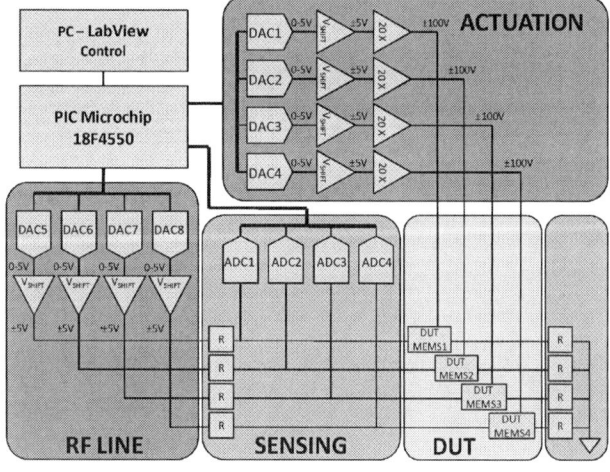

Figure. 3. Schematic view of the PCB utilized to stress the devices.

maintaining all these important restrictions we develop a microcontroller based board able to generate the triangular waveform during the "characterization period" and to stress the devices during the "long stress term period". A schematic view of the circuit is shown in Figure 3. We can distinguish four different areas: the "ACTUATION" part that generates the signal able to actuate the devices, the "RF LINE" part that generates the signal for the RF input of the device, the "SENSING" part that monitors the aperture and the closure of the switch and finally the "DUT".

The "ACTUATION" part is obtained from the series of the output port of the microcontroller that send the digitally voltage to a DAC converter, this generates an analog voltage that is shifted and amplified by two consecutive operational amplifier in order to obtain a voltage of ± 100 V. The "RF LINE" part is obtained from the series of the output port of the microcontroller with a DAC converter and an operational amplifier obtaining a voltage of 0-5 V. The "SENSING" part is composed of two resistor that limit the current flow over the RF path of the devices and an ADC (internal to the microcontroller) that acquire the voltage at the input port of the device. The acquired voltage is imposed by the microcontroller in case of open device (we choose arbitrarily 3.8 V) instead is obtained from the resistive partitioning in case of close device (1.9 V).

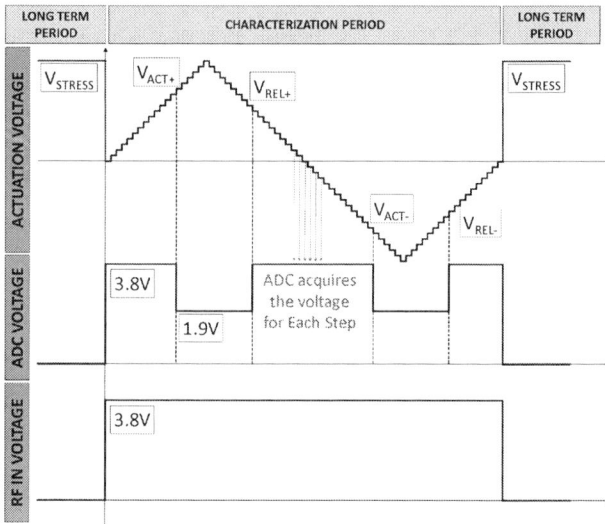

Figure. 4. Schematic representation of the stress procedure.

The four parts of the board are synchronized by the microcontroller that generates the right signals for each part of the circuit. During the "characterization period" the microcontroller generates an opportune staircase waveform and for each step acquires the information regarding the state of the switch (Figure 4). It can be seen that increasing the actuation voltage the ADC acquire 3.8 V whereas the device is open, after the actuation voltage (V_{ACT+}), the devices goes into down position and the acquired voltage (1.9 V) is obtained from the partition over the two resistors. In the same way the positive

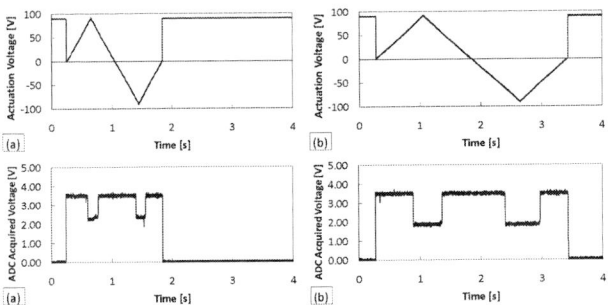

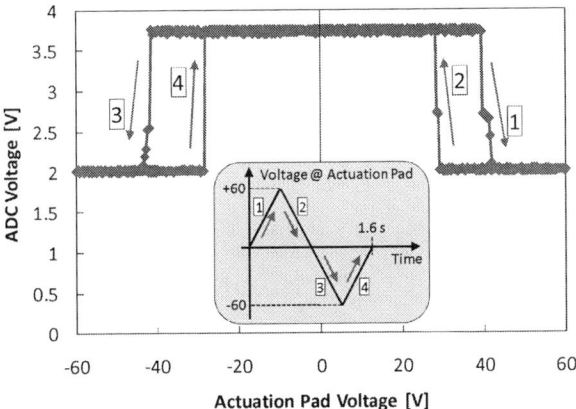

Figure. 5. Biasing voltages acquired with a DSO during a measurement process for two velocities: (a) 1.6 seconds measurement process and (b) 3.2 seconds measurement process.

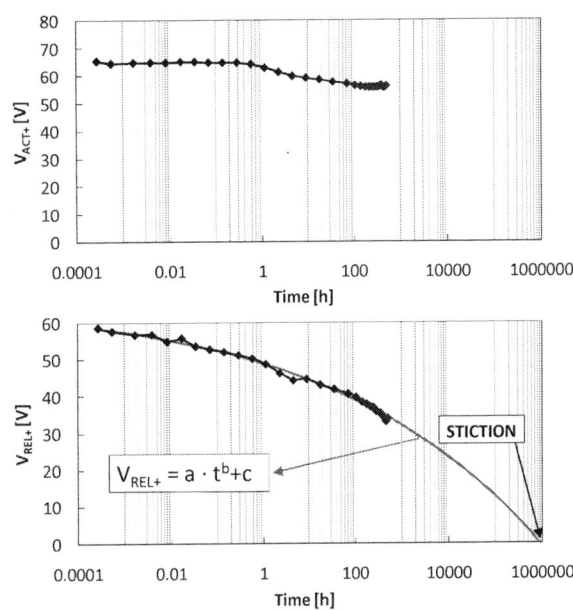

Figure. 6. Hysteresis graph obtained with the PCB during the "characterization period".

release voltage, the negative actuation voltage and the negative release voltage can be extracted.

The velocity of the measurement set up can be seen from Figure 5 where we show two different DSO acquisitions of the biasing voltages and the relative ADC acquired signal. The measures are referred to two different velocity: 1.6 s and 3.2 s. This parameter can be varied from a minimum of 0.8 s to a maximum of 28.8 s by acting on the microcontroller programming. In this work we consider a constant characterization time of 1.6 s that was considered sufficiently short to affect minimally the charge trapping phenomenon and long enough to not influence the dynamic mechanical response of the moving membrane.

With the acquired signal, we can reconstruct a graph like in Figure 6 where the positive and negative actuation and release voltages are easily visible. This is the classical hysteresis characteristic of a RF MEMS switch: increasing the voltage the positive actuation voltage is reached (point 1 in figure) and the switch goes in down position, then decreasing the voltage applied to the actuation pad, with a certain hysteresis, the positive release voltage is reached (point 2 in figure) and the switch goes in up position. The same behavior is visible for negative voltages (points 3 and 4 in figure). It can be clearly

noticed that the characterization is obtained monitoring the ADC acquired signal from the microcontroller.

IV. EXPERIMETAL RESULTS

The positive actuation and release voltages of two samples (Clamped-Clamped and cantilever), monitored during the stress, are visible in Figure 7 and in Figure 8. The stress times were 20 days for the Clamped-Clamped device and about two days for cantilever one. We use a power law fitting in order to

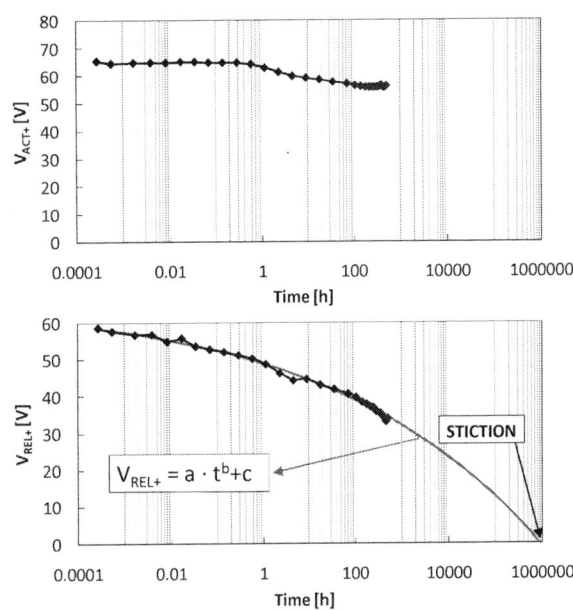

Figure. 7. Evolution of V_{ACT+} and V_{REL+} about a Clamped-Clamped device.

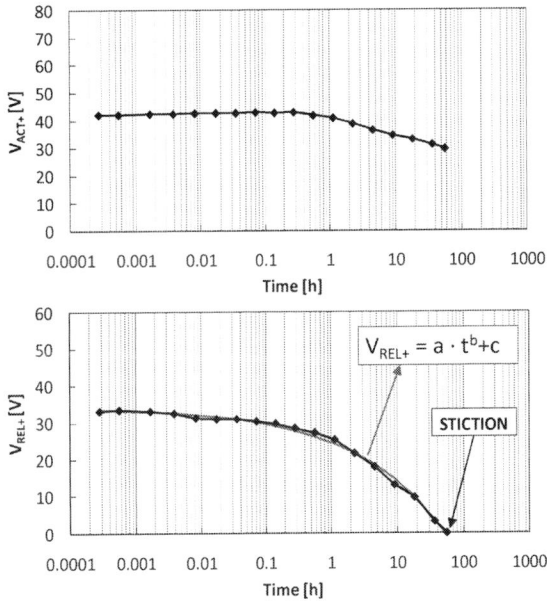

Figure. 8. Evolution of V_{ACT+} and V_{REL+} about a Cantilever device.

TABLE I. INITIAL AND FINAL EXPERIMENTAL DATA ABOUT A CLAMPED-CLAMPED DEVICE.

Time [h]	V_{ACT+} [V]	V_{REL+} [V]	V_{ACT-} [V]	V_{REL-} [V]
0.01	65.43	58.60	-62.95	-56.13
509 (21days)	59.22	35.97	-54.57	-32.26

TABLE II. EXTRACTED FITTING PARAMETER ABOUT THE CLAMPED-CLAMPED CONFIGURATION.

Fitting Parameters	a	b	c	R-square	Estimated lifetime
Value	-17.54	0.09735	66.43	0.9944	98.7 years

TABLE III. INITIAL AND FINAL EXPERIMENTAL DATA ABOUT A CANTILEVER DEVICE.

Time [h]	V_{ACT+} [V]	V_{REL+} [V]	V_{ACT-} [V]	V_{REL-} [V]
0.01	42.06	33.09	-32.47	-27.22
≈ 51	29.69	0	Stiction	Stiction

TABLE IV. EXTRACTED FITTING PARAMETER ABOUT THE CANTILEVER CONFIGURATION.

Fitting Parameters	a	b	c	R-square	Estimated lifetime
Value	-9.499	0.325	34.15	0.9954	51 hours

extrapolate the stiction time [11-12]:

$$V_{REL+} = a \cdot t^b + c \qquad (1)$$

We can see that the actuation voltage is minimally affected by the charge trapping mechanism (see Figures 7 and 8) while the release voltage is drastically decreased during the long actuation term period. This negatively affects the lifetime of the device and can bring to stiction occurrence. The phenomenon is reach after 51 hours of continuous biasing in cantilever configuration instead in Clamped-Clamped configuration the power law fitting estimates a stiction time of about 98 years.

The results are summarized in Table 1, 2, 3 and 4 for the two typologies respectively. The parameter c is related to the initial value of the release voltage (V_{REL+}) and it is not an important parameter to evaluate the robustness of the devices under test while b indicates the goodness to charge trapping issue because it is related to the slope of the curve: lower value of b is associated to a minor trend to accumulate charge. Clamped-Clamped design presents a b parameter that is one

order of magnitude lower respect to cantilever design. The fitting goodness is guarantee by the R-square parameter that is close to one in both cases. The Clamped-Clamped configuration presents better results due to its design based on a major restoring force.

In order to identify the deviation of release and actuation voltages we carried out an analysis like in [13] where the effect of the charge trapping is divides in narrowing and shifting of actuation and releases voltages. This kind of analysis allows us to better understand the entrapment mechanisms.

Let us define the difference between the positive and negative actuation and release voltages at the time step t_i respectively as:

$$\Delta V_{ACT}(t_i) = V_{ACT+}(t_i) - V_{ACT-}(t_i) \qquad (2)$$

$$\Delta V_{REL}(t_i) = V_{REL+}(t_i) - V_{REL-}(t_i) \qquad (3)$$

the narrowing of the actuation and release voltage windows can then be defined as:

$$\Delta V_{ACT\_NAR}(t_i) = \Delta V_{ACT}(fresh) - \Delta V_{ACT}(t_i) \qquad (4)$$

$$\Delta V_{REL\_NAR}(t_i) = \Delta V_{REL}(fresh) - \Delta V_{REL}(t_i) \qquad (5)$$

and the shifts of the center of symmetry of $\Delta V_{ACT}(t_i)$ and $\Delta V_{REL}(t_i)$ respectively as:

$$\Delta V_{ACT\_SHIFT}(t_i) = \frac{V_{ACT+}(0)+V_{ACT-}(0)}{2} - \frac{V_{ACT+}(t_i)+V_{ACT-}(t_i)}{2} \qquad (6)$$

$$\Delta V_{REL\_SHIFT}(t_i) = \frac{V_{REL+}(0)+V_{REL-}(0)}{2} - \frac{V_{REL+}(t_i)+V_{REL-}(t_i)}{2} \qquad (7)$$

The results presented in Figures 9 and 10 show that narrowing is more relevant respect to shift both in the actuation voltage and release voltage analysis (shifting is around 2 volts maximum instead narrowing is respectively 18 volts maximum about actuation voltage and 55 volts maximum about release voltage).This is in accordance with the dielectric-less design. In fact the use of an oxide layer over the actuator is compatible with shifting [14] instead dielectric-less switches present major narrowing of both actuation and release voltages because the trapped charges are dislocated respect to the actuation structure [15-16]. In this case the narrowing is the predominant effect respect to the shifting one. Results show that to obtain comparable narrowing of actuation and release voltages for both typologies, the stress time is one order of magnitude longer for Clamped-Clamped devices that are better candidates for long actuation periods.

The results show how a dedicated set up based on a microcontroller can be used to stress the device for long time. Future developments will allow us to analyze the problem as a function of temperature and then to extract any degradation kinetics related to temperature. This can be done easily with packaged devices and the set up presented in this work.

978-1-4673-4845-4/13 $31.00 © 2013 IEEE

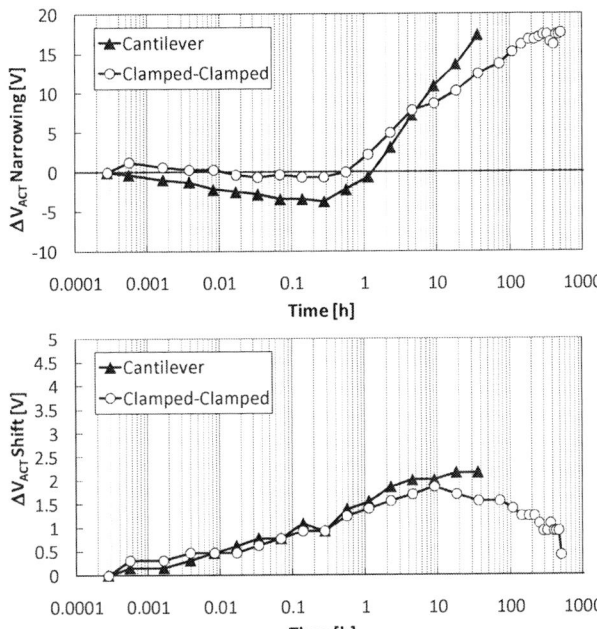

Figure. 9. ΔV_{ACT} Narrowing and ΔV_{ACT} Shift of the two typologies.

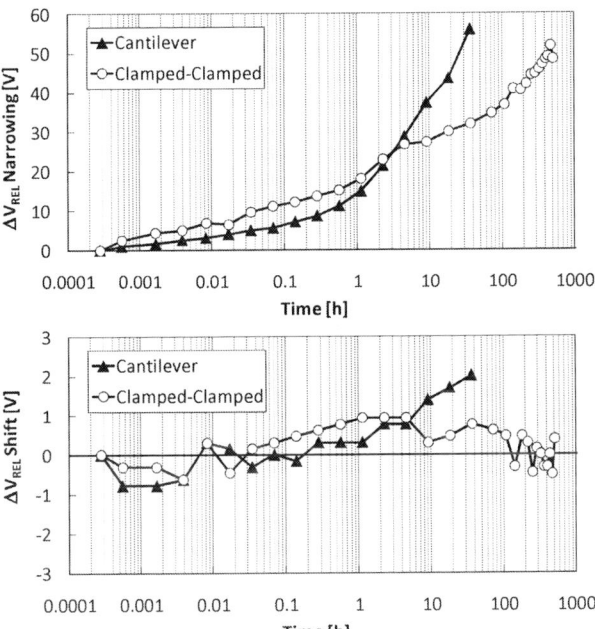

Figure. 10. ΔV_{REL} Narrowing and ΔV_{REL} Shift of the two typologies.

CONCLUSIONS

The analysis of charge trapping mechanism was presented with experimental results carried out for long time period (up to 20 days). The stress set up can be applied to packaged devices and the predicted lifetime is calculated with a fitting

procedure: a power law fitting is used to predict the stiction point. The results show that the choice of different designs (i.e. clamped-clamped and cantilever configurations) is a fundamental parameter to improve the reliability in term of charge trapping issue.

ACKNOWLEDGMENT

The authors would like to thank B. Margesin and Viviana Mulloni, from FBK-IRST Microsystems Division, Povo (TN), for the fabrication of the devices investigated in this work. This work was partially supported by the European Space Agency project ITT AO/1-5288/06/NL/GLC, "High Reliability MEMS Redundancy Switch" and in part by ENIAC project END "Models, Solutions, Methods and Tools for Energy Aware Design". The END project has received funding from the ENIAC Joint Undertaking under Grant Agreement No. 120214 and from the national programmes/funding authorities of Belgium, Greece, Italy, and Slovakia.

REFERENCES

[1] J. DeNatale, R. Mihailovich, "RF MEMS reliability", *TRANSDUCERS, Solid-State Sensors, Actuators and Microsystems, 12th International Conference on, 2003* , vol.2, pp. 943- 946, 8-12 June 2003.

[2] L. Xiaoguang J. Small, D. Berdy, L.P.B. Katehi, W.J. Chappell, D. Peroulis, "Impact of Mechanical Vibration on the Performance of RF MEMS Evanescent-Mode Tunable Resonators", *Microwave and Wireless Components Letters, IEEE* , vol.21, no.8, pp.406-408, Aug. 2011.

[3] S. Melle, C. Bordas, D. Dubuc, K. Grenier, O. Vendier, et al., "Investigation of Stiction Effect in Electrostatic Actuated RF MEMS Devices", *Silicon Monolithic Integrated Circuits in RF Systems, 2007 Topical Meeting on* , pp.173-176, 10-12 Jan. 2007.

[4] Y. Xiaobin P. Zhen J.C.M. Hwang, D. Forehand, C.L. Goldsmith, "Temperature Acceleration of Dielectric Charging in RF MEMS Capacitive Switches", *Microwave Symposium Digest, 2006. IEEE MTT-S International*, pp.47-50, 11-16 June 2006.

[5] N. Tavassolian, M. Koutsoureli, G. Papaioannou, B. Lacroix, J. Papapolymerou, "Dielectric charging in capacitive RF MEMS switches: The effect of electric stress", *Microwave Conference Proceedings (APMC), 2010 Asia-Pacific*, pp.1833-1836, 7-10 Dec. 2010.

[6] D. Mardivirin, A. Pothier, M. El Khatib, A. Crunteanu, O. Vendier, P. Blondy, "Reliability of Dielectric Less Electrostatic Actuators in RF-MEMS Ohmic Switches", *Microwave Integrated Circuit Conference, 2008. EuMIC 2008. European*, pp.490-493, 27-28 Oct. 2008.

[7] W.M. van Spengen, R. Puers, R. Mertens, I. De Wolf, "Experimental characterization of stiction due to charging in RF MEMS", *Electron Devices Meeting, 2002. IEDM '02. International* , pp.901-904, 8-11 Dec. 2002.

[8] J. Iannacci, F. Giacomazzi, S. Colpo, B. Margesin, M. Bartek, "A general purpose reconfigurable MEMS-based attenuator for Radio Frequency and microwave applications", *EUROCON 2009, EUROCON '09, IEEE*, pp. 1197-1205, 18-23 May 2009.

[9] A. Tazzoli, E. Autizi, M. Barbato, F. Solazzi, J. Iannacci, et al., "Impact of Continuous Actuation on the Reliability of Dielectric-less Ohmic RF-MEMS Switches", *MEMSWAVE 2009*, pp. 129-132, trento, Italy.

[10] X. Rottenberg, B. Nauwelaers, W. De Raedt, H.A.C. Tilmans, "Distributed dielectric charging and its impact on RF MEMS devices", *Microwave Conference, 2004. 34th European* , vol.1, pp.77-80, 14 Oct. 2004.

[11] D. Mardivirin, A. Pothier, A. Crunteanu, B. Vialle, P. Blondy, "Charging in Dielectricless Capacitive RF-MEMS Switches", *Microwave Theory and Techniques, IEEE Transactions on*, vol.57, no.1, pp.231-236, Jan. 2009.

[12] A. K. Jonscher, "Dielectric relaxation in solids", *J. Phys. D, Appl.Phys.*, vol. 32, pp. R57–R70, 1999.

[13] A. Tazzoli, E. Autizi, M. Barbato, G. Meneghesso, et al., "Evolution of Electrical Parameters of Dielectric-less Ohmic RF-MEMS Switches during Continuous Actuation Stress", *Solid State Device Research Conference, 2009. ESSDERC '09. Proceedings of the European*, pp.343-346, 14-18 Sept. 2009.

[14] S. Melle, D. De Conto, D. Dubuc, K. Grenier, O. Vendier, et al., "Reliability modeling of capacitive RF MEMS", *Microwave Theory and Techniques, IEEE Transactions on* , vol.53, no.11, pp. 3482- 3488, Nov. 2005.

[15] X, Rottenberg, I. De Wolf, B.K.J.C. Nauwelaers, W. De Raedt, H.A.C. Tilmans, "Analytical Model of the DC Actuation of Electrostatic MEMS Devices With Distributed Dielectric Charging and Nonplanar Electrodes", *Microelectromechanical Systems, Journal of* , vol.16, no.5, pp.1243-1253, Oct. 2007.

[16] R.W. Herfst, P.G. Steeneken, J. Schmitz, A.J.G. Mank, M. van Gils, "Kelvin probe study of laterally inhomogeneous dielectric charging and charge diffusion in RF MEMS capacitive switches", *Reliability Physics Symposium, 2008. IRPS 2008. IEEE International*, pp.492-495, April 27 2008-May 1 2008.

SESSION 2: TSV and 3D

978-1-4673-4845-4/13 $31.00 © 2013 IEEE

Test Structure and Analysis for Accurate RF-Characterization of Tungsten Through Silicon Via (TSV) Grounding Devices

Volker Blaschke and Hadi Jebory
TowerJazz, Newport Beach, California, USA
volker.blaschke@towerjazz.com

Abstract—We present an analysis on the extraction of the through silicon via (TSV) inductance from single port and two port S-parameter results. The test structure design is shown to significantly impact the extracted value and could cause inaccurate results and subsequently errors in the Spice model if not accounted for. We will show that an analytical model of the return circuit loop that the TSV forms with the test structure, does provide a useful assessment of the accuracy of the measured results. This analysis further provides important input for test structure design and when to use single port or two port test structures for TSV measurement.

Index Terms—S-Parameter, through silicon via (TSV), inductance, grounding device, thru-wafer via (TWV), R-L-C resonator, shunt resonator, return circuit loop, SiGe power amplifier, common emitter configuration.

I. INTRODUCTION

The TSV described in this paper is a tungsten filled rectangular conductor that connects the lowest interconnect metal layer (M1) to the back of the wafer. In a grounding configuration the wafer backside is metallized completing a low impedance path to ground for transistors in common emitter configuration and for shunt capacitors or inductors in matching networks in the design of SiGe power amplifiers for the wireless application space. The TSV as a grounding device must have an impedance as low as possible and meet stringent requirements on the accuracy of particularly the modeled inductance. Any inaccuracy in the inductance value will degrade the efficiency and linear output power of the SiGe power amplifier. A reliable and accurate design process hence requires scalable models for the TSV that correctly model the inductance and resistance of arrays of TSV of variable sizes and are silicon validated.

In this work, the inductance and resistance of the TSV was extracted using S-parameter measurements on single port shunt TSV and two port series R-L-C resonator test structures. The test structure layout and extraction methodology are explained in section II, followed by measurement and extraction results in section III and conclusions in section IV.

II. TEST STRUCTURE DESCRIPTION AND EXTRACTION METHODOLOGY

A. 1-Port Test Structure

A single port test structure was used to extract resistance and inductance over frequency from the S-parameters. The test structure forms a return circuit of parallel conductors of 1, 2, 4, 8 and 16 going TSV and 48 return TSV. A short structure is used for de-embedding of contact and interconnect parasitic leading to the TSV arrays (Fig.1). The resistance and inductance is obtained from the S-parameters using equations:

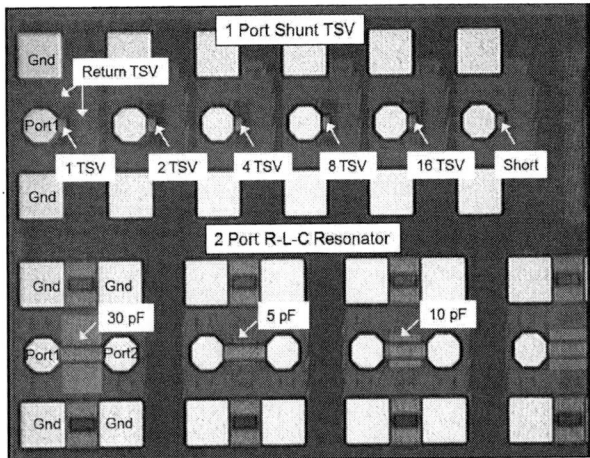

Fig. 1. RF test structures for S-Parameter measurement

$$Z_{11} = \frac{1 + S_{11}}{1 - S_{11}} \cdot 50 \tag{1}$$

$$L_{total} = \frac{imag(Z_{11})}{2 \cdot \pi \cdot f} \tag{2}$$

$$R_{total} = real(Z_{11}) \tag{3}$$

The extracted result will be for the entire return circuit loop of going TSV, return TSV, mutual coupling between the TSV's, spreading inductance / resistance from the backside metal plane and the front-side interconnect. From the inspection of a simplified schematic for this return circuit (Fig. 2), it can be seen that the device of interest (going TSV) cannot be extracted through measurement de-embedding alone. Hence to arrive at an estimate for the inductance of the TSV device, an analytical model needs to be developed that accounts for the components of the return circuit.

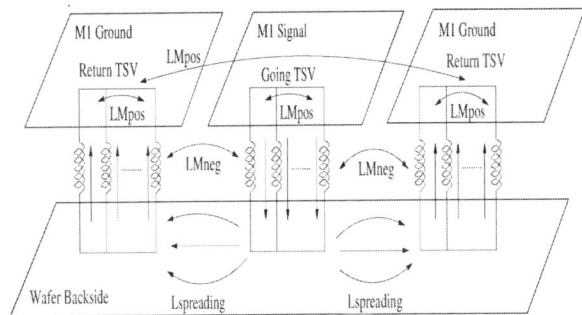

Fig. 2 Schematic for 1-port TSV test structure return circuit

The inductance of the circuit is the summation of the self-inductances L_1 of going TSV, L_2 of return TSV, $L_{spreading}$ for the backside plane spreading inductance, $L_{interconnect}$ for the wafer front-side interconnect inductance, minus twice the mutual inductance M_{12} between going and return TSV arrays:

978-1-4673-4845-4/13 $31.00 © 2013 IEEE

$$L_{total} = L_1 + L_2 - 2 \cdot M_{12} + L_{spreading} + L_{interconnect} \quad (4)$$

The self-inductances of the rectangular conductor TSV and mutual inductance between TSV's can be computed over the partial inductance matrix using Grover's equations [1]. An estimate for the spreading inductance on the wafer backside is obtained using the Biot-Savart law [2]. The interconnect inductance can be directly extracted from the short structure.

B. 2-Port Resonator Test Structure

A two port test structure in T-network configuration was used to indirectly extract inductance and resistance from the S-parameters on a series R-L-C resonator circuit (Fig.1). This method employs a capacitor that is shunted to ground through the TSV. Similar to the 1-port test structure, this circuit also forms a return circuit loop with going and return TSV, however the extraction process is greatly simplified as the series R-L-C resonator can be derived from the impedance Z_{12} of the T-network [3,4,5]. To facilitate the analysis and comparison between 1-port and 2-port results, the location and number of going TSV's and return TSV's was kept constant in both designs. The capacitor was implemented using a 2 fF/um^2 mimcap device and sized to span a capacitance range from 5 pF to 30 pF. This method requires one de-embedding structure for each capacitor size to determine the value of the capacitance in the resonator.

The impedance of the series R-L-C resonator of capacitor and TSV going array is equal to Z_{12} of the T-network:

$$Z_{12} = j\omega L + R + \frac{1}{j\omega C} \quad (5)$$

It's resonance frequency f_o is the frequency at which the phase of Z_{12} changes sign. Knowing resonance frequency and the capacitance value, the equivalent series inductance ESL of the resonator is calculated with:

$$ESL = \frac{1}{\left(f_o \cdot 2 \cdot \pi\right)^2 \cdot C} \quad (6)$$

The equivalent series resistance ESR is the minimum impedance value occurring at the resonance frequency.

III. RESISTANCE AND INDUCTANCE MEASUREMENT RESULTS

A. Resistance and Inductance Measurement on 1-Port RF Test Structure

The resistance and inductance over frequency for array sizes of 1, 2, 4, 8 and 16 TSV was extracted from the measured S-parameters on the 1-port RF test structure and de-embedded by subtracting the impedance of the short structure. The RF dependence of the resistance (Fig. 3) is expected as the width of the tungsten TSV is larger than the skin depth. The inductance (Fig. 4) shows an initial droop due to the skin effect reducing the internal self-inductance, followed by a flat curve up to the 20 GHz end frequency of the S-parameter sweep. This result however includes the return path parasitic as described above. To quantify the contribution of the return path, an analytical model of the test structure was computed and the effect of the placement of the

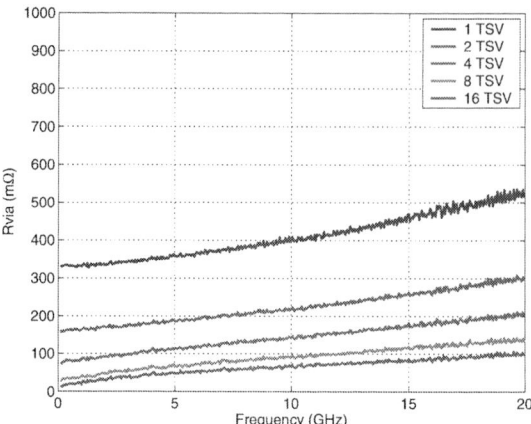

Fig. 3. Measured resistance over frequency for arrays of 1,2,4,8 and 16 TSV

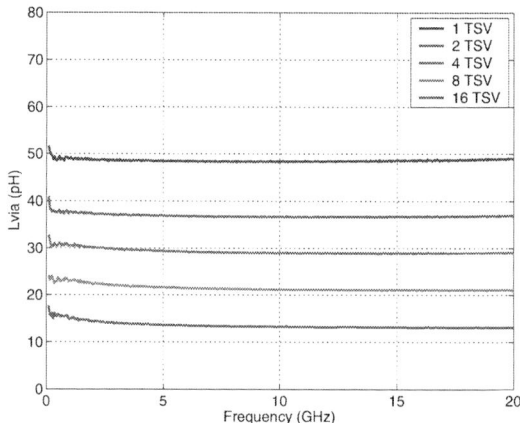

Fig. 4. Measured inductance over frequency for arrays of 1,2,4,8 and 16 TSV

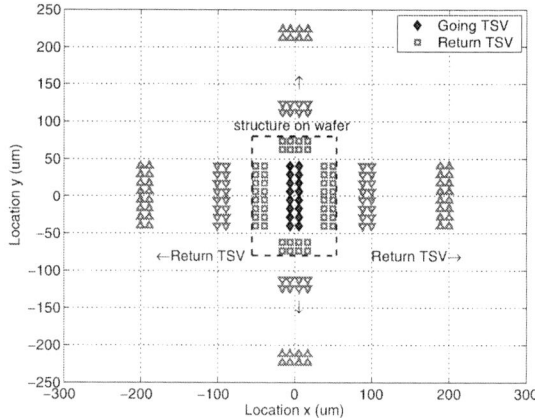

Fig. 5. Location of return TSV's (hollow symbols) in analytical analysis

return TSV investigated (Fig. 5). The analytical model will yield the DC inductance value. For zero offset corresponding to the test structure as measured, the going TSV is 22 pH, return TSV 12.6 pH, negative mutual coupling between TSV arrays is 8 pH and spreading inductance is 3.4 pH (Fig. 6). Since the spreading inductance computation assumes a closed loop path with the front-side of the wafer (interconnect plane in M1), the interconnect path was not de-embedded to prevent over-accounting of the parasitic inductance. The total inductance obtained by summation of the components is 30 pH and agrees fairly well with a measured value of 32 pH at 100 MHz for the array of 16 TSV, allowing an assessment of

978-1-4673-4845-4/13 $31.00 © 2013 IEEE

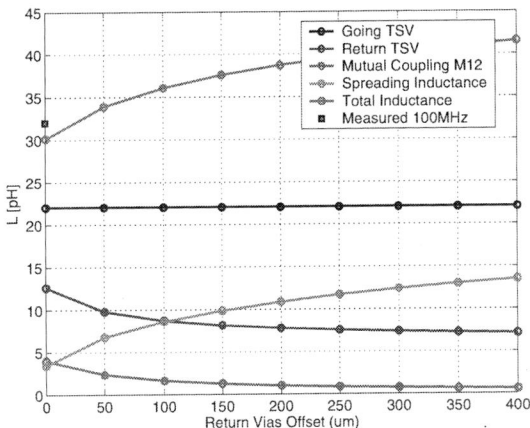

Fig. 6. Inductance components in return circuit path for a 16 TSV array as function of distance of the return TSV's.

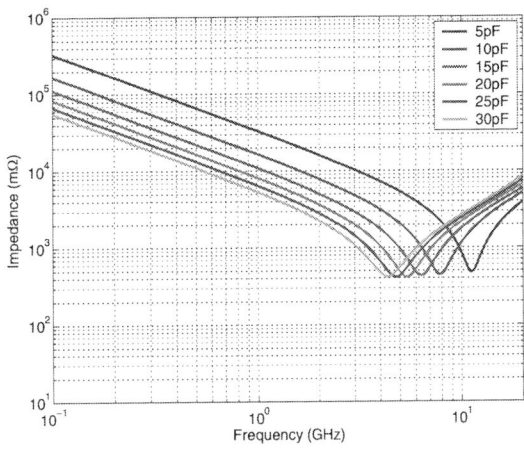

Fig. 7. Impedance for R-L-C resonator of 1 TSV with 5-30 pF capacitance.

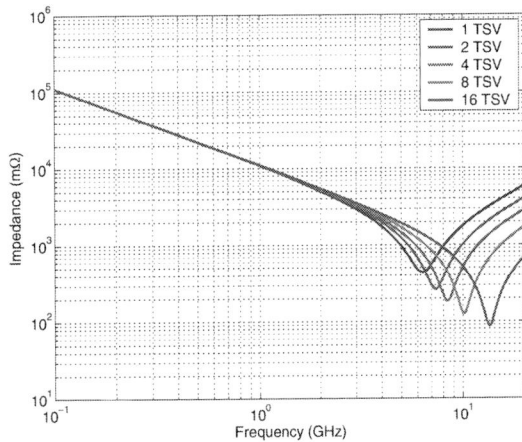

Fig. 8. Impedance of R-L-C resonator of 1-16 TSV with 15 pF capacitance

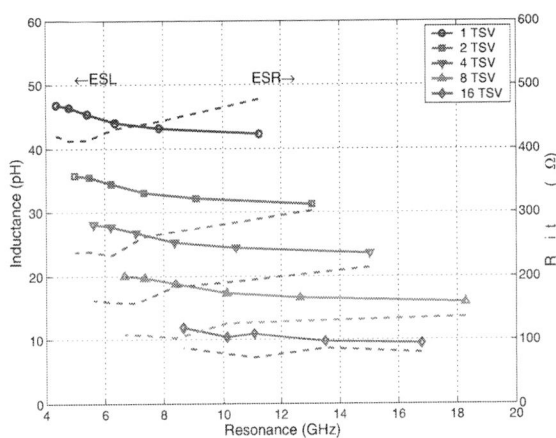

Fig. 9. ESL/ESR of R-L-C resonator of 1-16 TSV over resonance frequency

the effect of return TSV placement. Moving the return TSV out in steps of 50 μm decreases the inductance of the return TSV arrays due to less positive mutual coupling between the shunt currents of the 4 arrays. The negative mutual coupling M_{12} between going and return TSV decreases also. This is offset by an increase in the spreading inductance for the backside plane, leading to an overall increase of the total inductance as return TSV's are moved further away from the going TSV's. Hence this analysis shows that an "ideal" location of the return TSV such that the effect of confounding the measurement results is avoided, does not exist. The least error in inductance is obtained by placing the return TSV's close to the going TSV (zero offset in Fig. 6) to minimize the overall inductance of the loop. The single-port test structure therefore is not suitable to extract an accurate inductance value and does require an analytical "co-de-embedding" process to arrive at an estimate for the inductance of the TSV device. However due to its small footprint it is an attractive choice for process control and device specs monitoring, if calibrated with the proper analytical analysis.

B. Resistance and Inductance Measurement on 2-Port RF Test Structure

The impedance characteristics of the series R-L-C resonator are shown in Fig. 7 for a constant inductance of 1 TSV with varying capacitances and in Fig .8 for a constant capacitance of 15 pF with varying inductance values from arrays with 1, 2, 4, 8 and 16 TSVs. The resonance frequency decreases with the increase of capacitance and inductance. As the resonance frequency is decreased with a larger capacitor, the equivalent series resistance ESR is reduced due to a lower metal plate resistance and larger dielectric area for the displacement current (Fig. 9). The inductance value follows the opposite trend and increases with lower resonance frequency due to the increasing self-inductance of a larger capacitor plate. The delta in extracted inductance for resonators with different capacitor sizes is constant over the arrays of 1, 2, 4 or 16 TSVs supporting this assumption. Hence the most accurate value for inductance is obtained with the smallest available capacitor of 5 pF in this experiment.

978-1-4673-4845-4/13 $31.00 © 2013 IEEE

The inductance measurement results for the 2 methods are summarized in table I.

TABLE I
SUMMARY OF INDUCTANCE RESULTS

Inductance [pH]	1 TSV	2 TSV	4 TSV	8 TSV	16 TSV
P1 (data)	66	55	47	38	32
P1 (model)	63	55	49	41	30
P1 (model TSV)	**48**	**40**	**35**	**29**	**22**
P2 (data 5pF)	42	31	24	16	9
P2 (data + M12)	**43**	**32**	**26**	**20**	**18**

TABLE II
SUMMARY OF RESISTANCE RESULTS

Resistance [mΩ]	1 TSV	2 TSV	4 TSV	8 TSV	16 TSV
P1 (data raw)	397	225	144	98	82
P1 (data - short)	330	160	80	30	15
P2 (ESR)	420	238	161	108	87

The analytical model for the single port test structure matches the raw measurement data within 3 pH, confirming the assumptions of the model. The inductance value in table I for the 5 pF R-L-C resonator was de-embedded for the negative mutual coupling effect between going and return TSV, since this effect will lower the equivalent series inductance as obtained from Z_{12}. After de-embedding, the values from the resonator circuit are approximately 4-9 pH lower than the modeled values. This difference is likely due to a return loop circuit effect that is not accounted for in Z_{12} and requires further investigation.

The DC resistance from single-port measurements and the ESR value from the R-L-C resonators with the largest shunt capacitor of 30 pF are summarized in table II. The equivalent series resistance from the resonator circuit is in close agreement with the non-de-embedded resistance value of the 1 port test structure. Any differences are less than the measured probe contact resistance of 70 mΩ per port.

IV. CONCLUSION

The extracted inductance of the through silicon via (TSV) is strongly dependent on the design of the test structure. A single-port structure, while convenient in size, is not suitable to yield an accurate measured inductance value if not combined with an analytical analysis of the entire test structure return circuit loop. As inductance occurs only in loops of currents, the measurement result is for the entire return loop between signal and ground pads of the GSG probes. The two port resonator structure offers the advantage of a T-network configuration which provides a more direct result of the TSV going array through impedance Z_{12}. It is still confounded with return circuit loop effects that require further investigation and optimization of test structure. For both structures the dilemma of the TSV characterization is that for front-side wafer probing, a return circuit with going and return TSV needs to be formed using the actual device under test that is to be extracted, leading to confounding of the test results.

ACKNOWLEDGMENT

The authors wish to thank Dr. Volker Mühlhaus from Dr. Mühlhaus Consulting & Software GmbH for extensive Sonnet EM simulations to help analyze and interpret the data and Dr. Kai Kwok from Skyworks for valuable discussions on the analysis of the resonator measurement results. Further Dr. David Howard from TowerJazz is recognized for managerial support on the TSV program.

REFERENCES

[1] F.W. Grover, "Inductance Calculations," Van Nostrand Co. N.Y., 1946, Dover Edition, 2004..

[2] S. Weir and T. Dagostino, "PCB Power Delivery Optimizations for the Cost Driven Era," DesignCon 2009, TecForum Slides pp. 1-16, Feb 2009.

[3] M.J. Brophy, T. Saeger, W. Mickanin, "Resistance and Inductance of Through-Wafer Vias: Measurement, Modeling, and Scaling," Compound Semiconductor MANTECH 2005, Digest pdf 5.4, April 2005.

[4] V. Blaschke and R. Zwingman, "On-Wafer Inductance and Resistance Characterization of Sub-5pH Deep Silicon Via (DSV)," ICMTS 2010 Proceedings, pp. 136 - 139, March 2010.

[5] L.D. Smith, "MLC Capacitor Parameters for Accurate Simulation Model," DesignCon 2005, TecForum TF7 Slides pp. 38-58, Feb 2005.

978-1-4673-4845-4/13 $31.00 © 2013 IEEE

Test Structures for Electrical Evaluation of High Aspect Ratio TSV Arrays Fabricated Using Planarised Sacrificial Photoresist

R. Zhang, Y. Li, J. Murray, A.S. Bunting, S. Smith, C.C. Dunare,
J.T.M. Stevenson, M.P. Desmulliez[†], A.J. Walton
SMC, Institute of Integrated Micro and Nano Systems
School of Engineering,
part of the Institute of Integrated Systems,
The University of Edinburgh,
Edinburgh, EH9 3JF, UK
[†] School of Engineering & Physical Sciences, MiSEC
part of the Institute of Integrated Systems,
Heriot Watt University
Edinburgh EH14 4AS, UK
Y.Li@ed.ac.uk; Anthony.Walton@ed.ac.uk

Abstract—**An improved bottom-up electroplating technique has been successfully developed for the fabrication of TSV arrays with 9.5:1 aspect ratios. 125,500 TSVs have been fabricated in an area of 6×6 cm with a horizontal and vertical pitch of 240 µm. A method of visually inspecting the via yield is presented, and Kelvin test structures and contact chain test structures have been fabricated to electrically evaluate single and multiple TSVs respectively. The average resistance of the Cu vias was measured as of 9.1 mΩ using the Kelvin contact resistance structures.**

Keywords—*TSV, Kelvin structure, Daisy chain, electroplating, photoresist CMP*

I. INTRODUCTION

Through silicon vias (TSVs), are a key enabling technology for three-dimensional (3D) interconnection between integrated circuits (ICs) themselves and other technologies such as microsystems in a stacked system in package [1-5]. This technology is required to meet the increasing demand for smaller geometry electronic devices with improved performance and more functionality. High yield of TSVs filled using copper electroplating, and copper overburden removal are two key issues related to the delivery of the technology, which are the major contributors to the cost of TSVs.

When compared to the mature conformal Cu plating approach, the technique of bottom-up Cu plating is more suitable for filling void free, high aspect-ratio TSVs, as seed layers are not required on the sidewalls of vias. Test structures for characterisation of "via middle" process, where TSVs connect the lower metal level interconnect of the top die with the topmost metal wiring of the bottom die, have been previously reported [6]. The TSVs in that study were fabricated with a depth of only tens of microns, which requires wafer grinding to complete the process as well as a very thin substrate.

Clearly, the ability to fabricate vias without the requirement to thin the wafer is of potential interest [7]. This paper briefly presents an improved bottom-up plating technique developed by using planarised photoresist as sacrificial bridge for fabrication of TSVs [8]. It then describes a test chip design and the electrical characterization of the via arrays using Kelvin structures to measure contact resistance and via contact chain test structures for estimating the initial yield.

II. TEST CHIP DESIGN

In order to develop the via technology a 75mm test wafer was designed with an array of 125,500 TSVs with diameters of 40µm. Figure 1 shows a photograph of a test wafer after the vias have been filled with copper. The layout for each test chip is based upon this TSV array which consists of two diagonally offset arrays of vias. One array (typically connected to 120×120µm probe pads) is arranged with a 240µm pitch in both the horizontal and vertical axes. The other array, with the same pitch, is offset diagonally as shown in figure 2 and provides the electrical vias being characterised.

Figure 1. Photograph showing the layout of the array of 125,500 TSVs with diameters of 40µm.

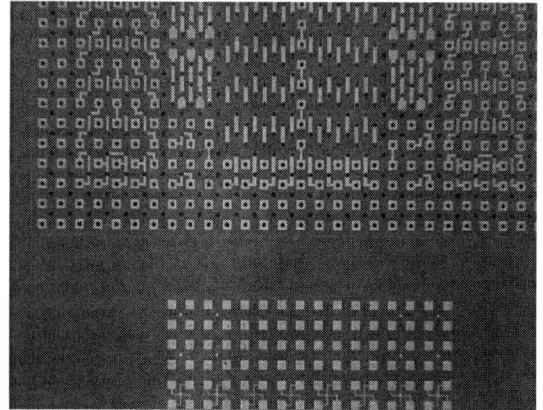

Figure 2. Photograph of a section of the layout of the test chip showing TSV via array (black dots), probe pads and interconnect for electrical test structures.

The test structures are built around these via arrays and the design can be probed using a 2×N probe arrangement with a 240μm pitch.

The entire via array on the wafer is divided into 36 die (see figure 3(a)) with two different die designs shown in figures 3(b) and 3(c), These consist of three types of test structures.

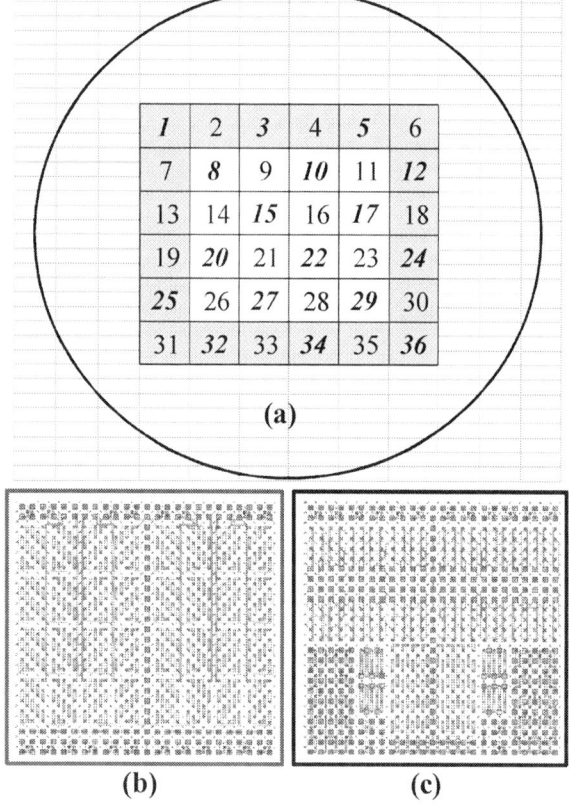

Figure 3. Chip layout of die on via array. (a) The yellow coloured dies identify the chips on the perimeter of the array. The italic numbered dies use (b) design 1, and other dies use (c) design 2.

In each die, there are Kelvin test structures (figure 4), TSV via chain structures (figure 5) and multiplexed yield structures (both single-TSV and serial-TSV-chain multiplexed designs) (figure 6) [9]. Die design 1 (die numbers in italics) contains 20 stand-alone TSV Kelvin test structures and two serial-TSV-chain multiplex test structures. Die design 2 (die numbers in standard font) contains 16 stand-alone Kelvin test structures, 4 TSV chains and 12 single-TSV multiplex test structures. In addition to the TSV related interconnect the chip also has structures for measuring the sheet resistance of the aluminium interconnect, located in the lower portion of figure 2. Clearly a key component of the test structures used to characterise TSVs is the requirement for interconnect on both sides of the wafer.

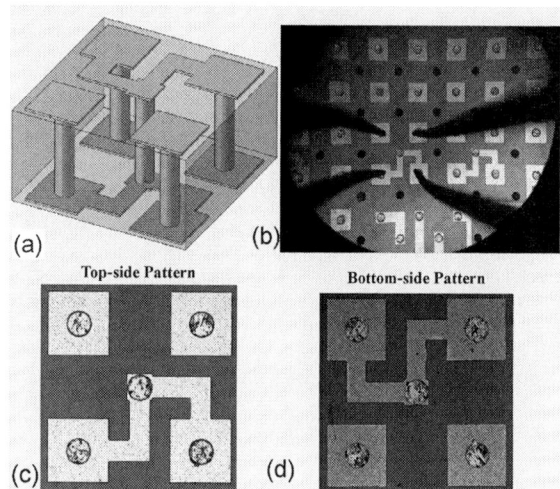

Figure 4. Kelvin test structure to measure the resistance of a single TSV (a) 3D illustration of the structure (b) Manual probing of the structure, (c) View of front and, (d) view of the back side of wafer.

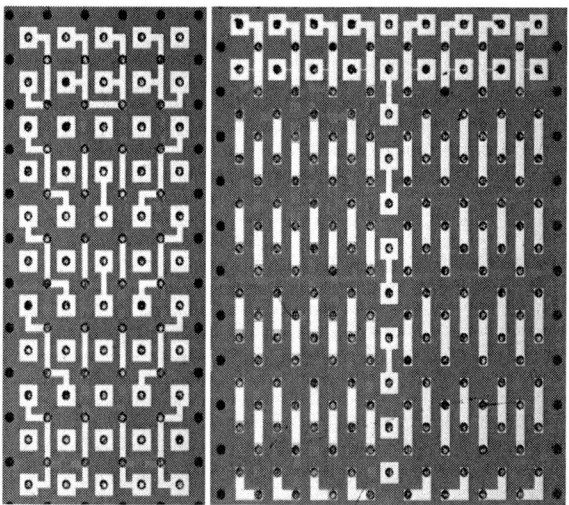

Figure 5 Photograph showing the layout of TSV chains on test chip. The layout on the backside of the chip has been added to the photograph (red tracks).

978-1-4673-4845-4/13 $31.00 © 2013 IEEE   38

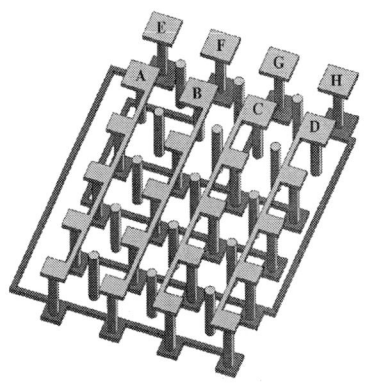

Figure 6. Multiplex test structure for measuring individual via connections.

Figure 4(a) shows a 3D illustration of the Kevin structure on the test chip which is used to measure the via resistance. The via contact chain is another electrical test structure to evaluate the yield of the TSVs. As with the Kelvin structure, the via chain requires interconnect patterning on both the front and back of the wafer. Figure 4 (c) and (d) shows this for the Kevin contact resistance structure with figure 5 showing contact chain interconnect (note the copper coloured backside wire has been manually added to the photograph). The main use of these two structures is to measure the via resistance and to verify the electrical continuity of a chain of TSVs. The front side wiring is connected through TSVs to the backside wiring so as to form a contact chain pattern. The test chip design has two different TSV daisy chain structures with 18 and 38 TSVs respectively. The probe pads on both sides of the wafer are also $120 \times 120 \mu m$ (with a pitch of $240 \mu m$) and the TSVs in the contact chains are connected by $50 \mu m$ wide Al tracks.

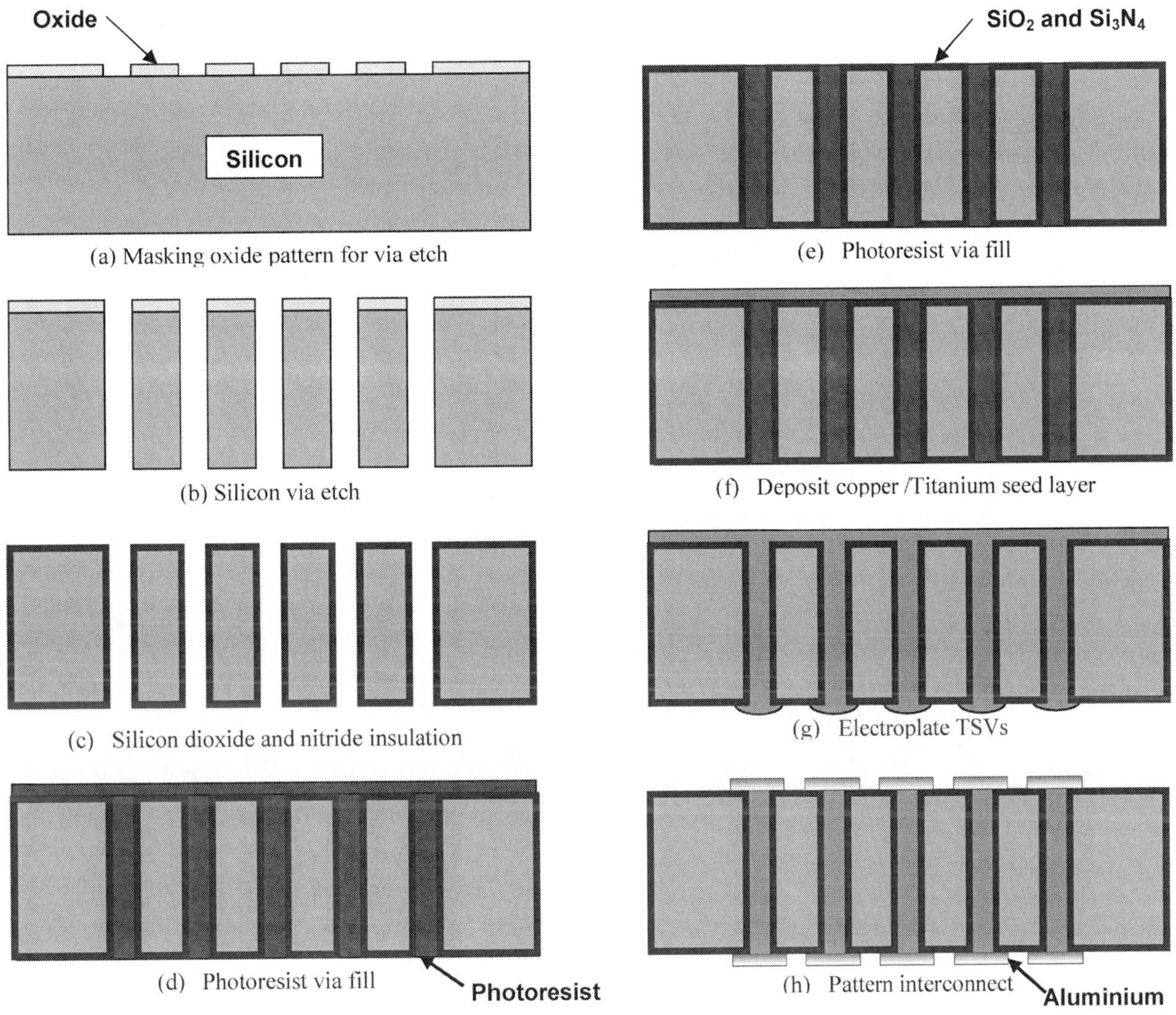

Oxide

Silicon

(a) Masking oxide pattern for via etch

(b) Silicon via etch

(c) Silicon dioxide and nitride insulation

(d) Photoresist via fill   **Photoresist**

SiO$_2$ and Si$_3$N$_4$

(e) Photoresist via fill

(f) Deposit copper /Titanium seed layer

(g) Electroplate TSVs

(h) Pattern interconnect   **Aluminium**

Figure 7. Schematic process flow for the fabrication of the TSV technology being characterised. In this process aluminium was used for the interconnect and contact pads.

III. FABRICATION PROCESS

The fabrication process, shown schematically in figure 7, [8] starts with a double-sided polished 75mm diameter silicon wafer, with a nominal thickness of 380μm. The first step is to etch the array of vias into the silicon using a deep reactive ion etch process (figure 7(b)). The Bosch process is used for this with SF_6 as the silicon etchant and CF_4 as the passivation gas. The vias are insulated with a thermal oxidation step followed by LPCVD silicon nitride. The vias are then filled with photoresist (figure 7(d)) and the excess photoresist removed using chemical mechanical planarization (CMP) (figure 7(e) and 8(a)). A copper seed layer is then deposited by sputtering (figure 7(f)) before the sacrificial photoresist is removed using a wet chemical strip (figure 8(b)). This allows the vias to be filled by bottom-up electroplating from the copper seed layer (figure 7(g)). As a consequence, the surface with the copper/titanium seed layer remains planarised. This approach with the thin bridge layer avoids the time-consuming and expensive polishing process to remove the thick overburden copper TSV hole sealing (normally thicker than 20μm) [10]. Aluminium interconnect is used in this process so the thin deposited copper electroplating seed layer and the over-plating

domes are first removed before the aluminum is sputtered and patterned (figure 7(g)). Alternatively, a copper damascene process could be employed to create interconnect.

It should be noted that current crowding at the perimeter of the TSV array during electroplating will result in the vias being filled more rapidly than those in the central portion, especially those in the corners of the array. However, the non-uniformity that can be observed in the top left-hand corner of the via array in figure 1 has been caused by this portion of the array being too close to the edge of the wafer holder used during the electroplating process.

Open-failures are one of the most critical faults which affect TSVs. It is important to note that even the presence of a single open-failure can potentially kill the functionality of a product chip. The larger the number of interconnects, the higher the probability of a chip failure due to a faulty TSV. Several phenomena can cause an open-failures associated with a given interconnect architecture. In this technology, the occurrence of these failures is most probably due to the trapped air bubbles inside of TSVs during electroplating. In addition to open-failures, poor contacts between the TSVs and

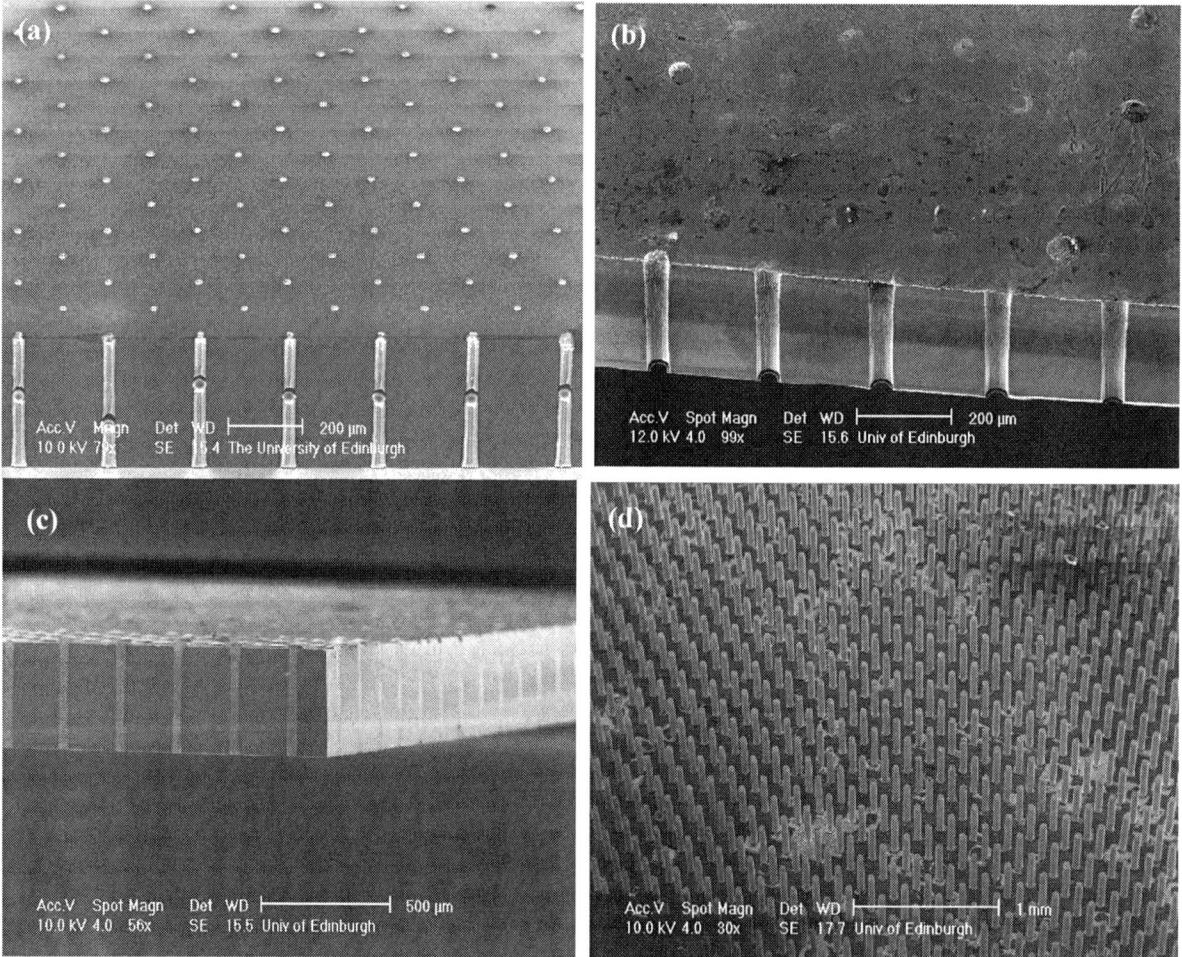

Figure 8. (a) Cross-section of vias filled with photoresist and planarised, (b) Cross-section of copper seed layer after photoresist removed from vias, (c) Cross-section of Copper filled via, (d) Optical visualisation of yield of Copper via array.

the testing pads will also degrade the performance of the via array, reducing process yield. This later problem results from issues with the polishing process that removes the small copper bumps on the front side of the wafer after the electroplating step.

IV. TEST STRUCTURES AND CHARACTERISATION

The quality of the Cu via fill and the yield of the TSVs were first visually evaluated by examining vertical cross-sections through the vias (Figure 8(c)). An alternative method to view the TSVs is shown in Figure 8(d), where a wet KOH etch has been used to remove the substrate and leave behind the copper via plugs. In this test chip the 125,500 TSVs are fabricated over a large 6×6 cm square area which presents a significant challenge to determine the overall yield. The substrate removal method is very effective in process development as it provides a rapid visual inspection that confirms whether every via has been filled. The main drawback is that small voids within a via will not necessarily be identified. The main advantages of this method compared with electrical structures are that:

- Close inspection can readily identify gross defects and the physical location of failed elements (provided the defect/void is not totally internal to the copper plug).

- No electrical interconnect is required.

Apart from sectioning vias another method to identify voids is through electrical measurement. The best structure to identify voids is the Kelvin contact resistance structure as the parasitic resistances associated with contact chains with large numbers vias mean that they are less likely to detect voids that do not totally occlude a via.

Single TSVs have been characterised using Kelvin contact resistance test structures, which are available on the test chip. When making these measurements it is important to remember that there is metallisation on both sides of the wafer and so insulation is required between the backside interconnect and the chuck. Another factor that provides a challenge is that the average resistance of the copper plugs is typically less than $10\,m\Omega$, while the aluminium interconnect used for the pads and connections between vias has been measured to have a relatively high sheet resistance of 60 $m\Omega/\square$. This makes it difficult to identify a high resistance via with a small void when using contact chain measurements and this is discussed in more detail later.

Initial, manually probed electrical tests on 436 of the Kelvin contact resistance structures gave an average via resistance value of 9.1 $m\Omega$, with a standard deviation for each set of vias being 3.4%. Figure 9 shows the distribution of these measurements, which appears to be bi-modal. Figure 10 shows the measured via resistance for structures on the 16 die which were not on the perimeter of the array (see figure 3) and it would appear that these vias have a much tighter distribution and lower resistance. The causes of this are not clear, but may be due to very small voids being formed in the perimeter vias as a result of increased bubble formation related to higher electroplating currents.

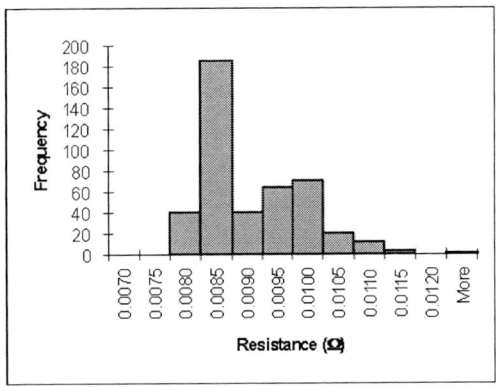

Figure 9. Resistance of vias measured using Kelvin structures on all 36 chips

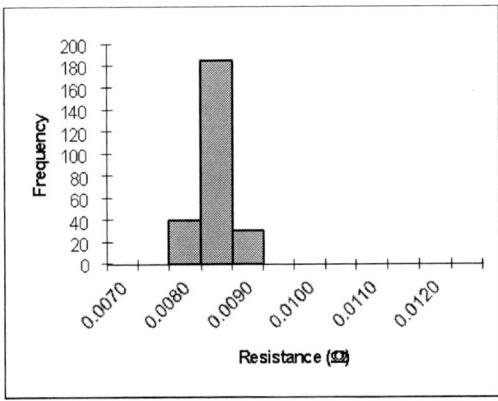

Figure 10. Resistance of vias measured using Kelvin structures on the 16 chips in the centre of the die layout shown in figure 3(a).

Further measurements and analysis are required in order to identify if there is any other systematic cross-wafer variation. Using some assumptions regarding via dimensions and copper resistivity, a theoretical value of 6 $m\Omega$ can be derived. Some of the difference between the measured and theoretical value is likely to be the additional contact resistance between the TSVs and the aluminium layers used to form the test structure. The rest may be related to differences in material properties between standard figures for bulk copper resistivity and this microscale electroplated copper and/or potential errors in the via geometry.

Resistance measurements on the TSV contact chains were performed using a PA200 semi-automatic probe station and HP4062-UX instrumentation. Contact chains have been measured with resistances of averaging 8.65±1.06Ω for 18 TSVs chains and 12.44±1.93Ω for those structures with 38 vias. As would be expected the chain resistance values divided by the number of vias also includes a share of the total track resistance. As a result these values are significantly larger than the Kelvin measurements of a single via. It should be noted that for this technology the via resistance is small compared with the interconnect (60 $m\Omega/\square$) which makes it difficult to detect high resistance vias with a partial void.

Open circuit discontinuities were observed in five out of sixteen chains and at the time of writing the cause of these failures has not been confirmed. However, as a result of handling during electrical characterisation there was severe scratching on the unprotected backside of the wafer, which together with the good results observed from the optical inspections, suggest open circuit problems are probably more likely to be related to broken interconnect tracks.

For yield prediction, TSV chains typically need to have large numbers of vias to permit the measurement of very low failure rates. However, as the number of series-connected elements increases, the chain resistance increases as well. If there is in a huge chain containing 100,000 or more TSVs, there is an even more severe challenge in detecting the presence of a partial void, then for the smaller chain lengths reported here.

Figure 6 shows a passive multiplexed test structure which requires only two levels of interconnect. The details of the structure, and the methods of measurement used, are described fully in [9]. This structure potentially enables the via yield to be tested in an efficient manner, provided that the individual via resistances are significantly higher than the interconnect metal tracks. However, with the 40μm TSVs reported in this paper this is not the case and so "to increase the resistance" designs were also included that replaced the single contact shown in figure 6 with a 23 via serial-TSV-chain. Each of the 23 via serial-TSV-chains has an estimated resistance around 6Ω. The resistance of the tracks associated with the electrical connection to each TSV-chains range between 7 to 20 Ω. Hence, while it was possible to identify open circuits using these structures they were also not able to identify soft failures.

V. Conclusions

This paper has described a technology for fabricating 40μm diameter electroplated Cu TSVs. A test chip has been designed and fabricated to help characterise a new process associated with the bottom-up electroplating of high aspect ratio through silicon vias. This includes test structures for the measurement of single TSVs as well as contact chains to aid the characterisation of the yield of via fabrication and the identification of failed vias.

For the technology reported, the resistance of the vias (~10mΩ) is at least 20 times smaller than the interconnect tracks. If a via has a partial void the increase in resistance is a small percentage of the total which makes it very difficult to electrically identify there is a vias with a partial voids in the contact chain. Clearly reducing the via diameter, which is very desirable in itself, is unlikely to help, especially if the resistance increase caused by the void remains similar. It is also possible to reduce the track resistance by electroplating thick Cu interconnecting tracks and this will raise the percentage increase in resistance caused by a partial void.

However, it should be remembered there is a practical limit as to thickness of the copper tracks.

In summary there is a challenge in electrically characterising the yield of the TSVs reported in this paper. The electrical Kelvin structures show a bimodal distribution that correlates with the via location and identifying the source of this is the subject of future work.

Acknowledgments

The authors would like to acknowledge financial support from the Edinburgh Research Partnership in Engineering and Mathematics (ERPem), CSC, the EPSRC (Engineering and Physical Sciences Research Council) IeMRC Smart Microsystems project (FS/01/02/10) and Texas Instruments' contribution to electroplating equipment and consumables.

References

[1] S. Q. Gu, P. Marchal, M. Facchini, F. Wang, M. Suh, D. Lisk, M. Nowak, "Stackable memory of 3-D chip integration for mobile applications", Proc. IEDM, 2008, pp. 1–4.

[2] D. Velenis, M. Stucchi, E. J. Marinissen, B. Swinnen, E. Beyne, "Impact of 3-D design choices on manufacturing cost", Proc. IEEE Int. Conf. 3-D Syst. Integr., Sep. 2009, pp. 1–5.

[3] U. Kang, H.-J. Chung, S. Heo, S.-H. Ahn, H. Lee, S.-H. Cha, J. Ahn, D.-M. Kwon, J. H. Kim, J.-W. Lee, H.-S. Joo, W.-S. Kim, H.-K. Kim, E.-M. Lee, S.-R. Kim, K.-H. Ma, D.-H. Jang, N.-S. Kim, M.-S. Choi, S.-J. Oh, J.-B. Lee, T.-K. Jung, J.-H. Yoo, C. Kim, "8 Gb 3-D DDR3 DRAM using through-silicon-via technology", Proc. Int. Solid-State Circuit Conf., 2009, pp. 130–131, paper 7.2.

[4] H. Yoshikawa, A. Kawasaki, I. Tomoaki, Y. Nishimura, K. Tanida, K. Akiyama, M. Sekiguchi, M. Matsuo, S. Fukuchi, K. Takahashi, "Chip-scale camera module (CSCM) using through-silicon-via (TSV)", Proc. Int. Solid-State Circuit Conf., 2009, pp. 476–477, paper 28.5.

[5] V. Suntharalingam, R. Berger, S. Clark, J. Knecht, A. Messier, K. Newcomb, D. Rathman, R. Slattery, A. Soares, C. Stevenson, K. Warner, D. Young, L. P. Ang, B. Mansoorian, D. Shaver, "A 4-side tileable back-illuminated 3-D-integrated mpixel CMOS image sensor", Proc. Int. Solid-State Circuit Conf., 2009, pp. 38–39, paper 2.1.

[6] Stucchi, M., Perry, D., Katti, G., Dehaene, W., Velenis, D., "Test Structures for Characterization of Through-Silicon Vias", IEEE Trans Semiconductor Manufacturing, Vol 25 , no 3, pp 355 – 364, Aug. 2012.

[7] P. Dixit, and J. Miao, "Aspect-Ratio-Dependent Copper Electrodeposition Technique for Very High Aspect-Ratio Through-Hole Plating", Journal of The Electrochemical Society, Vol. 153, No. 6, pp. G552-G559, 2006.

[8] R. Zhang, Y. Li, C.C. Dunare, A.S. Bunting, J.T.M. Stevenson and A.J. Walton, "Planarised Photoresist as Bridge for Fabrication of High-Aspect Ratio TSVs Array", Presented at 38th International Conference on Micro and Nano Engineering (MNE2012), Toulouse, France, 2012.

[9] A.J. Walton, W. Gammie, D. Morrow, J.T.M. Stevenson, and R.J. Holwill, "A novel approach for reducing the area occupied by contact pads on process control chips", Proc. IEEE ICMTS, San Diego, USA, Vol. 3, pp. 75- 80, 1990.

[10] J.H. Lai, H.S. Yang, H. Chen, C.R. King, J. Zaveru, "A 'mesh' seed layer for improved through-silicon-via fabrication", J. Micromech. Microeng., 20 (2010) 025016.

A novel silicon interposer for measuring devices requiring complex two-sided contacting

Jaber Derakhshandeh, Negin Golshani, Loek A. Steenweg, Wim van der Vlist, Lis K. Nanver

DIMES, Delft University of Technology,
Feldmannweg 17, 2628 CT, Delft, The Netherlands
j.derakhsh@gmail.com

Abstract— **The design and fabrication process is presented for a novel silicon-based interposer suitable for dies where it is necessary to place multiple contact pads on the both sides of wafer. This interposer transfers all contacts to the same side of the wafer so that the measurement can be done using conventional probe stations for one-sided probing.**

Keywords—Double sided probing; Silicon DRIE, Silicon interposer; Photodiode detector

I. INTRODUCTION

Probing systems for double-sided probing are being used for electrical characterization in many applications involving structures such as vertical interconnects, detectors for emission measurement purposes and optical and MEMS devices. Measurement systems that can handle two-sided measurements are very expensive and complex and not generally available in electrical measurement laboratories [1,2,3,4].

In this paper we present a silicon-based interposer to transfer the backside contacts of the wafer to the front side. The idea is to glue the device under test (DUT) to this interposer and then wire bond the front-side and backside contacts to the interposer. The designed package is illustrated in Fig. 1.

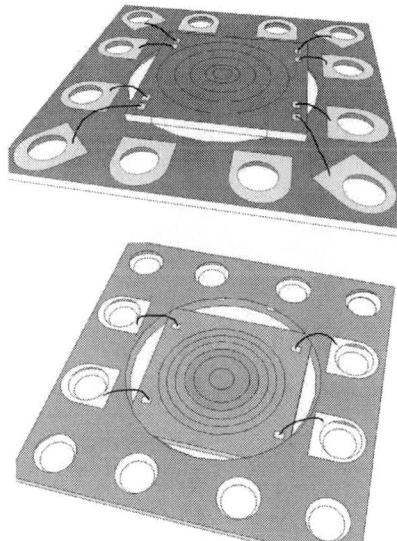

Figure 1. Illustration of the front- and backside of the designed package for the dies with two-sided contacts.

Using this package it would be possible to measure the devices on conventional probe stations.

There are several advantages of using this package. First of all the package is made on silicon wafers using conventional integrated circuits fabrication methods which makes it compatible with CMOS processing. The formation of through silicon Vias (TSVs) is now well-developed in silicon technology which gives the possibility of designing packages with small dimensions. Moreover, this package has advantages such as thermal mismatch improvement, better thermal conductance and stability of dimensions because both the die and the package are made of silicon.

In some applications such as the electron and x-ray detectors, studied in connection with this work, external circuitry is required to amplify the signal level. With this package it is possible to integrate some of the electronic devices such as JFETs and capacitors on the package itself. Therefore it would be possible to enhance the performance of the final product by using such a package.

II. DESIGN PROCEDURE

The exact layout for interposer depends on the number of the contact pads on the front- and the backside of DUT and its physical dimensions. In our wafer-scale layout there are 9 packages with different via sizes as displayed in Fig. 2. there were 9 packages with different VIA sizes. It consists of two masks: one for via etching and one for metallization.

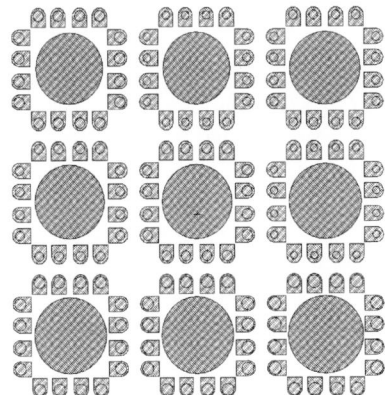

Figure 2. The designed wafer-scale layout for silicon interposers containing 9 packages with different via sizes.

The via mask has been used to etch through the silicon wafer and the metallization mask has been used for metal patterning and etching on both sides of the wafer. As it can be seen in Fig. 2, each package has 16 small holes for potentially contacting 16 pads and one big hole in the middle of the package where the DUT can be mounted on one side of the package while still having access to the other side for bonding purposes. The size of this opening depends on the size of the DUT. The device under test should have sufficient overlap with the interposer to allow gluing it securely to the package.

The etching profile needs to be sloped for metallization purposes. It means that we should calculate the on-mask size of the vias based on the final desired via sizes. Fig. 3 shows a simple drawing of the etching profile dimensions used to calculate the size of vias for the layout design. The etching angle is around 63.5° and the thickness of the wafer is 560 μm as shown in the drawing. To have a diameter of 1.78 mm for the central holes, which is the size of the hole to hold the DUT, we need to have a diameter of 2.86 mm diameter in the design. The smallest via size used in the design was 560 μm which resulted in a 5 μm opening on the front side of interposer.

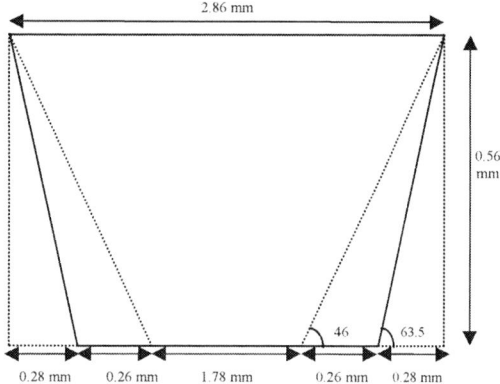

Figure 3. Calculation of via and hole sizes in the design.

III. FABRICATION PROCESS

As mentioned before we used a <100> double-side polished wafer with a thickness of 560 μm. After making alignment markers (zero layers) on the wafer, a 6 μm thick PECVD oxide was deposited on both sides of wafer to act as a mask for silicon etching. Then using the designed via mask, we patterned the resist followed by dry etching of the oxide mask. To etch through the silicon we used a Bosch-process deep-reactive-ion-etching (DRIE) system called Adixen. The etching temperature was -10° C to have better selectivity to oxide and we used a combination of O_2, SF_6 and C_4F_8 gases. Oxygen is used to make the profile similar to KOH wet etching. The optimized recipe gives an angle around 60° for etching profile. By adjusting the oxygen flow-rate we can obtain different angles. Without oxygen the etching will proceed vertically straight down resulting in an angle of 90°. The etching time

was 2 hours resulting in complete silicon etching and stopping on the oxide layer.

As shown in Fig. 4, after through-silicon etching we removed the oxide mask in a BHF solution and then cleaned the wafer in a nitric acid bath for 10 min followed by DI water washing.

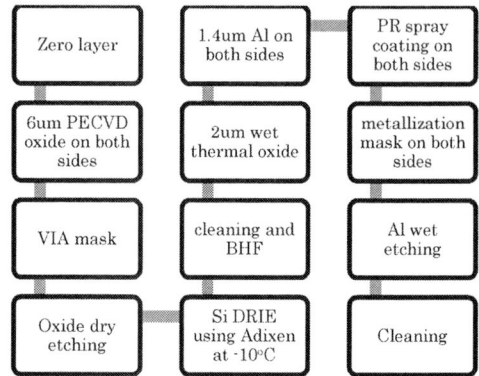

Figure 4. Process flow for fabricating the designed package.

The next step was to isolate the vias from the silicon package. We used wet oxidation for 9 hrs at 1100°C to achieve 2 μm thick oxide inside the holes.

After that 1.4 μm Al was deposited on both sides of the wafer. For better step coverage and contacting of the metal on both sides, we deposited the Al at 350° C using a pulsed DC sputtering system. Using the next mask we etched the Al. In this step photoresist has to be spray coated since we have holes in the wafer. We spray coated the resist on both sides of the wafer and then we did exposure using the same mask on both sides of the wafer. After development and baking the resist, we used wet etching to remove the Al. The final step was cleaning in acetone and nitric acid followed by DI water rinsing.

Fig. 5 shows a graphical view of front- and backside of package. The Al contacts are visible in this figure.

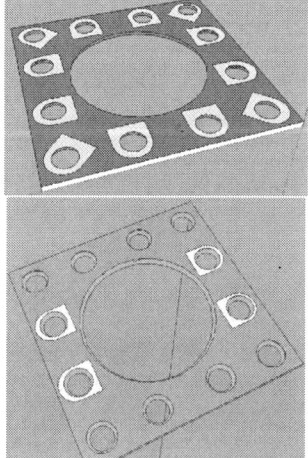

Figure 5. Top and bottom sides of designed package.

A top view photo of the fabricated wafer is displayed in Fig. 6. The front-side view of vias, holes and contact pads are visible in this image. The DUT is to be placed in the middle of big holes and glued to this side of the package.

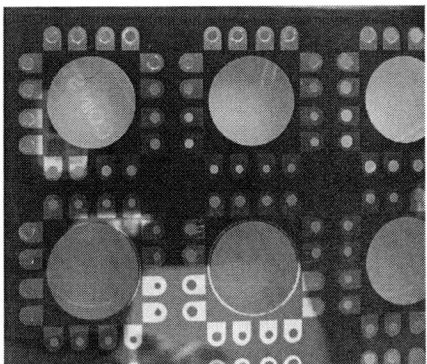

Figure 6. Front side image of the fabricated package.

A SEM image of a contact via taken from the frontside is shown in Fig. 7. The layer surrounding the hole is Al which is extended for wire bonding and measurement needle purposes.

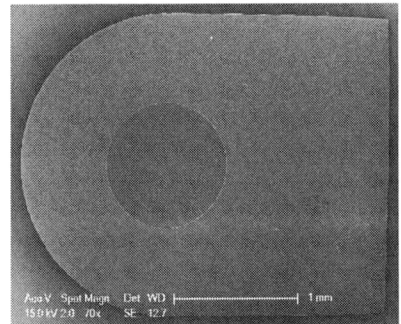

Figure 7. SEM image of the fabricated contact hole

In Fig. 8 a backside view of the fabricated package is displayed. The location of the bond pads on the backside of the DUT has to be visible through the big holes so that wire bonding can be done through this hole. A special tool was made in order to wire bond the contact pads through this hole. The whole package is isolated from the vias by the 2 μm thick thermal oxide.

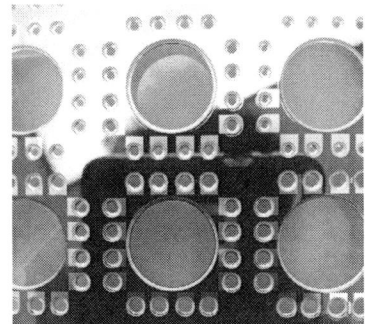

Figure 8. Backside image of the silicon interposer.

Fig. 9 is a magnified SEM image of the etched via showing the pseudo-KOH etching profile. This profile ensures having electrical contact from the one side of the wafer to the other.

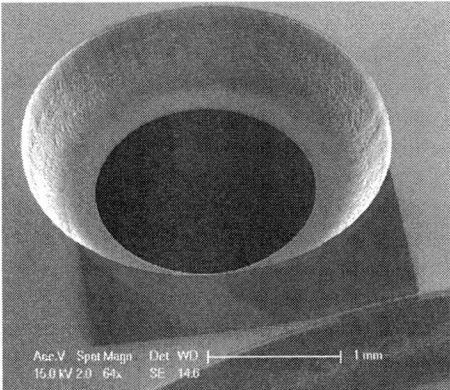

Figure 9. SEM image of a contact hole showing the pseudo-KOH etching of the vias

In Fig. 10 two vias next to the central big hole is shown. The dark regions are Al contact pads for wire bonding to the DUT contact pads.

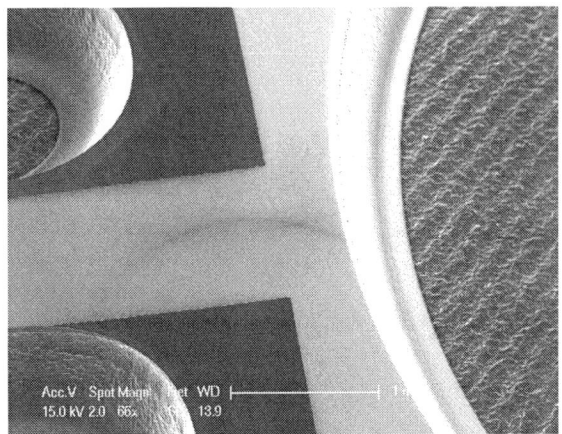

Figure 10. SEM image of two vias and the large central hole

IV. MEASUREMENT RESULTS

For the measurement of devices the Cascade Microtech Summit 12000 probe station was used. The DC measurements were performed on the parameter analyser Agilent 4156A/4156C and for CV measurements a HP 4284 LCR meter was used.

The final packaged device is shown in Fig. 11. We developed a Teflon container to prevent any damage to the backside wire bonds. All contact pads are available for measurement from the top side of the complete package. The device that we mounted on the silicon package/interposer had several bulk p-n diodes with two contacts on the backside and several contacts on front side of the wafer.

978-1-4673-4845-4/13 $31.00 © 2013 IEEE

Figure 11. Mounted p-n diodes on the package with access to front and backside contact pads.

Using this developed package we could do all DC and CV measurements on the probe station successfully.

Fig. 12 shows a DC current voltage measurement between two neighbouring vias. As can be seen in this plot, even when applying a voltage of -100 V between two neighbouring contact pads in the package, there is no leakage current between them.

The measured resistance between top and bottom contacts of one via was in the order of $0.8485m\Omega$ which is negligible compared to the series resistance found in most devices.

CONCLUSION

In this paper we demonstrated a silicon-based interposer enabling the measurement of dies with contact pads on both sides of the wafer using conventional Cascade probing systems. The fabricated package has successfully been used to test p-n diodes under development for use as radiation detectors.

ACKNOWLEDGMENT

We would like to express sincere thanks to all DIMES clean room staff, in the DIMES technology center in TU Delft for their assistance in the preparation of these devices. Also special thanks to Kees Kooijman and Patrick Vogelsang from FEI Company for fruitful discussions.

REFERENCES

[1] T. S. Tarter and N. T. Do, "Method and apparatus for electrical characterization of an integrated circuit package using a vertical probe station," U.S. Patent 6396296, May 2002.

[2] T. Burcham, P. McCann, and R. Jones, "Double-sided probing structures," U.S. Patent Application 2008, 0265925, Oct. 30, 2008.

[3] Kuan-Chung Lu et al, "Vertical Interconnect Measurement Techniques Based on Double-Sided Probing System and Short-Open-Load-Reciprocal Calibration", 2011 Electronic Components and Technology Conference

[4] Daniel Schurz, Warren W. Flack, Robert L. Hsieh, "Dual Side Lithography Measurement, Precision and Accuracy", SPIE 2004 #5375-127

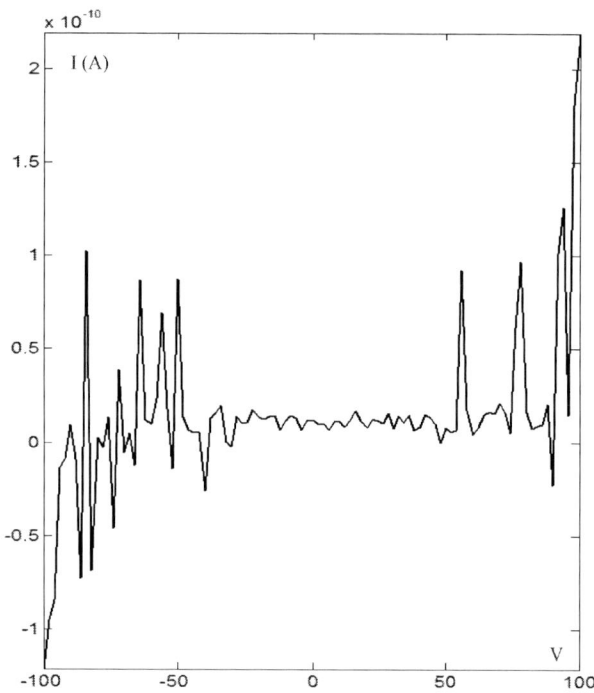

Figure 12. Electrical measurement of the isolation between two contact pads.

SESSION 3: Capacitance

978-1-4673-4845-4/13 $31.00 © 2013 IEEE   48

Characterization of Capacitance Mismatch Using Simple Difference Charge-Based Capacitance Measurement (DCBCM) Test Structure

Ken Sawada[1], Geert Van der Plas[2], Yuichi Miyamori[3], Tetsuya Oishi[4],
Cherman Vladimir[2], Abdelkarim Mercha[2], Verkest Diederik[2], and Hiroaki Ammo[4]

[1] Sony Corporation to IMEC, imec vzw, Kapeldreef 75, B-3001 Leuven, Belgium;
[2] IMEC, imec vzw, Kapeldreef 75, B-3001 Leuven, Belgium;
[3] Sony Semiconductor Corporation,
4001-1 Haramizu, Kikuyo-machi, Kikuchi-gun, Kumamoto, 869-1102 Japan;
[4] Sony Corporation, 4-14-1 Asahi-cho, Atsugi-shi, Kanagawa, 243-0014 Japan
Ken.Sawada@jp.sony.com

Abstract—**We propose a test structure named difference charge-based capacitance measurement (DCBCM) for measuring matching of MOM capacitance with better than 10 atto-farad (aF) accuracy and MOS capacitance with few tens of aF accuracy. The test structure is a derivative of the Charge-based Capacitance measurement (CBCM) technique [1]. In the structure two matched (or intentionally mismatched) capacitors are charged with alternating voltages on one side and on the other side the charges are alternated between two output nodes. We can eliminate parasitic leakage and charge injection components and extract the capacitance difference from the resulting output current that is proportional to the capacitance difference. It is found that mismatch of 20fF MOM capacitances with intentionally 100aF offset can be measured with 7.2aF absolute accuracy. With an adequate input pulse scheme, we also demonstrated a measurement of 100-200fF MOS capacitance mismatch with bias voltage dependence which showed sensitivity of σ = 0.06%. The proposed DCBCM technique is suitable for evaluating small capacitance mismatch for beyond 20nm node.**

Keywords— *Capacitance matching; CBCM; difference charge-based capacitance measurement; DCBCM; atto-farad*

I. INTRODUCTION

The scaling of technology has resulted in a reduction of capacitance values that are of practical use in circuits. Measurement of mismatches between such a small capacitances becomes more important in both process characterization and precise analog circuit design. Measuring capacitance is used in many test structures. For example, a charge based capacitance measurement method [2] is known for absolute capacitance measurement. There are also many derivative methods like CIEF-CBCM [3] [4] to increase measurement accuracy by cancelling charge injection components. Obtaining mismatch from absolute capacitance measurements requires accuracy on the absolute capacitance to be well below the expected mismatch, a stronger requirement than needed for absolute capacitance characterization. To realize simple and accurate method we

propose a new test structure to measure mismatch characteristics specifically. Current methods to assess the matching of capacitance consist of somehow complicated circuit [5] and it is perceived to be difficult to measure accurately the difference of very small capacitances (100fF and well below), as well as measure capacitance difference of non-linear capacitors, such as Varactors and moscaps .

In this work we designed and evaluated a simple test structure named DCBCM for measuring capacitance mismatch which consists of only two FETs in its core circuit and we also confirmed this technique with aF order accuracy.

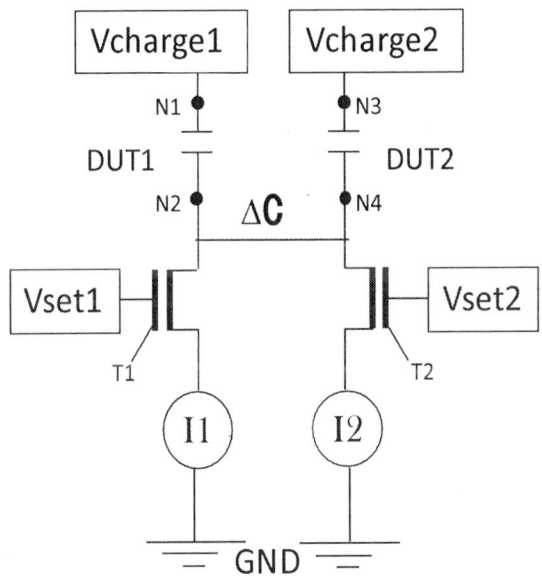

Figure 1. DCBCM core circuit.

II. DESIGN FOR DCBCM TEST STRUCTURE

Fig. 1 shows a schematic of DCBCM core circuit. Fig. 2 shows the timing chart of control and measurement of the test structure of Fig. 1.

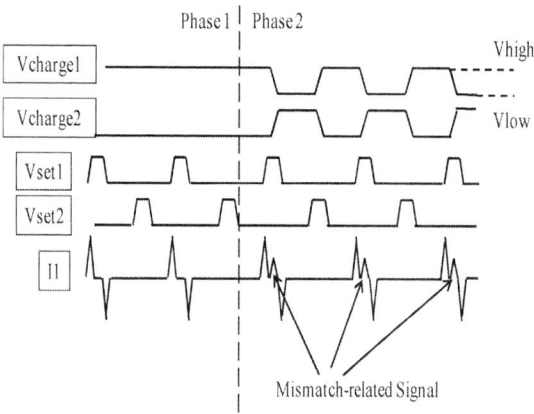

Figure 2. Timing chart of control and measurement of the test structure of Fig. 1.

In the structure two matched capacitors are charged with alternating differential voltages on one side and on the other side the charges are alternated between two output nodes using a pair of switch FETs. The switching (Vset) and charge (Vcharge) signals are appropriately timed to ensure accurate cancellation of leakages and charge injection. The switching signals are non-overlapping, as is the case in the standard CBCM method, the charging signals change when switches are on, causing the charging currents to run to the output nodes without causing voltage changes on the nodes (except voltage drop across switches). The resulting output current is proportional to the capacitance difference and inversely proportional to the voltage swing and frequency of alternation

$$\Delta C = \Delta I / (\Delta V * f) . \qquad (1)$$

where ΔI is the difference between the DC currents measured during the first phase and the second phase (i.e. the peaks of mismatch-related signals in Fig. 2.), ΔV is the voltage difference between Vhigh and Vlow and f is the frequency of the alternations of the input pulses.

Vcharge blocks do not affect measurement if they have sufficiently low output impedance. Impedance should be substantially low such that settling of Vcharge nodes is much faster than period of measurement. This corresponds to

C_vcharge_node * R_vcharge

$$\ll \text{period of measurement.} \qquad (2)$$

where it is needed at least 10 times smaller for CR time constant than period of measurement. For example in 10ns (100MHz) measurement condition, this is at most 1ns, or for 10pF parasitic less than 100Ohm. Relaxing frequency relaxes this impedance proportionally.

Coupling from any Vcharge node to central node of DCBCM is critical. Since any parasitic coupling adds to difference we carefully made the test structure with symmetrical/balanced design.

We designed suitable sizes of low leakage 1.8V NMOSFETs (L=0.3um, W=0.025um /DUT_fF) for switch FETs. Fig. 3 shows results of ΔC obtained from SPICE simulations of our DCBCM circuit. The DUTs of pair capacitances are 20fF and 20fF+ΔC (ΔC=10-500aF intentionally mismatched) MOMcap. The frequency and the voltage difference of input pulses are set to 25MHz and 1V. At the 100aF mismatch point, the accuracy of the ΔC simulation value shows 3.5% (3.5fF) to that of intentionally designed value.

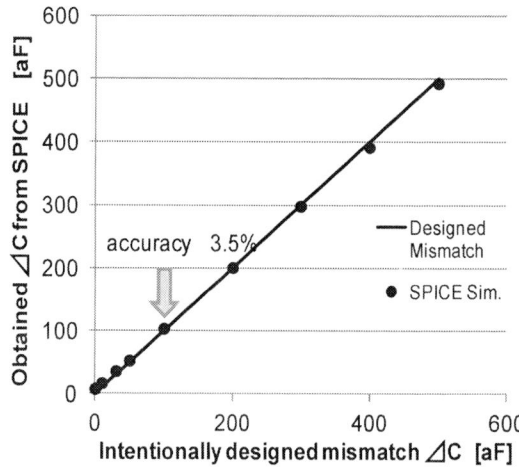

Figure 3. Comparison of ΔC from SPICE simulation vs. designed mismatch.

III. MEASUREMENT RESULTS

Fig. 4 shows an example of the test chip fabricated by a 40 nm foundry process. TABLE I shows DUTs of pair capacitances fabricated in DCBCM test structures. Fig. 5 reports measurement results of capacitance mismatch between two elements of 20fF and 20fF+ΔC (ΔC=100-500aF intentionally mismatched) MOMcap. In the measurement, the frequency and the voltage difference of input pulses are changed from simulated ones for purposes of increasing signal to noise ratio and set to 625kHz and 5V.

capacitances in the Pelgrom plot of Fig. 6, and the measurement sensitivity as small as σ = 0.05%.

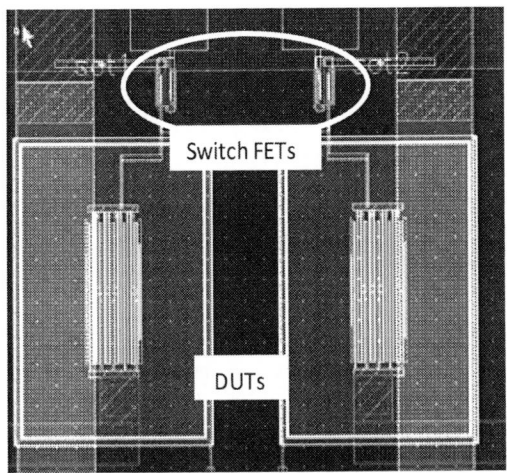

Figure 4.   An example of the test chip .

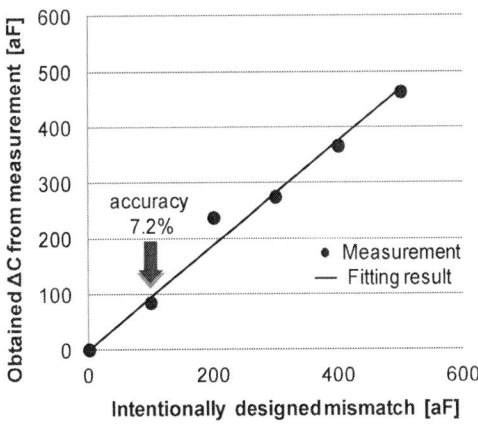

Figure 5.   Comparison of ΔC from measurement vs. designed mismatch.

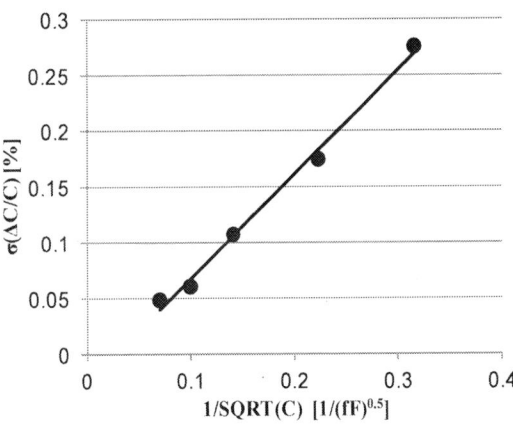

Figure 6.   Pelgrom plot for measured capacitance mismatches vs. capacitance.

TABLE I.   DUTs OF PAIR CAPACITANCES FABRICATED IN DCBCM

Capacitor Structure	Pair Capacitance [fF]	Switch NFETs (L/W) [um]
MOMCAP	20/20	0.27/0.5
	20/20.1	0.27/0.5
	20/20.2	0.27/0.5
	20/20.3	0.27/0.5
	20/20.4	0.27/0.5
	20/20.5	0.27/0.5
	10/10	0.27/0.32
	20/20	0.27/0.5
	50/50	0.27/1.25
	100/100	0.27/2.5
	200/200	0.27/5.0
MOSCAP	100/100	0.27/2.5
	200/200	0.27/5.0

At the 100aF mismatch point (at the smallest intentionally mismatch condition), the accuracy of the ΔC measurement fitting line shows 7.2% (7.2aF) to that of intentionally designed value. This gives same accuracy order as we simulated (Fig. 3). To plot the measurement mismatches versus 1/sqrt(capacitance), we measured 27 pairs of 10fF-200fF DUTs. We obtain clear linear dependence in a wide range of

When considering using DCBCM technique to measure MOScap, we face a trade-off between measurement accuracy and the voltage difference of input pulses (V_high-V_low). If the voltage difference is increased (decreased), the variability of C in ΔC/C (error component) is increased (decreased) because of the voltage dependence of MOScaps. At the same time the current from capacitance difference is increased (decreased). To avoid this trade-off, an improved control and measurement timing chart is introduced (Fig. 7). Comparing Fig.2 and Fig. 7, waveforms in phase 1 is replaced with 180 degree inverted phase 2 waveforms. This doubles effectively the voltage applied. For example 0.4V of the voltage difference (V_high/V_low=0.4V/0.0V) can be used in Fig. 7 instead of 0.8V of that (V_high/V_low=0.4V/-0.4V) in Fig.2.

978-1-4673-4845-4/13 $31.00 © 2013 IEEE

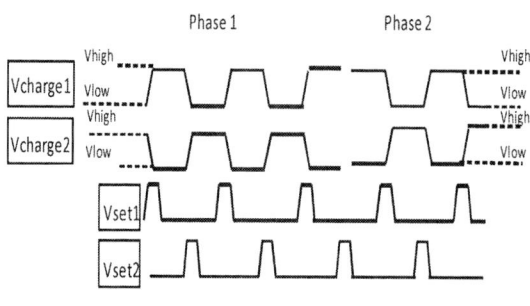

Figure 7.   Timing chart of control of the input pulses of Fig. 1.

In MOScap mismatch measurement bias voltage for MOScap were set to -1V, 0V and +1V as indicated in Fig. 8. Fig. 9 indicates $\Delta C/C$ as a function of bias and sizing of MOScap. We can see a reasonable mismatch dependency of bias voltages. 200fF MOScap shows smaller mismatches than that of 100fF MOScap and the mismatch is highest at 0V bias due to steep slope characteristics of CV curve of MOScap. In this case the measurement sensitivity is as small as $\sigma = 0.06\%$.

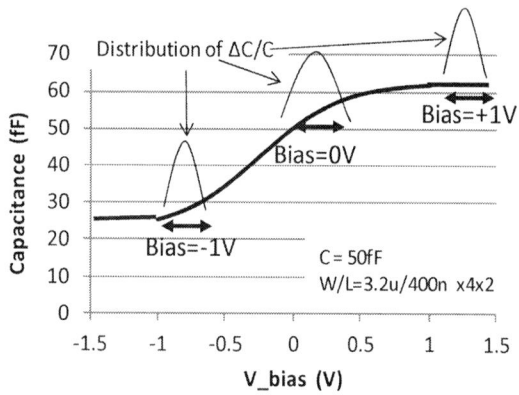

Figure 8.   C-V characteristic of a designed MOS capacitance.

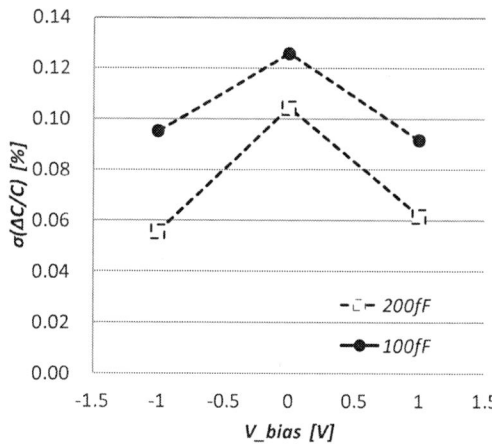

Figure 9. $\Delta C/C$ as a function of bias and sizing of MOS capacitance.

IV.   CONCLUSIONS

This paper has presented a simple difference charge-based capacitance measurement (DCBCM) test structure and technique to measure capacitance mismatches of tens of fF pairs. Thanks to alternate charging and discharging biasing scheme, we can eliminate parasitic components in the measurement current and achieved aF order accuracy. DCBCM gives a simple and accurate measurement method suitable for evaluating small capacitance mismatch for beyond 20nm node.

REFERENCES

[1]   J. C. Chen, B. W. McGaughy, D. Sylvester, and C. Hu, "An on-chip, attofarad interconnect Charge-Based Capacitance Measurement (CBCM) Technique," International Electron Device Meeting, pp. 3.4.1-3.4.4., 1996.

[2]   D. Sylvester, C. Chen, and C. Hu, "Investigation of Interconnect Capacitance Characterization Using Charge-Based Capacitance Measurement (CBCM) Technique and Three-Dimensional Simulation," IEEE J. Solid-State Circ., vol. 33, no. 3, pp. 449-453, Mar. 1998.

[3]   Y. W. Chang, H. W. Chang, T. C. Lu, Y. C. King, W. Ting, Y. H. Joseph Ku, and C. Y. Lu, "Interconnect Capacitance Characterization Using Charge-Injection-Induced Error-Free (CIEF) Charge-Based Capacitance Measurement (CBCM)," IEEE Trans. Semicon. Manufacturing, Vol. 19, No. 1, pp. 50-56, Feb. 2006.

[4]   E. Baruch, S. Shperber, R. Levy, Y. Weizman, J. Fridburg, and R. Marks, "A Simple System for on-die Measurement of atto-Farad Capacitance," Proc. 24th ICMTS, pp.19-21, Apr. 2011.

[5]   S. W. Sin, H. G. Wei, U. F. Chio, Y. Zhu, S. P. U, R. P. Martins, and F. Maloberti, "On-Chip Small Capacitor Mismatches Measurement Technique using Beta-Multiplier-Biased Ring Oscillator," IEEE Asian Solid-State Circ., pp. 49-52, Nov. 2009.

978-1-4673-4845-4/13 $31.00 © 2013 IEEE

Comparison of C-V measurement methods for RF-MEMS capacitive switches

Jiahui Wang, Cora Salm, and Jurriaan Schmitz

MESA+ Institute for Nanotechnology, University of Twente
Enschede, the Netherlands
Fax: +31 534891034. Email: j.wang-1@utwente.nl

Abstract—**The applicability of several capacitance-voltage measurement methods is investigated for the on-wafer characterization of RF-MEMS capacitive switches. These devices combine few-picofarad capacitance with a high quality factor. The standard quasistatic and high-frequency measurements are employed, as well as the recently introduced very-low-frequency method. S_{11} is measured by a network analyzer to calculate the capacitance of the device from radio-frequency measurements. Significant differences are found around the pull-in and pull-out voltages.**

Keywords—RF-MEMS switch; capacitance; parasitics; measurement

I. INTRODUCTION

RF-MEMS capacitive switches and varactors have penetrated the high-volume handset market in 2011 [1]. These new devices combine a high tuning ratio with a good quality factor (see e.g. [2]), making them an attractive choice for tunable impedance matching at the antenna. For process development and process control, electrical testing at wafer level of these components is essential, but is confronted with three major issues: the device is very sensitive to the ambient before packaging, it has a relatively low total capacitance of only a few picofarad, and its mechanical behavior has a very long time scale compared to other microelectronic components, in the millisecond range [3].

In this paper we will present a measurement approach which effectively tackles the ambient issue with a standard wafer-level probe station, and show and compare capacitance-voltage (*C-V*) measurements on RF-MEMS capacitive switches using the quasistatic, very-low-frequency, high-frequency and radio-frequency *C-V* measurement techniques.

II. EXPERIMENTAL

The RF-MEMS capacitive switches were fabricated on a Si substrate. Fig. 1 shows the top view of the RF-MEMS capacitive switch designed for ground-signal-ground (GSG) probing. The large perforated square (300 x 300 μm²) is the moving electrode, connected to ground; the underlying firm electrode is connected to the signal (S) pad. The movable top electrode is connected to four anchors by the curved springs. All metallization is aluminum. The devices (and open/short de-embedding structures) are contacted with two micromanipulators, with the exception of the RF-CV measurement where a GSG probe was used.

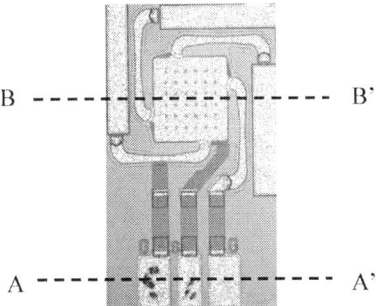

Fig. 1 Top view microscope image of the RF-MEMS capacitive switch under study. The A-A' and B-B' dashed lines indicate the cross-sections of figures 2 and 3.

A Keithley 595 Quasistatic meter (QSCV) was used for the quasistatic *C-V* measurement. Both an HP 4284A LCR meter (LCR) and a Keithley SCS 4200 (K4200) were used for the high-frequency *C-V* measurement.

The very-low-frequency (VLF) measurements [4] were conducted using the Keithley SCS 4200 with two pre-amplifiers. In the set-up of the VLF measurement, starting values for capacitance and resistance should be given. Only with well-chosen values (from previous measurements or calculated estimates) we obtain good convergence towards the expected capacitance.

A ZVB20 network analyzer [5] was used for 1-port radio-frequency *C-V* (RF-CV) measurement. Open/short/load calibration is carried out on a standard calibration pattern. The test structure parasitics are consecutively eliminated by de-embedding (as described later). The device impedance is calculated from S_{11} by the following equations [6]:

$$Z_m = 50\Omega \times \frac{1 + S_{11}}{1 - S_{11}}$$

$$Z_{DUT} = \frac{Z_o \times (Z_{mDUT} - Z_s)}{Z_o - Z_{mDUT} + Z_s} \qquad C_{DUT} = \frac{-1}{2\pi f \times \text{Im}(Z_{DUT})}$$

where Z_{DUT} is the impedance of the device under test; Z_o is open compensation; Z_s is the short compensation; Z_{mDUT} is the measured impedance of the device under test; Z_m is the measured impedance.

To remove any moisture, which is the main cause of stiction failure, a 10 minutes dry air flush at room temperature is employed followed by a 10 minutes heating step in dry air at

150 °C. All measurements were at room temperature with dry air flow. The measured capacitance kept stable when we repeated the *C-V* measurement. This indicates that no change of the device occurred during the measurement.

III. EXPERIMENTAL RESULTS

A. Device and system parasitics

A careful de-embedding is needed to get the accurate capacitance of the device under test (C_{DUT}), because the parasitic capacitance is of the same order as the DUT capacitance. The schematic cross sectional view of the bond pad area (A-A') is shown in Fig. 2 (a). Parasitic capacitance exists between the bond pads, through air and through the silicon substrate. Given that the outer pads are both grounded, the equivalent circuit in the bond pad area A-A' can be simplified as shown in Fig. 2 (b).

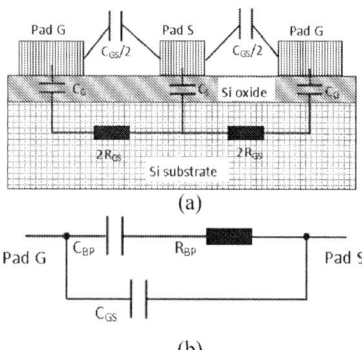

(a)

(b)

Fig. 2 (a) The schematic cross sectional view of the device from A-A' of Fig. 1, (b) the small-signal equivalent circuit of the parasitics.

The schematic cross sectional view of the device at B-B' is shown in Fig. 3 (a). Besides the device capacitance under study C_{DUT}, parasitic capacitance is found around the switchable electrode via the substrate. Given the symmetry, again the parasitic equivalent circuit can be simplified as the sub-circuit of C_{BE+} and R_{BE}, as shown in Fig. 3 (b).

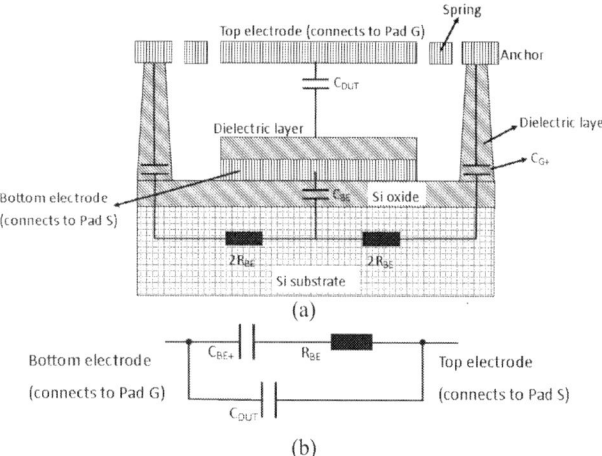

(a)

(b)

Fig. 3 (a) The schematic cross sectional view of the device from B-B' of Fig. 1, (b) the small-signal equivalent circuit of the parasitics.

Apart from the parasitics shown in Fig. 2 and Fig. 3, there are cable inductance (L_{SYS}), pad-to-probe resistance (R_{SYS}) and open capacitance (C_{SYS}) outside the test structure; and there are short inductance (L_s) and resistance (R_s) introduced by the connection lines between the pads and the device. A total equivalent circuit is shown in Fig. 4 by combining all parasitics.

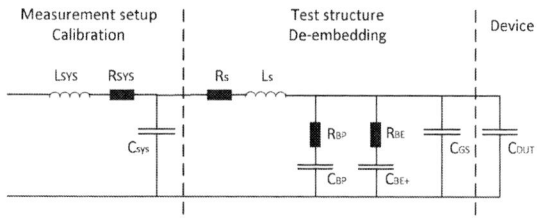

Fig. 4 The small-signal equivalent circuit of parasitics of the device and measurement system.

The device is larger at B-B' than at A-A'. As a consequence, the RC-times of the parasites $R_{BP}C_{BP}$ and $R_{BE}C_{BE+}$ are significantly different. In a device study across a large frequency range, this may lead to different open capacitance corrections for different regimes.

The open–correction is capacitive at our measurement frequency range from quasi-static to 1 GHz. Several reference structures were available on-wafer for open de-embedding, but no structure was designed such that an accurate open correction could be obtained for all measurements. Hence, one DUT was sacrificed to the purpose: the top electrode was mechanically removed from the device, and the damaged structure was then used as the open reference [7]. This approach does disconnect the two ground pads, so we divided the obtained open impedance by two when the devices were probed with two micromanipulators (for QSCV, VLF and high-frequency measurements). For radio-frequency measurements, the two ground pads are connected using the GSG probe, so the obtained "open" capacitance is correct.

The short compensation is only important in radio-frequency measurement. We found a significant difference between the device's parasitic inductance and that of the provided "short" reference structure on the wafer. By estimation of the induction-loop areas of the DUT and the "short" test structure, L_s as obtained from the short measurement could be corrected manually (multiplying L_s with the loop area ratio). With the corrected L_s value, the capacitance became frequency independent.

Exemplary capacitance-frequency curves of the device, obtained by the five measurement techniques, are shown in Fig. 5. In this figure, obtained capacitance values in the "up" and "down" state of the switch are shown from all five measurement approaches. As the quasistatic measurement does not associate with a measurement frequency, the QSCV values are indicated with dashed lines. High-frequency measurements (using LCR and K4200) below 10 kHz were inaccurate, as the device capacitance is too low. Hence these data are not shown here.

978-1-4673-4845-4/13 $31.00 © 2013 IEEE   54

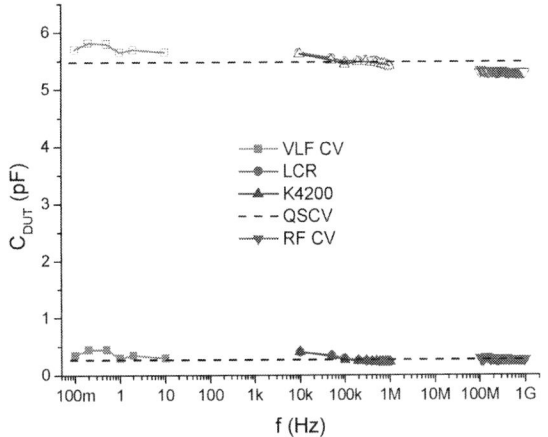

Fig. 5 The device capacitance as a function of frequency, as determined by the five measurement approaches. The capacitance is almost independent of frequency. The open-symbol measurements are obtained under DC bias of -15 V, while the solid symbols represent measurements at 0 V bias.

The obtained average value of C_{DUT} is 5.65 pF and 0.3 pF at down-state and up-state, respectively, almost independent of frequency. This indicates an accurate de-embedding. However, in the full C-V curve we did observe distinct differences. These are treated in the next subsection.

B. C-V Comparison measured by five approaches

Fig. 6 shows full C-V curves measured by these five approaches. Upon closer inspection, the C-V curves measured by different instruments exhibit some differences, in particular near the pull-in and pull-out voltages (V_{pi}, V_{po}). In Fig. 6, the voltage region between -15 V and -8 V shows distinctly different curves for HF and RF capacitance (gradually going down) on one hand, and QSCV and VLF capacitance (remaining constant, then going up) on the other hand.

This gradual capacitance decrease with reducing electric field in the downstate is commonly and consistently observed on RF MEMS capacitive switches. It is attributed to the flattening of the dielectric-metal interface or zipping effects [8-12].

However, in the QSCV and VLF measurements, the C_{DUT} at down-state does not decrease with the absolute value of DC bias voltage close to the pull-out voltage. Instead, C_{DUT} increases when the DC bias is close to V_{po}, so either the measured values are incorrect, or the commonly accepted contact model is at least incomplete.

It should be noted that the quasistatic measurement is error-prone upon the sudden switching of the device, where the assumption of quasistatic conditions does not hold. (At pull-out, an infinitesimal change in DC voltage leads to a large capacitance change.) This however does not explain any discrepancies seen several measurement points away from V_{po}.

The VLF and QSCV measurements typically have a slower voltage sweep than the high-frequency methods. We investigated whether the different measurement time scale is the cause for this effect, by artificially slowing down the high-frequency C-V sweeps. The measurement time does not influence the C-V curve in these high-frequency measurements even with 4 seconds hold time and 4 seconds wait time.

A more detailed comparison of the C-V curves near V_{po} is shown in Fig. 7. In RF, K4200 and LCR measurements, C_{DUT} decreases with an average rate of 0.025 pF/V between -20 V and -11 V. The decrease is somewhat faster between -11 V and V_{po} (see Figs. 7a–c). In the VLF measurements (Fig. 7d), C_{DUT} decreases with a slower rate of 0.02 pF/V from -20 V to -11 V (independent of the frequency). Between -11 V and V_{po}, C_{DUT} increases with DC bias voltage. The QSCV measurement is similar (Fig. 7e), be it somewhat obscured by the lower measurement accuracy.

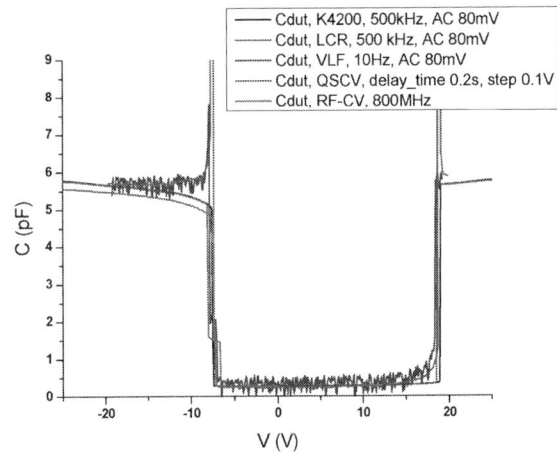

Fig. 6 Comparison of the C-V curves measured by five approaches.

This C-V behavior has been observed consistently on various devices under test. Further, no measurement-induced device changes have been observed on any of the measurement techniques or devices.

From our experiments reported above, we can rule out the measurement time and measurement-induced device changes as causes for the observed discrepancy between high-frequency and low-frequency measurements in the voltage range between -11 V and V_{po}. But otherwise, the origin if this discrepancy is still unclear.

978-1-4673-4845-4/13 $31.00 © 2013 IEEE

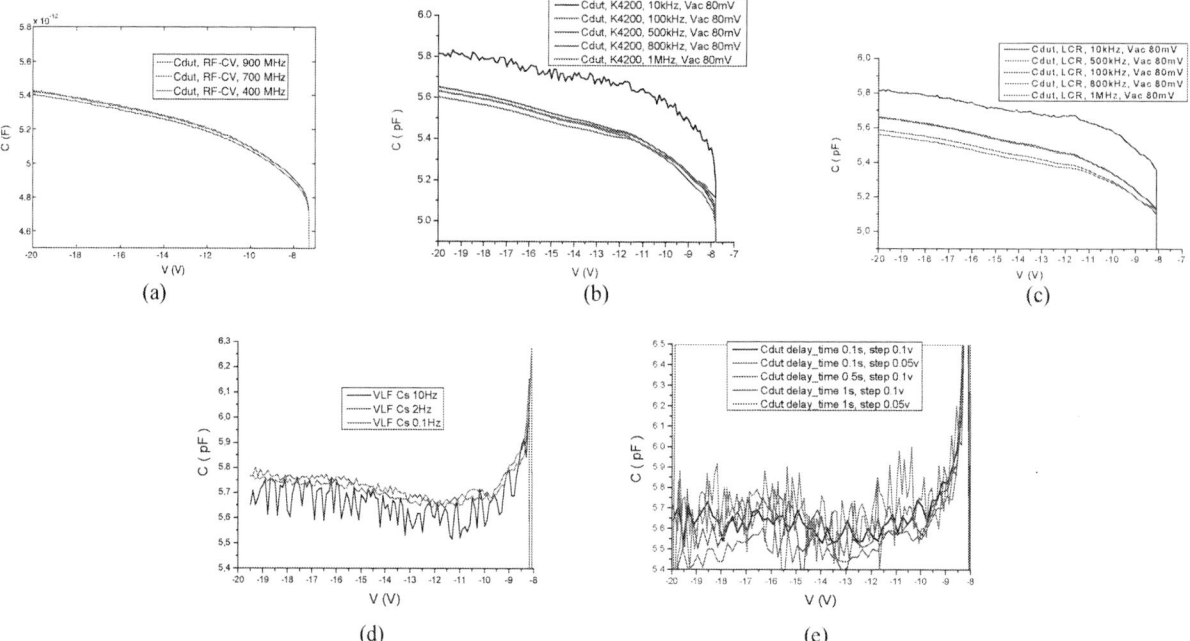

Fig. 7 Comparison of the capacitance of the device in the down-state, (a)RF-CV, (b) K4200, (c) LCR, (d) VLF, (e) QSCV measurements.

C. Intermediate capacitance state

We measured an 'intermediate state' between the down-state and the up-state (cf. Fig 6). Fig. 8 zooms in on the intermediate state of the *C-V* curve. The capacitance of this state is 1.5 pF in K4200, LCR and RF-CV measurements and 2 pF according to QSCV and VLF measurements.

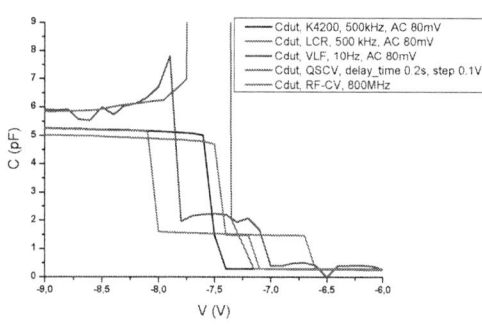

Fig. 8 *C-V* curves measured by five techniques, around the intermediate state observed at pull-out.

To further investigate this intermediate state, the topography of the device was measured by optical means using an MSA-400 Micro-system-analyzer. Fig. 9 (a) shows the topography of the device in the down-state. The red line drawn in the figure runs from the center of the movable top electrode to the fixed end of one anchor. We quantified the height of the device along this line as a function of the applied DC bias. As shown in Fig. 9 (b), the center of the top electrode moves up

slightly before the whole electrode comes up, likely corresponding to a lower capacitance than the full down-state. One possible explanation of this behavior is that the top electrode is not flat. The center part is higher than the edge part as a result of the fabrication process. A second factor may be that the electric field at the edge of the electrode is larger than in the center, so the electric force at the edge is larger.

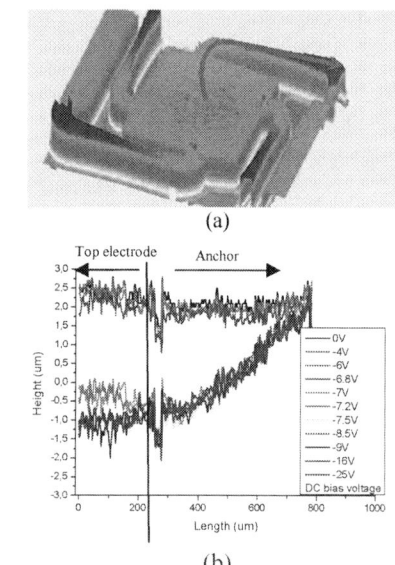

Fig. 9 Dynamic topography of the device, (a) the device in down-state, (b) the height of the surface of the device along the red line in (a) under different DC bias voltages.

According to the measurement results of more than 10 devices, the intermediate capacitance is always there but differs somewhat from device to device. It also changes slightly when we repeat the V-sweep measurement. The different capacitance values observed in the intermediate state among different measurements may then be caused by the stiffness of the device.

D. Pull-in voltages

The V_{pi} values measured by the different techniques do not coincide, as shown in Fig. 10. For the RF-CV measurement, V_{AC} was not determined and therefore V_{pi} is indicated with a dashed line. The power of the network analyzer we used for the RF-CV measurement is 0 dB (1 mW). In QSCV, V_{pi} is hard to estimate from the obtained curve; consecutive measurements lead to values randomly differing up to 0.2 V. Hence QSCV values of V_{pi} are not shown in this graph.

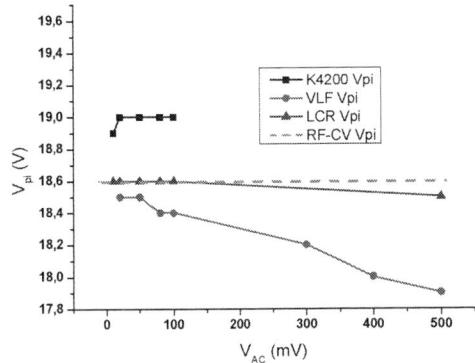

Fig. 10 Pull-in voltage measured by four methods, as a function of the applied ac test signal amplitude.

The observed differences among the pull-in voltages measured by different techniques may be related to device drift. (Note that these are unpackaged devices.) When the test signal has a larger amplitude, the instantaneous electric field may be significantly higher that the DC bias alone, which can lead to earlier pull-in. This is particularly observed with VLF CV measurements (with an almost linear dependence), where the measurement frequency is much lower than the device switching speed.

IV. DISCUSSION

A comparison of the measurement results of these five instruments is made in Table 1 (next page). High-frequency measurements are faster and have better accuracy than QSCV and VLF measurements. The RF-CV and high-frequency measurements, using LCR and K4200 instruments, show similar C-f (cf. Fig 5). The C-V measurement with the K4200 shows somewhat higher noise in the measurement than the LCR meter (cf. Fig 7b and 7c) but offers multiple functions like DC voltage list sweep, frequency sweep and time sweep.

The VLF and QSCV measurements exhibit higher noise, but do reveal slightly different behavior of the RF MEMS switch which calls for further investigation. Because of the measurement principle, QSCV measurements show unphysical overshoots during pull-in and pull-out. VLF measurement is slower but it can be used for precise measurements on devices with high series resistance (not shown here); it also has multiple functions like high-frequency measurements on the K4200.

V. CONCLUSIONS

The C-V curves of RF-MEMS switches measured by five different methods are compared. An equivalent circuit of the parasitics is given. After de-embedding, the device capacitance is almost independent of frequency across 10 orders of magnitude. The device capacitance appears different in the down-state near the pull-out voltage; quasistatic C-V and very low frequency measurements yield different results from RF-CV and high-frequency C-V measurements. The root cause of this difference is still under study. An intermediate capacitance state was observed at pull-out and associated with the stepwise raising of the moving electrode.

ACKNOWLEDGEMENT

This work has been performed in the EPAMO project, which is funded by public authorities of participant countries as well as by the ENIAC Joint Undertaking.

REFERENCES

[1] J. Bouchaud, "RF MEMS switches, varactors finally penetrate mobile handsets", IHS iSuppli MEMS Market Brief, Vol. 4 issue 12, December 2011.

[2] M. P. J. Tiggelman et al., On the trade-off between quality factor and tuning ratio in high-frequency capacitors, IEEE Trans. El. Dev. 56 (9) 2128 (2009).

[3] P. G. Steeneken et al., *Dynamics and squeeze film damping of a capacitive RF MEMS switch*, J. Micromech. Microeng. 15 (2005) 176–184.

[4] *Performing very low frequency capacitance-voltage measurements on high impedance devices using the model 4200-SCS semiconductor characterization system*, Keithley application note 3140.

[5] J. Schmitz et al., *RF capacitance-voltage characterization of MOSFETs with high leakage dielectrics*, IEEE Electron Device Letters, 24 (1) (2003) 37-39.

[6] Thomas H. Lee, *The design of CMOS radio-frequency integrated circuits*, Cambridge University Press, 2004.

[7] As suggested by R. W. Herfst, private communication, June 2012.

[8] H. M. R. Suy et al., *The static behavior of RF MEMS capacitive switches in contact*, Proceedings of Nanotech MSM, June 2008.

[9] Lifeng Wang et al., *Capacitance characterization of dielectric charging effect in RF MEMS capacitive switches under different humidity environments*, MEMSYS, Jan. 2012.

[10] A. Hariri et al., *Modeling of dry stiction in micro electro-mechanical systems (MEMS)*, J. Micromech. Microeng. 16 (2006) 1195-1206.

[11] A. B. Yu et al., *Effect of surface roughness on electromagnetic characteristics of capacitive switches*, J. Micromech. Microeng. 16 (2006) 2157-2166.

[12] H. M. R. Suy et al., *A Compact Scalable Circuit Model for RF MEMS Switches*, Nanotech MSM, May 2007.

TABLE I.   COMPARISON OF THE MEASUREMENTS OF A CAPACITANCE-VOLTAGE SWEEP FROM -20 V TO 20 V (STEP OF 0.1 V) BY FIVE TECHNIQUES

Instrument	QSCV	VLF	K4200	LCR	RF
Typical meas. time	> 1 minute	>> 20 minutes	0,2-2 minutes	1-4 minutes	< 5 seconds
Typical accuracy	≈ 0.1 pF	0.01 pF to 0.1 pF	< 0.001 pF at frequency from 50 kHz to 1 MHz	< 0.001 pF at frequency from 10 kHz to 1 MHz	Calculated from S_{11}
Range of DC voltage sweep	From -20 V to 20 V	From -20 V to 20 V	From -30 V to 30 V	From -40 V to 40 V	From -30 V to 30 V
Range of frequency	N.A.	0.01 Hz to 1 Hz	1 kHz - 10 MHz	20 Hz-1 MHz	80 MHz – 1 GHz
Ease of calibration	Difficult (small compensation noise)	Difficult (small compensation noise)	Easy	Easy	Easy
Artefacts	Yes (near V_{pi} and V_{po})	Yes but small (near V_{pi} and V_{po})	No	No	No
Sweep functions	V-sweep	V-sweep, f-sweep	V-sweep, DC voltage list sweep, f-sweep and time sweep.	V-sweep	V-sweep, f-sweep

Effective Channel Length Estimation Using Charge-Based Capacitance Measurement

Katsuhiro Tsuji and Kazuo Terada

Faculty of Information Sciences, Hiroshima City University
3-4-1, Ozuka-Higashi, Asa-Minami-Ku, Hiroshima, 731-3194, JAPAN
tsuji@hiroshima-cu.ac.jp

Abstract—**An effective channel length is estimated from the capacitance-voltage (C-V) curves of actual size MOSFETs which are measured using charge-based capacitance measurement (CBCM). To evaluate the accurate capacitances between the gate and the channel of sample MOSFETs, their parasitic capacitances are removed by using the test MOSFETs having various channel size and special test structure. A good linear relation between the gate-channel capacitance and the design channel length is obtained and then, the effective channel length is estimated from it. It is found that the obtained effective channel length is shorter than that extracted by the conventional channel resistance method.**

Keywords—effective channel length; charge-based capacitance measurement; MOSFET;

I. INTRODUCTION

An effective channel length (L_{EFF}) is an important parameter to determine MOSFET electric characteristics. However, as miniaturization of MOSFETs is progressed, the channel structure has become complex, and it becomes difficult to extract L_{EFF} accurately using channel resistance method (CRM). The reason is that the conventional CRM uses the drain current model for the MOSFET having the uniform channel [1]. L_{EFF}-extraction methods using gate capacitance measurement have also been proposed [2, 3]. While a channel resistance cannot be determined by these methods, the measurement is free from the ambiguities introduced by the non-uniformity of the channel resistance and the gate voltage-dependent mobility. On the other hand, these methods need to measure small capacitance which is difficult to measure or the capacitance of MOSFET having especially wide channel to reduce influences of parasitic capacitances. We have reported that capacitances of actual size MOSFETs can be measured by CBCM method [4, 5]. In this study, by using the techniques, we attempt to estimate L_{EFF} from C-V curves of actual size MOSFETs.

II. MEASUREMENT OF CAPACITANCES

A. CBCM Test Structure

The test circuit is structured by DMA (Device Matrix array) which consists of 64 cells (= 8 rows x 8 columns), row decoder and column decoder. Fig. 1 shows an equivalent circuit of the unit cell. It consists of NMOS (N-channel MOSFET) transfer gates and CBCM part. The both decoders and NMOS transfer gates are designed with 0.6-μm channel and 3-V supply technology. For the CBCM part, they are

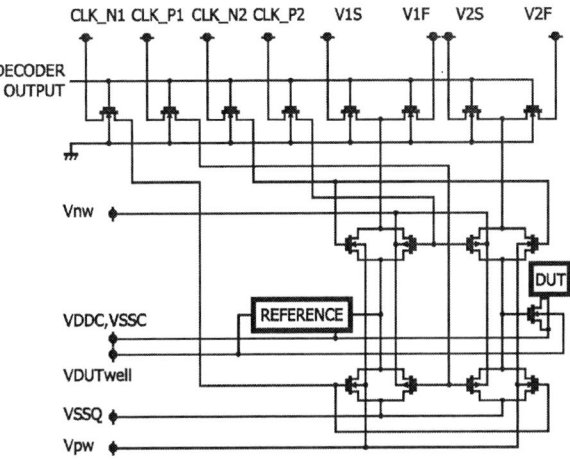

Figure 1. An equivalent circuit of unit cell structure for measurement of gate to substrate capacitance.

TABLE I. THE MEASURED CHANNEL STRUCTURE AND CHANNEL DIMENSIONS.

channel structure	with halo regions, and $T_{OX} = 2$ nm (T_{OX}: gate oxide thickness)
channel width W	0.6, 1.0 μm
channel length L	0.12, 0.25, 0.6, 1.0 μm

designed with 65-nm channel and 1.2-V supply technology. DUT MOSFET and REFERENCE are connected to the middle node of between two CMOS transmission gates, respectively. Table 1 shows the channel structures and the channel dimensions. A total of 8 combinations of channel dimensions for NMOS are used in this study. The capacitance of DUT MOSFET (C_{DUT}) is measured by the charge-injection-induced-error-free CBCM (CIEF-CBCM) method [6]. The method is based on twice DC current measurements. Table 2 shows the measurement conditions. Fig. 2 shows the non-overlapping pulses applied to test circuit. In the first step (STEP1), the current (I_{STEP1}), which contributes to charge and to discharge whole capacitances involved in the current path, is measured at V1F or V2F terminal. In the second step (STEP2), the current (I_{STEP2}) without charge and discharge to

TABLE II. The Measurement Conditions for Fig. 1.

$VDDQ$ = V1F, V2F	0 ~ 1.2 V, 0.05 V step
$VSSQ$	0 V
$VDUTwell$, $VSSC$	STEP1 : Fig. 2(e) STEP2 : Fig. 2(f)
Frequency	0.45 MHz

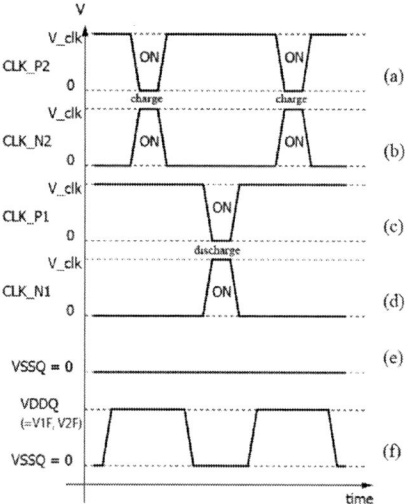

Figure 2. Timing chart of non-overlapping pulses.

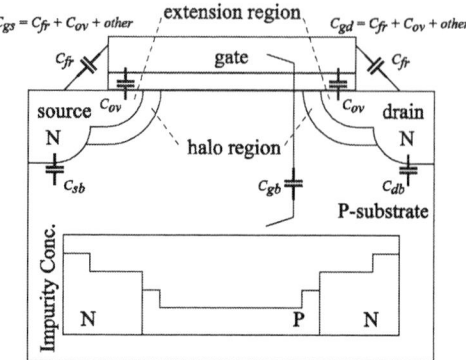

Figure 3. Schematic view of extrinsic parasitic capacitances, where C_{fr} and C_{ov} are fringe capacitance and overlap capacitance, respectively.

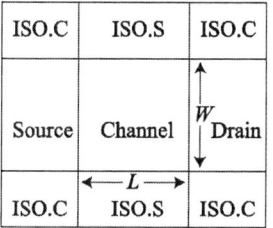

Figure 4. Top view of MOSFET structure, where ISO.C and ISO.S are isolation of corner and isolation of side, respectively.

DUT MOSFET is measured. Net current contributing to C_{DUT} is obtained by subtracting I_{STEP2} from I_{STEP1}, then, C_{DUT} is expressed by

$$C_{DUT} = \frac{1}{f} \cdot \frac{d(I_{STEP1} - I_{STEP2})}{dVDDQ}, \qquad (1)$$

where f is the frequency of non-overlapping pulses, and $VDDQ$ is the applied voltage to V1F and V2F terminals in Fig. 1.

B. Elimination of Parasitic Capacitaces

Parasitic capacitances (C_{gs}: fringe C_{fr}, overlap C_{ov} and other capacitances in between gate and source, C_{gd}: C_{fr}, C_{ov} and other capacitances in between gate and drain and C_{gb}: gate-body capacitance) are involved in the obtained C_{DUT}, as shown in Fig. 3. Also, junction capacitances C_{sb}, C_{db} in Fig. 3 are one of the parasitic capacitances, however, C_{sb} and C_{db} are negligible, because same voltage is applied to $VDUTwell$ and $VSSC$ terminals (that is: substrate and drain/source of DUT MOSFET) during the above measurement. To estimate L_{EFF} accurately, above parasitic capacitances need to be removed.

Fig. 4 shows a top view of MOSFET structure, where ISO.C and ISO.S are an isolation region of corner and an isolation region of side, respectively. Additional parasitic capacitances other than them shown in Fig. 3 exist in these

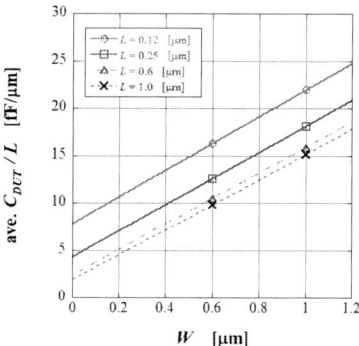

Figure 5. Plots of the average C_{DUT}/L vs. W. The parasitic capacitances per unit length for each L are obtained from these y-intercepts.

isolation regions. Then, the measured C_{DUT}s involve their capacitances. Therefore, these parasitic capacitances also need to be removed to increase estimation accuracy of L_{EFF}.

C. Removal of Gate to Body Capacitance

Fig. 5 shows plots of C_{DUT}/L versus W, where L and W are channel length and channel width, respectively, and C_{DUT} is average of the data for 24 test MOSFETs having the same L

978-1-4673-4845-4/13 $31.00 © 2013 IEEE

and W. Then, C_{DUT}s shown in Fig. 5 are measured at where DUT MOSFETs are in strong inversion. C_{DUT} values at $W = 0$, the capacitances which are extracted from y-intercept in Fig. 5 [7], is considered to be parasitic capacitances in four ISO.C and two ISO.S regions in Fig. 4. Thus, because the obtained capacitance is a parasitic capacitance between gate and substrate along the channel length direction, outside the immediate channel area, it is considered that these capacitances obtained from Fig. 5 are corresponding to the components of four ISO.C and two ISO.S regions in C_{gb} in Fig. 3. Thus, by performing the procedure shown in Fig. 5 for each measurement voltage, parasitic capacitances along L for each measurement voltage are obtained, and are subtracted from C_{DUT}.

D. Measurement of Gate to Drain/Source Capacitances

The parasitic capacitances C_{gd} and C_{gs} are evaluated by the test circuit in Fig. 6. The connection of DUT MOSFET in this circuit is different from that of Fig. 1. As shown in Fig. 6, drain and source of DUT MOSFET are connected to the middle node of between CMOS transmission gates. Thus, the gate-drain/source capacitances can be measured by this test structure. The measurement condition is represented in Table 3. The measurement of current (I_{STEP1}) of STEP1 is the same way as obtaining C_{DUT}. The measurement of current I_{STEP2} of STEP2 is measured by applying the pulse of Fig. 2(f) to $VSSC$ terminal, that is: the gate of DUT MOSFET becomes the same as the voltage of the middle node of between CMOS transmission gates. On the other hand, $VDUTwell$ terminal is keeping the condition of STEP1. Consequently, the current which charge and discharge to both C_{gd} and C_{gs} is not measured. Therefore, the sum of $C_{gd} + C_{gs}$ is extracted from Eq. (1).

Fig. 7 shows the measured ($C_{ge} + C_{gs}$), which are average values of 8 test MOSFETs, for each channel dimension. It is found that those values are almost independent of both the applied voltage and L. On the other hand, it is also found that those values do not increase in proportion as the channel width W, because the obtained capacitances involve components that are independent of W. To extract more accurate gate-drain/source capacitances, it is needed that their components are removed from the obtained capacitances. Fig. 8 plots the average values of all measurement points of Fig. 7 for each channel dimension as a function of W. This figure is plotted by regarding the measured values shown in Fig. 7 as constant value which is independent of voltage. The straight lines are the regression lines for each L. The components that are independent of W are obtained from their y-intercepts. It is considered that they are capacitances in ISO.C regions in Fig. 4. Fig. 9 plots the corrected average capacitances per unit W, where the independent components of W obtained from Fig. 8 are removed. Because the values in Fig. 9 represent the independence of W, it is considered that they show accurate ($C_{gd} + C_{gs}$) per unit W. In this study, average of these values is used as the extrinsic parasitic capacitance of the channel width direction. The value is 0.42 fF/μm. To obtain more accurate gate capacitance, the value is subtracted from C_{DUT} which are removed parasitic capacitances along L.

TABLE III. THE MEASUREMENT CONDITIONS FOR FIG. 5.

$VDDQ$ = V1F, V2F	0 ~ 1.2 V, 0.05 V step
$VSSQ$	0 V
$VSSC$	STEP1 : Fig. 2(e) STEP2 : Fig. 2(f)
$VDUTwell$	STEP1, STEP2 : Fig. 2(e)
Frequency	0.45 MHz

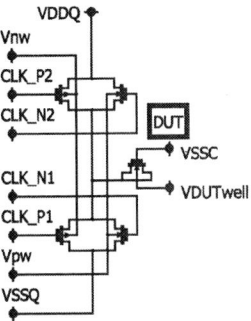

Figure 6. DUT MOSFET connection to the middle node of between CMOS transmission gates for measurement gate to drain/source capacitances.

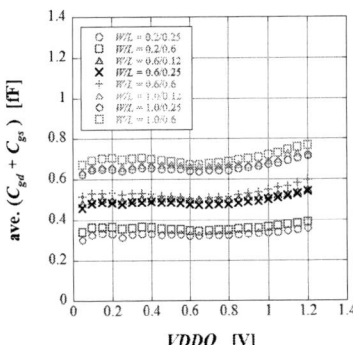

Figure 7. Measured $C_{gd} + C_{gs}$, where are average values 8 test MOSFETs, for each channel dimension.

III. L_{EFF} EXTRACTION

Fig. 10 shows C_G per unit W as a function of $Vgwell$, where C_G and $Vgwell$ are, respectively, the corrected gate capacitance which are removed the parasitic capacitances from C_{DUT} and the voltage which is defined as "effective voltage between gate and substrate of DUT MOSFET", that is: $Vgwell = VDDQ - VDUTwell$. It is noted that the represented capacitances are the average value per unit width of 24 test MOSFETs, for each channel dimension. It is considered that the parasitic capacitances are eliminated from the measured C-V curves.

978-1-4673-4845-4/13 $31.00 © 2013 IEEE

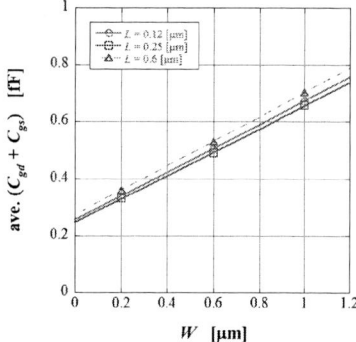

Figure 8. Plots of average values for each channel dimension as a function of W.

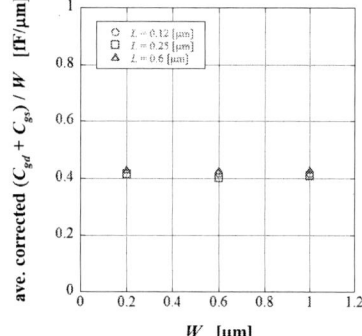

Figure 9. Plots of corrected average values which are removed the components obtained from Fig. 8, where the plots are the capacitance per unit W.

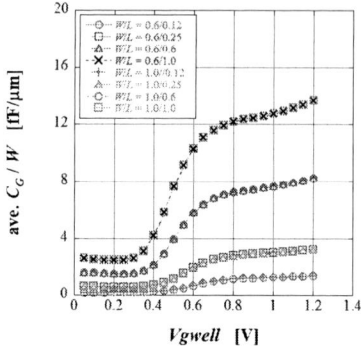

Figure 10. C_G per unit W as function of $Vgwell$, where C_G is the corrected gate capacitance which are removed the extrinsic parasitic capacitances from C_{DUT}.

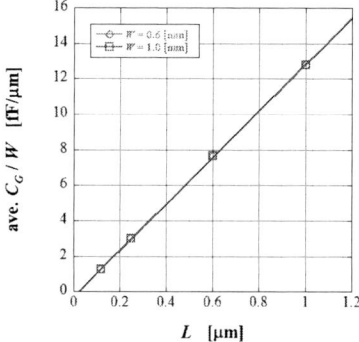

Figure 11. Plots of average C_G/W when $Vgwell = 1.0$ V. ΔL is obtained from the x-intercept.

Assuming that C_G in strong inversion region nearly equals to the gate oxide thickness capacitance C_{OX}. C_G/W can be expressed as follow:

$$\frac{C_G}{W} = \frac{\kappa_{OX}\varepsilon_0}{T_{OX}} \cdot L_{EFF} = \frac{\kappa_{OX}\varepsilon_0}{T_{OX}} \cdot (L_{DES} - \Delta L), \qquad (2)$$

where κ_{OX}, ε_0 and T_{OX} are oxide dielectric constant, permittivity in vacuum and oxide thickness, respectively. ΔL represents the design channel length $L_{DES} - L_{EFF}$, where L_{DES} is equal to L in this study. Fig. 11 shows average C_G/W, which are obtained from Fig. 9, as a function of L when $Vgwell = 1.0$ V. The straight lines are the regression lines. The both lines coincide. A good linear relation between the gate-channel capacitance and the design channel length is obtained. ΔL is obtained from the x-intercept, then, the value is around 0.02 µm. Thus, it is found that the extracted L_{EFF} becomes shorter than L_{DES}.

IV. CONCLUSION

For MOSFET having complex channel structure, it has been known that L_{EFF} extracted from the conventional CRM becomes abnormally longer than L_{DES}. It has also been obtained that L_{EFF} extracted from the improved CRM becomes slightly longer than L_{DES}. However, in this study, the extracted L_{EFF} becomes shorter than L_{DES}. The reason is considered that the extension regions in drain/source regions are regarded as the MOSFET channel in CRM. On the other hand, in this study, the extension regions are regarded as a part of parasitic capacitances C_{gd} and C_{gs}, and are removed. Therefore, it is considered that the extracted L_{EFF} in this study becomes shorter than L_{DES}.

REFERENCES

[1] K. Terada and H. Muta, "A new method to determine effective MOSFET channel length," Jpn. J. Appl. Phys. **18** (1979) pp. 953-959.

[2] J. C. Guo, S. S. Chung, and C. C.Hsu, "A new approach to determine the effective channel length and drain-and-source series resistances of miniaturized MOSFET's," IEEE Trans. ED., vol.41, no.10, October 1994, pp. 1811-1818.

[3] D. Fleury, A. Cros, K. Romanjek, D. Roy, F. Perrier, B. Dumont, H. Brut, and G. Ghibaudo, "Automatic extraction methodlogy for accurate measurements of effective channel length on 65-nm MOSFET technologyand below," IEEE Trans. Semiconductor Manufacturing 21 (2008) 504.

[4] K. Tsuji, K. Terada, R. Kikuchi, T. Tsunomura, A. Nishida, and T. Mogami, "Evaluation of MOSFET C-V Curve variation using test structure for charge-based capacitance measurement," Proc. Int. Conf. on Microelectronic Test Structures, pp. 8-12. April 2011.

[5] K. Tsuji, K Terada, R. Takeda, T. Tsunomura, A. Nishida, and T. Mogami, "Threshold voltage variation extracted from MOSFET C-V curves by charge-based capacitance measurement," Proc. Int. Conf. on Microelectronic Test Structures, pp. 82-86. March 2012.

[6] Y. W. Chang, H. W. Chang, C. H. Hsieh, H. C. Lai, T. C. Lu, W. Ting, J. Ku, and C. Y. Lu, "A novel simple CBCM method free from charge injection-induced errors," IEEE Elec. Dev. Lett., vol.25, no.5, May 2004, pp. 262-264.

[7] J. C. Guo, C. C. H. Hsu, P. S. Lin, and S. S. Chung, "An accurate "Decoupled C-V" method for characterizing channel and overlap capacitances of miniaturized MOSFET," Proc. VLSI-TSA1993, 1993, pp. 256-260.

978-1-4673-4845-4/13 $31.00 © 2013 IEEE

A new Ultra-Fast Single Pulse technique (UFSP) for channel effective mobility evaluation in MOSFETs

Z. Ji[1*], J. Gillbert[2], J. F. Zhang[1], and W. Zhang[1]

[1] School of Engineering, Liverpool John Moores University
Liverpool, UK
[2] Keithley Instruments, UK

*Corresponding author contact: Z.Ji@ljmu.ac.uk Tel: +44 1512312383

Abstract— **A new technique is proposed for mobility evaluation to overcome the shortcomings of conventional techniques. By measuring Id and Cgc simultaneously within 3 μs, it removes adverse impact of Vd on mobility, avoids cable-switching, and minimizes charge trapping. Furthermore, it can work on highly 'leaky' devices without special RF structure. It is shown that mobility can be extracted with gate leakage current density as high as 40 A/cm^2. The sources of error are then systematically analyzed. This technique can be easily implemented in Keithley 4200 semiconductor analyzer with two 4225-PMUs and therefore it can serve as a simple and robust tool for accurate mobility exaction for material selection during technology development.**

Keywords—Mobility measurement, MOSFET, Ultra-fast measuremnt

I. INTRODUCTION

Channel carrier mobility is a key parameter for selecting materials and process development. Conventionally, it is extracted from inversion charges, Q_i, measured by split C-V technique and channel conduction current, I_{ch}, from I_d-V_g measurement through (1) [1].

$$\mu_{eff} = \frac{L}{W} \cdot \frac{I_{ch}}{V_d \cdot Q_i} \tag{1}$$

This technique suffers from a number of disadvantages. Since V_d=0 is applied for split C-V measurement and non-zero V_d cannot is applied for I_d-V_g measurement, Q_i from split C-V is larger than the real inversion charge for I_{ch}, leading to an under-estimation of mobility [2]. With the downscaling of gate oxide and the introduction of new dielectric materials, gate leakage [3] and charge trapping [4] are severe and can cause further significant errors in mobility evaluation. Moreover, it requires changing cables during measurement, limiting measurement speed. Although many attempts [2-8] have been made to improve the technique, there is no technique that can overcome all these shortcomings at present. In this work, Ultra-Fast Single Pulse (UFSP) technique is proposed to provide a complete solution to solve all the problems mentioned above.

II. DEVICE AND EXPERIMENTAL SETUP

A. Device

One control pMOSFET with 2.3nm SiO_2 dielectric was selected to calibrate the UFSP technique since it exhibits negligible gate leakage and fast trapping. This device has a p$^+$

poly-Si gate. To demonstrate the applicability of UFSP on devices with considerable charge trapping, one pMOSFET with HfSiON/SiON gate stack was used. The equivalent oxide thickness is 1.65nm and has a TiN gate. Finally, one ultra-thin nMOSFET with HfSION/SiON gate stack is used to demonstrate the applicability to leaky devices. The equivalent oxide thickness is 1.28nm and has a TiN gate. Unless specified, the samples have a channel length and width of 10 μm x10 μm.

B. Experimental setup

The UFSP technique is based on transient Current-Voltage measurement by using two Keithley 4225-PMUs, as shown in Fig. 1. Four Keithley 4225-RPMs are used to reduce cable capacitance effect and achieve accurate measurement below 100 nA in μs. For UFSP measurement, a single pulse (shown schematically in inset of Fig. 2) is applied to the gate. The device is switched on and then off. The corresponding currents at source and drain is measured, i.e. I_d^{on}, I_s^{on}, I_d^{off} and I_s^{off}. The measurement time (=edge time) in this work is set at 3 μs. All the measurements are controlled by Keithley KTEI software.

III. ULTRA-FAST SINGLE PULSE TECHNIQUE (UFSP)

Firstly, one pMOSFET with 2.3 nm SiO_2 is used to demonstrate the technique. Four measured currents are plotted in Fig. 2. Their difference can be examined by analyzing the current flow in the channel during the transient measurement as shown in Fig. 3. Three types of current are present: channel conduction current, I_{ch}, displacement current between gate and source/drain, I_{dis_s} and I_{dis_d}, and the leakage current between gate and source/drain, I_{g_s} and I_{g_d}. When device is switched off-to-on, the direction of I_{dis_s} and I_{dis_d} is toward the channel, same direction as I_{ch} at the source, but in opposite direction to I_{ch} at the drain. This leads to $I_s^{on} > I_d^{on}$. When device is switched on-to-off, I_{dis_s} and I_{dis_d} change direction, but I_{ch} does not. This leads to $I_d^{off} > I_s^{off}$. I_{g_s} and I_{g_d} are independent of V_g sweep direction and always flow from the source and drain towards gate under negative V_g. Based on above analysis, channel current, I_{ch}, gate current between gate and channel, I_g, displacement current between gate and channel, I_{dis} can be separated by using (2) - (4). C_{gc} can be calculated using (5).

$$I_{ch} = \frac{I_d^{on} + I_d^{off} + I_s^{on} + I_s^{off}}{4} \tag{2}$$

$$I_g = I_{g\_s} + I_{g\_d} = \frac{I_s^{on} + I_s^{off} - I_d^{on} - I_d^{off}}{4} \tag{3}$$

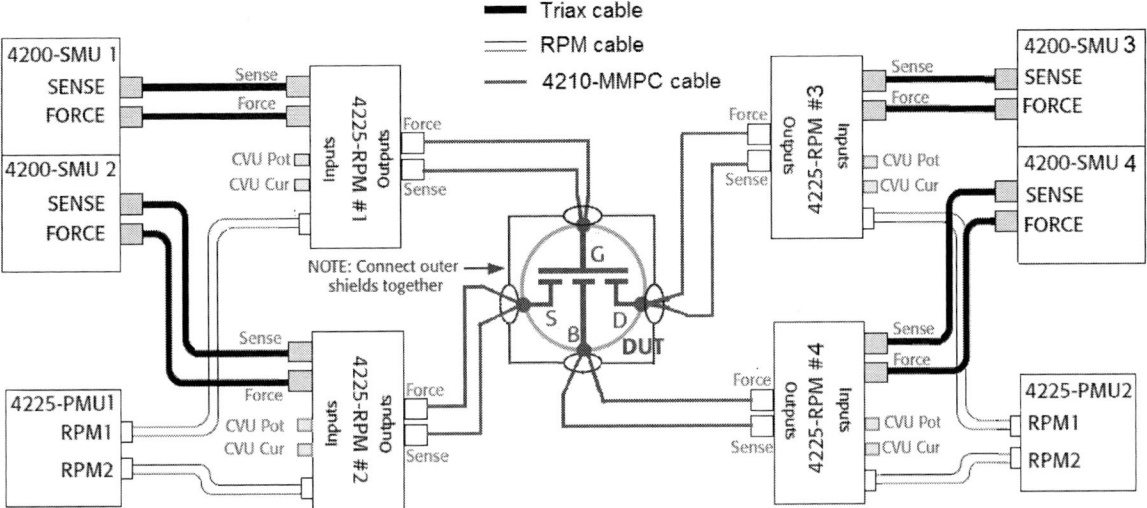

Fig 1: Experiment setup for Ultra-fast Single Pulse (UFSP) technique. Two Keithley dual-channel 4225-PMUs are used for performing transient measurements. Four Keithley 4225-RPMs are used to reduce cable capacitance effect and achieve accurate measurement below 100 nA. Moreover, these 4225-PRMs also facilitate the comparison between DC and transient measurement by automatically switching between different configurations.

$$I_{dis} = I_{dis\_s} + I_{dis\_d} = \frac{I_d^{off} - I_d^{on} + I_s^{on} - I_s^{off}}{4} \quad (4)$$

$$C_{gc} = I_{dis} / [dV / dt] \quad (5)$$

Fig. 4 plotted the extracted transient I_{ch}, I_g and C_{gc} and compared them with DC measurement. Good agreement has been obtained. As expected, this pMOSFET with relatively thick oxide has negligible I_g compared with I_{ch}. Once C_{gc} and I_{ch} are evaluated accurately, Q_i can be obtained by integrating C_{gc} against V_g. With the corresponding I_{ch}, channel effective mobility, μ_{eff}, is calculated through (1). Fig. 5 compares μ_{eff} evaluated by UFSP technique under 3 different V_d. Good agreements are obtained indicating the errors induced by V_d from the conventional techniques has been removed.

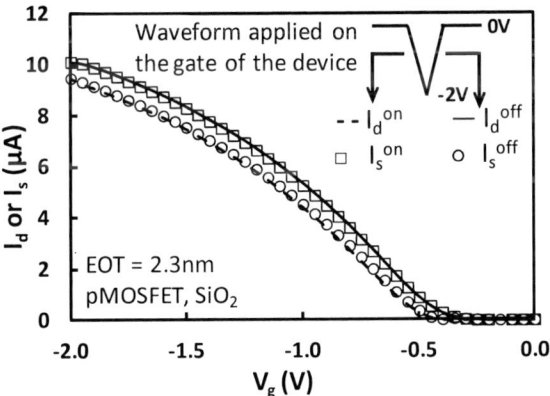

Fig 2: Four currents measured from source and drain corresponding to the off-to-on and on-to-off V_g sweep. The applied drain bias, Vd, is -100mV. Schematic V_g waveform is shown in inset.

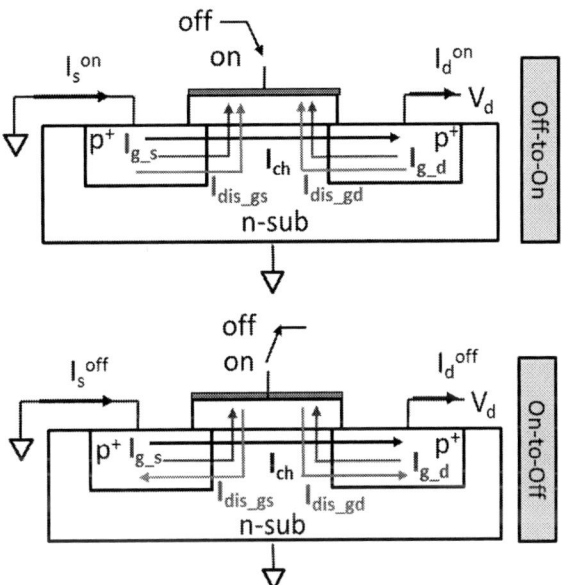

Fig 3: Schematic diagram of current flow during transient measurement. The measured currents at source and drain include three components: conduction current I_{ch} (black line), displacement currents I_{dis_gs} or I_{dis_gd} and gate leakage current Ig_s or Ig_d.

Nowadays, the oxide thickness of the device shrinks significantly. Therefore, in order to be a useful technique, it must have the capability to work on ultra-thin devices. High leakage current will affect both I_d-V_g and C_{gc} measurements and in turn the mobility extraction. In order to minimize its impact, frequency up to Giga Hz was used and special RF structure with large area is required.

978-1-4673-4845-4/13 $31.00 © 2013 IEEE

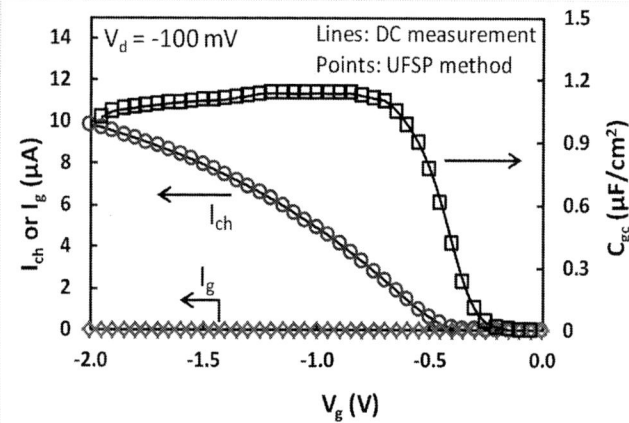

Fig 4: Ich ('o'), Ig ('◊') and Cgc ('□') extracted simultaneously from the currents in Fig.2 by using (2)-(5). Black lines are from the measurement using conventional Id-Vg and split C-V technique.

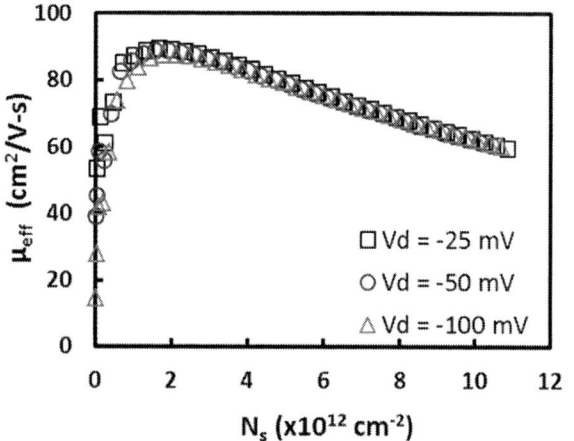

Fig 5: The effective channel mobility, μ_{eff}, extracted under 3 different V_d by using UFSP technique.

Unlike the existing techniques, UFSP can work very well with leaky devices of standard structure. one 'leaky' nMOSFET with EOT of 1.28nm is used. Four currents measured from the source and drain corresponding to the off-to-on V_g sweep and on-to-off V_g sweep are shown in Fig. 6 (a). By using (2)-(5), I_{ch} ('□'), I_g ('o') and C_{gc} ('x') are extracted and plotted in Fig. 6(b). Ig from DC measurement is also plotted for comparison in Fig. 6(b). Good agreement is obtained. The device has a leakage current density as high as 45 A/cm². Fig. 6 (c) shows that electron mobility can be reliably measured for this leaky device where Ig is as high as 45 A/cm². UFSP can tolerant much higher leakage density under higher measurement speed, thus special RF structure will not be necessary for mobility evaluation.

To demonstrate the applicability of UFSP to devices with significant charge trapping, one pMOSFET with HfO_2/SiO_2 stack was used. Large amount of traps locate very close to Si/SiO_2 interface in such dielectric stack and they can exchange charges with substrate rapidly. The conventional split C-V takes seconds, making them indistinguishable from channel mobile charges. As a result, inversion charges will be over-estimated and in turn the channel effective mobility will be underestimated. The UFSP technique only takes microseconds,

minimizing charge trapping effect. Fig. 7 compares the mobility extracted by these two techniques. It clearly shows that after suppressing the trapping, the mobility extracted from the UFSP ('□') is considerably higher than that by the conventional split C-V ('o').

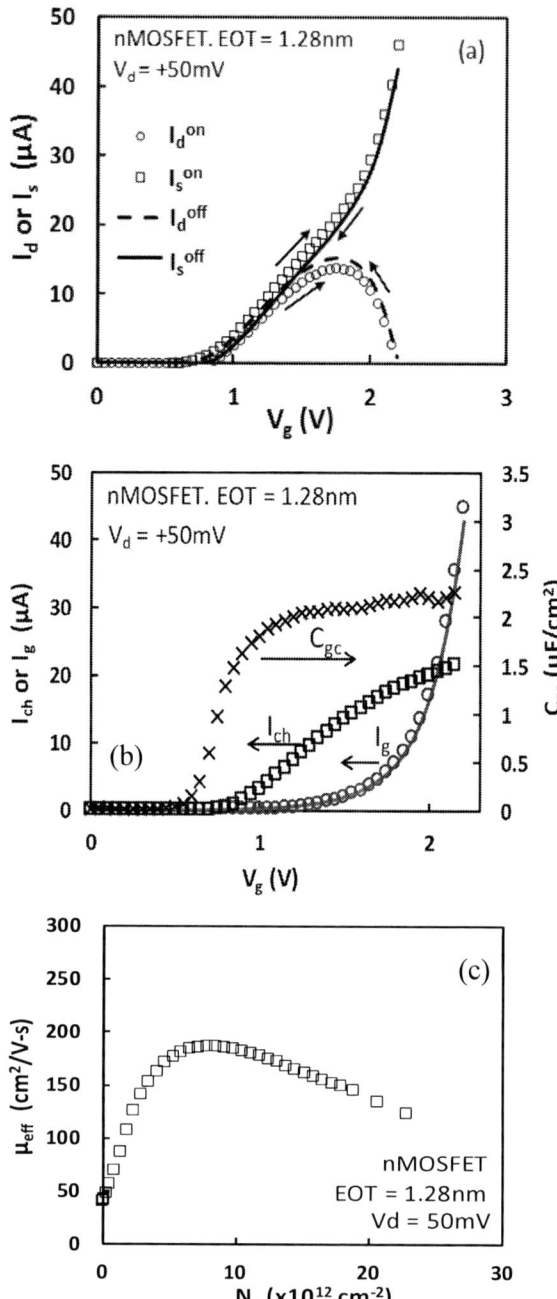

Fig 6: (a) Four currents measured from the source and drain corresponding to the off-to-on and on-to-off V_g sweeps by UFSP technique. An nMOSFET with EOT of 1.28nm is used. **(b)** I_{ch} ('□'), Ig ('o') and C_{gc} ('x') are extracted from the currents in (a) with (2)-(5). The blue line is the leakage current obtained by DC measurement. **(c)** Channel effective mobility, μ_{eff}, is calculated by using the extracted I_{ch} and C_{gc} with (1).

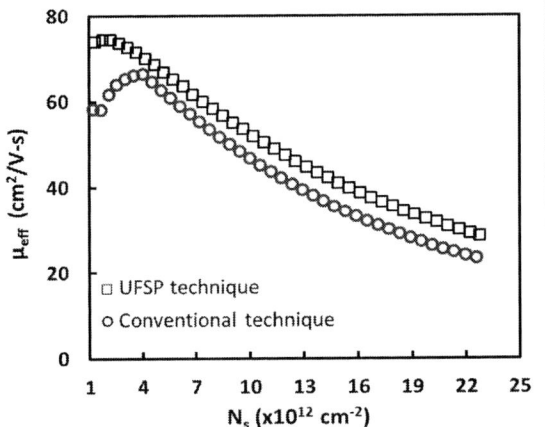

Fig 7: A comparison of mobility extracted by UFSP and conventional technique for a device with HfO₂/SiON dielectric of considerable fast trapping.

IV. SOURCES OF ERROR

A. Geometric impact

The results shown above are all with the long channel devices and therefore, potentially the geometric effect can become significant during the ultra-fast pulsed measurement. In order to estimate the impact of geometric effects in our conditions, simulation is carried out based on (6) given by [9]:

$$N(t) = \frac{8N(0)}{\pi^2} \exp(-\frac{\pi^2 D}{L^2} \cdot t) \qquad (6)$$

where N(0) is the starting value of charge density in the channel, N(t) is the average density at time 't' after switching-off, D is the diffusion coefficient for MOSFET and L is the gate length.

Fig 8 shows the channel charge decay calculated from (6) for pMOSFETs of different channel lengths. N(0) is set at 1×10^{13} cm⁻² for illustration. As expected, when the device has a channel length of 100 μm, significant channel charges remain in the channel after several microseconds, causing the geometric component. When the channel length reduces to 10 μm, however, almost all inversion charges in the channel can flow back to the source/drain within 100 ns.

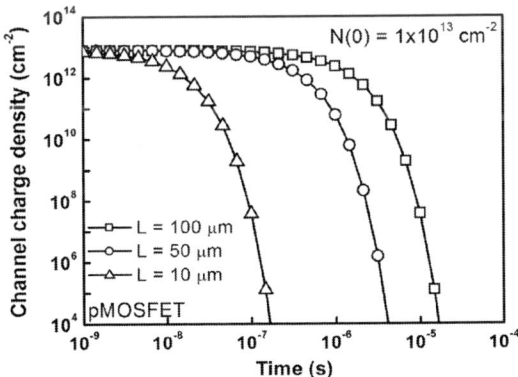

Fig. 8 Channel charge density as a function of time after switching off, calculated according to (6). Channel length varies from 100 μm to 10 μm. The initial charge density N(0) is set at 1×10^{13} cm⁻².

Fig 9 (a) and (b) shows the channel charge variation with time after switching off under different N(0) for both pMOSFET and nMOSFET of L=10μm. In both cases, almost all the inversion charges flow back to the source/drain after 100 ns. Since our measurement time is several microseconds, the impact of geometric component is negligible for the devices used in this work, i.e. L≤10μm.

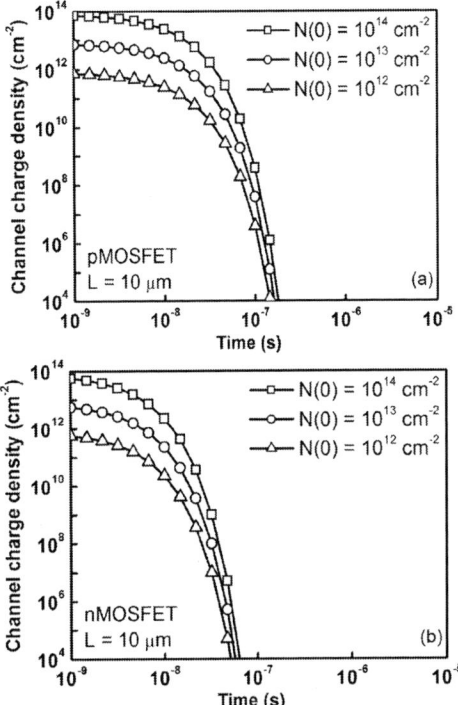

Fig. 9 Channel charge density as a function of time after switching off for pMOSFET (a) and nMOSFET (b). Channel length is fixed at 10 μm. The initial charge density N(0) varies from 1×10^{14} cm⁻² to 1×10^{12} cm⁻².

To further confirm that the geometric component is insignificant in our case, Fig. 10 shows that the measured C_{gc} from the 1.28nm nMOSFET is insensitive to the pulse edge time. A reduction of pulse edge time will increase the geometric component. If it were important, it should have an impact on the C_{gc}.

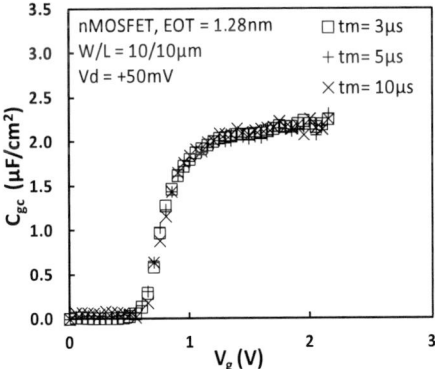

Fig. 10 A comparison of the Cgc measured under three different measurement time, tm, namely the pulse edge time. nMOSFET with EOT of 1.28nm is used. Device channel length and width is of 10 μm by 10 μm. The applied drain bias, Vd, is +50mV.

B. Parasitic capacitance and Series resistance

Parasitic capacitance of bond pads is another source of error. When parasitic capacitance is not negligible, it can raise the measured C_{gc} and cause an offset. However, this has not been observed in our measurement. As shown in Fig.4, the C_{gc} measured directly from the UFSP method shows no offset indicating the impact of parasitic capacitance on C_{gc} is negligible. This agrees with [10] that gate-to-channel capacitance, C_{gc}, is less sensitive to parasitic capacitance, when compared with gate-to-bulk capacitance, C_{gb}.

S/D series resistance can also affect both C_{gc} and I_{ch} and thus affect the extracted mobility. Without considering series resistance, Cgc is calculated using (5). The presence of series resistance at source and drain side, R_s and R_d, will change the effective gate voltage to $V_{geff\_s} = V_g - I_s * R_s$ on source and to $V_{geff\_d} = V_g - (V_d - I_d * R_d)$ on drain side. As a result, the real ramp rate dV_{geff}/dt varies between dV_{geff_s}/dt and dV_{geff_d}/dt.

The typical series resistance is reported being less than 100 Ω [11]. 2.3nm pMOSFET is used to assess the impact of series resistance on C_{gc}. In Fig. 11, dV_g/dt is replaced by either dV_{geff_s}/dt or dV_{geff_d}/dt to calculate C_{gc}, assuming $R_s = R_d = 100\ \Omega$. It is found that the impact of series resistance on C_{gc} measurement is negligible as shown in Fig. 11.

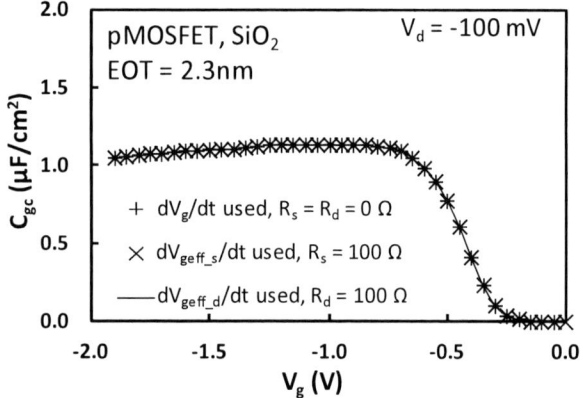

Fig. 11 Impact of series resistance on measured C_{gc}. pMOSFET with 2.3nm SiO₂ gate dielectric is used and the width and length is 10μmx10μm.

The presence of series resistance, R_{sd}, can also affect the channel current, I_{ch}. This is caused by the current-induced potential drop on R_{sd} and thus the real drain bias w.r.t. the source side, V_{ds}, is smaller than the applied drain bias, V_d. Like conventional mobility extraction methods, UFSP technique itself has not taken the effect into account. Channel current can be corrected using $I_{ch}/(V_d - I_{ch} * R_{sd}) * V_d$ if series resistance cannot be neglected. For a given technology, longer channel devices are less sensitive to series resistance than shorter channel devices due to their smaller I_{ch}.

To estimate the error introduced by R_{sd} on I_{ch} and the extracted mobility for the device used in the work, Fig. 12 (a) compares the channel currents with and without taking R_{sd} =100 Ω into account. The corresponding mobility with and without R_{sd} correction is shown in Fig. 12 (b) and the error is within 1%. As a result, the impact of typical series resistance

on the mobility extraction is negligible for a long channel device (L = 10 μm).

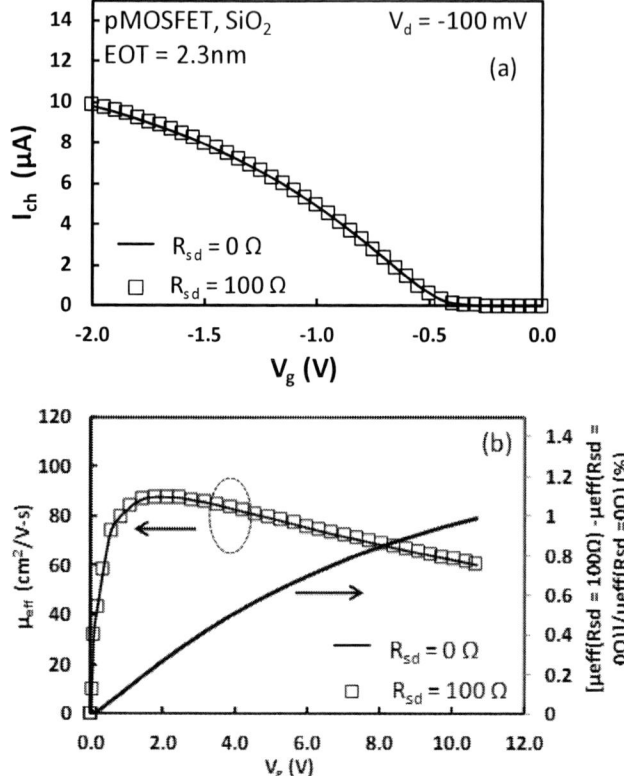

Fig. 12 Impact of series resistance on measured I_{ch} (a) and μ_{eff} (b). Device A of pMOSFET with 2.3nm SiO₂ gate dielectric is used and the width and length is 10μmx10μm.

V. TECHNIQUE LIMITATION

Although UFSP is an improved technique over the conventional split C-V for mobility extraction, it has its own limitations.

A. Device geometry

UFSP separated displacement current I_{dis}, and channel current I_{ch}, from the total measured current. However, this contains one implicit requirement that the magnitude of these two currents should be comparable. Under a certain dV_g/dt and V_d, I_{ch} is proportional to W/L and I_{dis} is proportional to W*L. Therefore, the ratio between I_{dis} and I_{ch} is proportional to L^2 and independent of W. This indicates that the selection of the channel length of the device is critical for UFSP technique. If the channel length of the device is selected too short, I_{dis} will become too small compared with I_{ch} and C_{gc} cannot be accurately determined. Therefore, UFSP technique is not suitable for mobility extraction on short channel devices. Fig. 13 shows that UFSP technique is applied on a set of 2.3nm pMOSFET with SiO₂ dielectric. They have the same channel length (10μm) but different channel widths (from 10μm to 0.25μm). Reasonable accurate C_{gc} has been obtained with the device area down to 2.5μm².

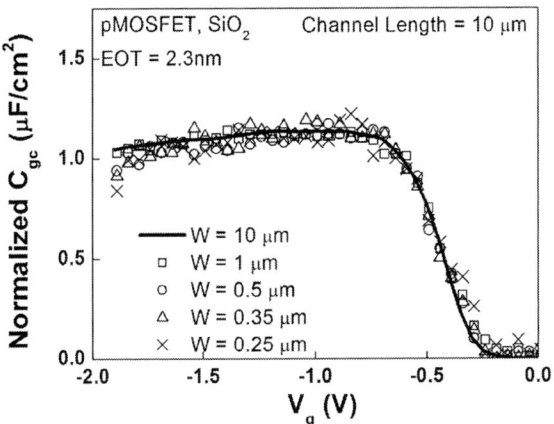

Fig. 13 Impact of channel width on measured Cgc. pMOSFET with 2.3nm SiO$_2$ gate dielectric is used. The length is 10µm and width varies from 10 µm down to 0.25µm.

B. Interface states

Another limitation of the UFSP technique is that it can only work on the device with negligible interface states. It is well-known that these interface states locate spatially at the Si/SiO$_2$ interface [12] and thus can communicate with the substrate within short time [13]. As a consequence, during the measurement, interface states cannot be frozen. The population of the mobile inversion charge and these immobile interface states will change simultaneously and they are virtually indistinguishable. In another word, channel inversion charges will be over-estimated if interface states are not negligible. This in turn leads to the under-estimation of the channel effective mobility.

CONCLUSION

In conclusion, a novel ultra-fast single pulse technique (UFSP) has been developed to achieve mobility extraction from only one measurement with fast speed. UFSP enables I$_d$-V$_g$ and C$_{gc}$-V$_g$ to be measured at the same time and the cumbersome change of cabling between these two measurements is not required. UFSP successfully reduced the measurement time to microseconds, more than 10^5 faster than the conventional technique. This fast speed helps to minimize charge trapping effects which is usually exhibited by devices with alternate gate stack materials. UFSP is shown to be tolerant to the interference from gate leakage current in ultra-thin oxide devices. Mobility is successfully extracted on the device with leakage current density as high as 40 A/cm^2.

REFERENCES

[1] S. Takagi, A. Toriumi, M. Iwase, and H. Tango, "On the universality of inversion layer mobility in Si MOSFET's: Part I-effects of substrate impurity concentration," IEEE Trans. Electron Dev., vol. 41, no. 12, pp. 2357-2362, 1994.

[2] C. L. Huang, J. V. Faricelli, and N. D. Arora, "A new technique for measuring MOSFET inversion layer mobility," IEEE Trans. Electron Dev., vol. 40, no. 6, pp. 1134-1139, 1993.

[3] L. Pantisano, L. Trojman, J. Mitard, B. Dejaeger, S. Severi, G. Eneman, G. Crupi, T. Hoffmann, I. Ferain, M. Meuris and M. Heynes, "Fundermentals and extraction of velocity saturation in sub-100nm (110)-Si and (100)-Ge," in VLSI Symp. Tech. Dig., pp. 52 – 53, 2008.

[4] A. Kerber, E. Cartier, L. Pantisano, R. Degraeve, T. Kauerauf, Y. Kim, A. Hou, G. Groeseneken, H. E. Maes and U. Schwalke, "Origin of the threshold voltage instability in SiO2/HfO2 dual layer dielectrics," IEEE Electron Device Lett., vol. 24, no. 2, pp. 87 – 89, 2003.

[5] S. M. Thomas, T. E. Whall, E. H. C. Parker, D. R. Leadley, R. J. P. Lander, G. Vellianitis, and J. R. Watling, "Accurate effective mobility extraction in SOI MOS transistors," in Ultimate Integration of Silicon, pp. 31-34, 2009.

[6] D. V. Singh, P. Solomon, E. P. Gusev, G. Singco, and Z. Ren, "Ultra-fast measurements of the inversion charge in MOSFETs and impact on measured mobility in high-k MOSFETs," in IEDM Tech. Dig., pp. 863-866, 2004.

[7] G. Liu, D.Guo, K. Xiu, W. K. Henson and P. J. Oldiges, "Intrinsic effective mobility extraction with extremely scaled gate dielectrics," Appl. Phys. Lett., vol. 97, no. 2, 023509, 2010.

[8] J. Koga, S. Takagi, A. Toriumi, Int. Conf. Solid State Devices and Materials, pp.895, 1994.

[9] G. V. D. Bosch, G. Groeseneken and H. E. Maes, "On the geometric component of charge-pumping current in MOSFET's," IEEE Electron Dev. Lett., vol. 14, no. 3, pp. 107-109, 1993.

[10] F. Lime, C. Guiducci, R. Clerc, G. Ghibaudo, C. Leroux and T. Ernst, "Characterization of effective mobility by split C(V) technique in N-MOSFETs with ultra-thin gate oxides," Solid State Electron., vol.47, pp. 1147 – 1153, 2003.

[11] K. Iniewski, A. Balasinski, B. Majkusiak, R. B. Beck and A. Jakubowski, "Series resistance in a MOS capacitor with a thin gate oxide," solid state electron., vol.32, pp. 137 – 140, 1989.

[12] J. F. Zhang, Z. Ji, M. H. Chang, B. Kaczer, and G. Groeseneken, "Real Vth instability of pMOSFETs under practical operation conditions," in IEDM Tech. Dig. , pp. 817-820, 2007.

[13] J. F. Zhang, C. Z. Zhao, A. H. Chen, G. Groeseneken, and R. Degraeve, "Hole traps in silicon dioxides. Part I. Properties," IEEE Tran. Electron Devices, vol. 51, no. 8, pp. 1267-1273, 2004.

978-1-4673-4845-4/13 $31.00 © 2013 IEEE

SESSION 4: Noise and RF

978-1-4673-4845-4/13 $31.00 © 2013 IEEE

Optical High Frequency Test Structure and Test Bench definition for on Wafer Silicon Integrated Noise Source characterization up to 110 GHz based on Germanium-on-Silicon Photodiode

S. Oeuvrard[1,2], J.-F. Lampin[2], G. Ducournau[2], L. Virot[1,3], J.M. Fedeli[3], J.M. Hartmann[3], F. Danneville[2], Y. Morandini[4], D. Gloria[1]

[1]STMicroelectronics, TR&D/TPS Lab., 850 rue Jean-Monnet, 38926 Crolles Cedex, France
[2]IEMN, Cité Scientifique Avenue Poincaré, 59652 Villeneuve d'Ascq Cedex, France
[3]CEA LETI, Minatec Campus, 17 Rue des Martyrs, 38054 Grenoble, France
[4]DOLPHIN INTEGRATION, 39 avenue du Granier, 38942 Cedex Meylan, France

Abstract— **A new Optical-High-Frequency test structure and dedicated test bench have been developed to characterize a Germanium-on-Silicon photodiode intended to be used as an integrated noise source, a first step to high frequency transistor noise figure on-wafer extraction. Continuous wave signals have been measured from these 1550 nm photodiodes, with RF power higher than -20 dBm at 109 GHz.**

Keywords—Silicon Photonics, Germanium Photodiode, Test Structure, Noise Source, Noise Measurement

I. INTRODUCTION

The outstanding transistors current gain cut-off frequency (Ft) achieved in CMOS and BiCMOS technologies, featuring Ft above 300GHz, is opening up new opportunities for THz applications using Silicon chipsets. A major consideration to design low Noise Figure circuit lies in the experimental extraction of transistor noise parameters: minimum noise figure NF_{min}, noise equivalent resistance R_n and optimum source impedance Γ_{opt}. This extraction is realized on wafer and modeled down to sub-millimeter wave range. Nevertheless, there are no commercial noise sources available over 170 GHz, avoiding the use of traditional characterization bench for noise parameters determination [1], [2]. An alternative solution could be the use of photodiode as white noise generator [3]. In this paper, we propose to study, for the first time, a silicon integrated photodiode, potentially able to be used as millimeter wave noise generator, through the design of dedicated Opto-Electrical test structure and bench for on-wafer measurement. The proposed solution enables the photodiode millimeter wave output spectrum characterization versus optical stimulus up to THz. The methodology has been evaluated on SOI technology and it turned out that the photodiode reached an output RF power as high as -20 dBm at 109 GHz, with an input optical power of 12 dBm reaching the input grating coupler. A level that is very promising for noise generator use at these frequencies, even if further measurements need to be done to determine the Excess Noise Ratio of this device.

This paper is organized as follows. First, the classical on wafer transistor Noise parameters extraction method is described. Then, the specific Opto Electrical test structure as the Opto Electrical Test bench and associated methodology

are detailed, and finally, the obtained results on SOI technology, described in [4], are reported.

II. TRANSISTOR NOISE PARAMETERS EXTRACTION CLASSICAL BENCH LIMITATIONS FOR THZ FREQUENCIES

It is well known in the theory of linear noisy networks that a complete characterization of noise in a linear two-port at one frequency requires the knowledge of the four noise parameters NF_{min}, R_n, $Real(\Gamma_{opt})$, $Imag(\Gamma_{opt})$. The measurement method used in this case to determine these parameters is the multi impedance generally based on the Y-factor [5]. This method relies on the measurement of the DUT noise factors for different impedances presented to the DUT, thanks to the equation below:

$$NF(Y_s) = NF_{min} + \frac{Rn}{\Re e\{Y_s\}}|Y_s - Y_{opt}|^2 \quad (1)$$

where Ys is the DUT input admittance.

The basic concept of a traditional noise setup enabling this experiment is presented on figure 1. It is composed of a Noise source, generating a white noise signal, the on-wafer Device Under Test (DUT) for which one wants to extract the four noise parameters, and the Noise Receiver with its associated amplifier.

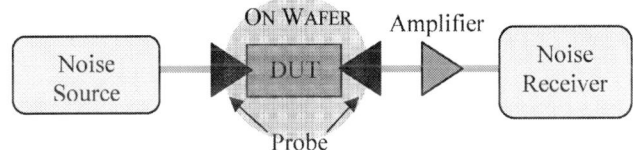

Figure 1. Usual Noise bench for on wafer transistor Noise parameters extraction.

Our study will focus on the photodiode, for the realization of a noise source, as it is the main limitation to carry out the sub-millimeter wave noise measurement. The main characteristic of interest for a noise source is its Excess Noise Ratio (ENR), expressed by the difference between the equivalent noise source hot and cold temperatures, divided by 290K. In the particular case of a photodiode use, the cold state will be when the device is not lighted, and the hot state when

978-1-4673-4845-4/13 $31.00 © 2013 IEEE

it is lighted by an optical white signal. In that work we focus in first step on the RF Output power measurement versus light one. In our case, this optical signal is delivered by an Erbium-Doped Fiber Amplifier. A good noise source will ensure enough noise power at the DUT input, and will allow remaining in the Noise Receiver sensitivity range.

III. OPTO-ELECTRICAL TEST STRUCTURE FOR ON WAFER PHOTODIODE CHARACTERIZATION

The photodiode characterization requires the design of a specific test structure, which includes an optical input and an electrical output. The optical part is made of a grating coupler that allows coupling the light on wafer, equivalent to the RF pad at the output (figure 2). The coupling of the light is done by an optical fiber, placed with an 11° to normal. Then, an optical integrated waveguide is used to feed the PIN junction. The injected light is converted into a photocurrent, which produces an output electrical signal, propagating through a transmission line, in the metallization levels above. Due to the reduced losses of the waveguide compared to the transmission lines, the photodiode is placed as near as possible to the RF output pad.

In order to maximize the coupling between the optical fiber and the grating coupler, the photocurrent of the photodiode is simultaneously measured. The fiber angle and its position above the coupler are meticulously adjusted until the maximum of photocurrent is reached. An accuracy in the range of 0.5 µm is requested in the fiber positioning in order to optimize the optical coupling.

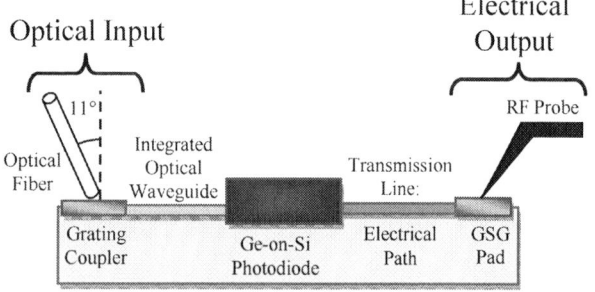

Figure 2. Schematic of the photodiode test concept : Optical input and Electrical output

A cross-sectional schematic of the PIN photodiode is shown on figure 3. The Ge detector is grown on top of a rib waveguide by epitaxial process. As the light propagates along the Si waveguide, it couples through the Germanium intrinsic region, where it is absorbed. The waveguide is based on rib architecture to minimize the losses and maximize the coupling. To increase the cut-off frequency of PIN photodiodes, the vertical geometry, stacking the different Germanium layers p-doped, intrinsic and n-doped, is preferred [6]. Nevertheless, it is hardly compatible with this particular way of leading the light, through a rib waveguide, as it decreases the coupling and then decreases the responsivity of the photodiode. The vertical geometry is also adding more steps in the process, that is why in the case, we focus on horizontal architecture.

The test structure contains PIN photodiodes with different widths and lengths of the intrinsic area, layered on a waveguide connected to the grating coupler (figure 4). The contacts above the P and N areas are connected to a GSG RF pad. The grating coupler is located 300 µm away from the RF pad, to avoid any contact between the optical fiber and the RF probes during the test. The GSG pad pitch is compatible with 100 µm pitch probes, allowing us to measure RF signal up to 220 GHz.

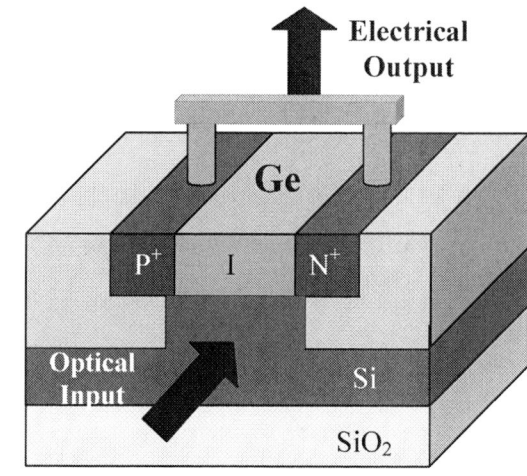

Figure 3. 3D schematic side view of the PIN Germanium Photodiode

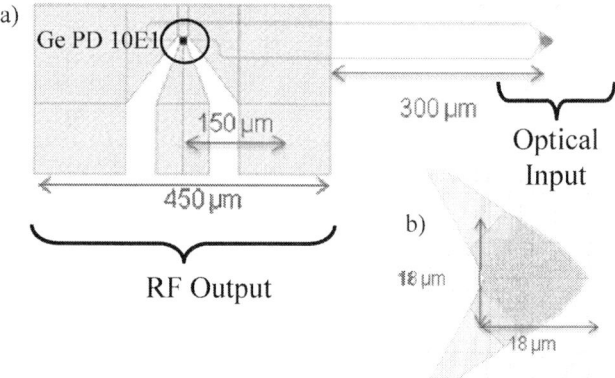

Figure 4. Test structure schematic (a) and zoom on the grating coupler (b).

The geometry of the photodiode lead to a 3 dB cut-off frequency of about 20 GHz, and a 10 dB cut-off frequency of 45 GHz. Nevertheless, its specific application in a noise source context allow us to use it far above the intrinsic bandwidth, as long as it still deliver enough output power to stay in the receiver sensitivity.

IV. OPTO-ELECTRICAL TEST BENCH AND METHODOLOGY DESCRIPTION

The developed photodiode bench measurement is described figure 5. The device is reversely biased through SMU (Source Monitor Units), and the optical input is lighted by two continuous lasers amplified by an Erbium Doped Fiber Amplifier (EDFA) at 1550 nm. The wavelength difference between lasers corresponds to the generated frequency inside the photodiode.

The output RF power is detected by a Spectrum Analyzer connected to the RF probes. However, the spectrum analyzer is only detecting up to 67 GHz. Then, the frequency response of the photodiode has been measured in two different band, from 5 to 67 GHz and from 75 to 110 GHz. For the W band (75-110 GHz), a mixer has been added to transpose the input signal below 67 GHz for detection. This mixer adds around 30 dB of conversion losses that have been taken into account for de-embedding. S parameters of the bench, including the cables, the probe, with and without the mixer, have been measured to de-embed the RF out power from the own equipment losses.

Photography of the bench, with the optical fiber, the camera and the RF probes, is shown on figure 6. PIN junction characterization can be made versus DC bias and Optical source power. In order to determine the optimal position of the optical fiber on the grating coupler, the photocurrent is continuously measured: the maximum of photocurrent corresponds to a maximum of output power. In the characterized technology, the fiber is oriented with an 11° angle to maximize the light coupling inside the grating. Softly changing this angle does not seem to have a significant impact on the output signal.

To visualize the position of each element on the wafer, a camera detecting the visible spectrum is placed with a 45° angle, and a LED in front. The camera detects the reflection of this LED light on the wafer, enabling the visualization of RF and optical pads, but cannot display the optical signal escaping through the optical fiber. Thus, the photocurrent measurement is the easiest way to determine the optimal position of the fiber on the grating coupler.

V. OBTAINED RESULTS ON SOI TECHNLOGY

A PIN diode (PD Ge 10E1), 10 µm long and 1µm large (width of the intrinsic part of the PIN) in a technology from CEA/LETI, has been characterized. The measurements have been realized in two frequency bands, from 5 to 67 GHz and from 75 to 110 GHz (W band). Open and short structures have been measured to de-embed the results (Optical and Electrical paths) and thus determine the actual optical power illuminating the photodiode. The insertion losses from the output of the EDFA to the grating coupler have been estimated to 10 dB.

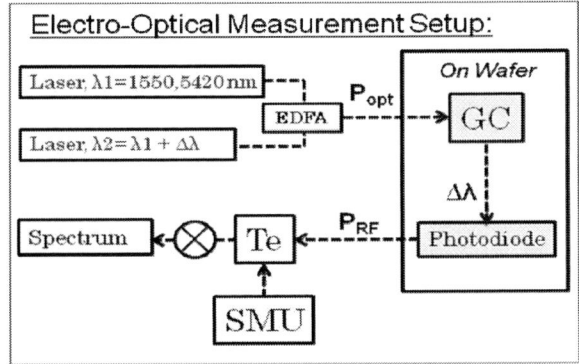

Figure 5. Schematic of the electro-optical measurement setup. GC stands for Grating Coupler, Te for bias Tee and SMU for Source Monitor Units.

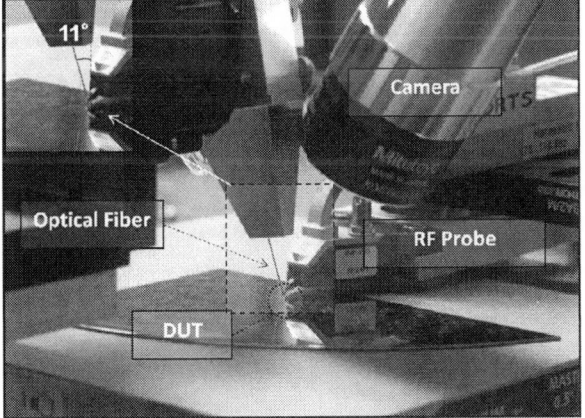

Figure 6. Photography of the test bench including : the DUT (center), the optical fiber (left), the camera (above) and the probe (below)

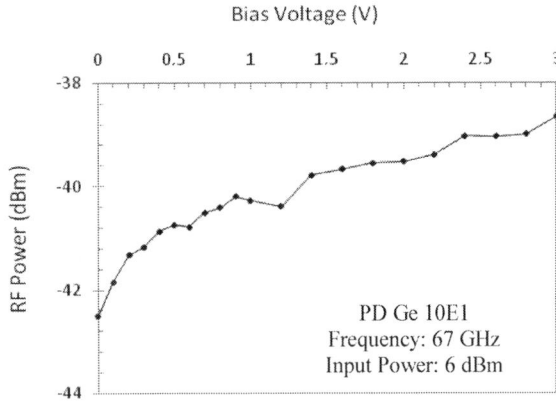

Figure 7. Bias effect on the Photodiode RF Output Power with 6 dBm Optical Input Power, at 67 GHz

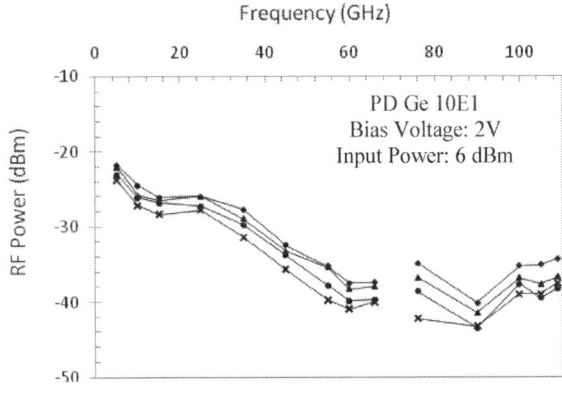

Figure 8. RF Output Power frequency dependance with 6 dBm Optical Input Power and 2V bias.

978-1-4673-4845-4/13 $31.00 © 2013 IEEE

The output RF power versus DC bias at 67 GHz and the output power versus frequency for a 2V reversed bias are shown in figure 7 and 8, respectively. Bias effect on the power delivered by the photodetector is limited after 2V. To avoid any avalanche effects, the authors reversely biased the photodiodes at 2V.

The discontinuity observed between both frequency bands is mainly due to the used different setups. As the spectrum analyzer is only detecting RF power up to 67 GHz, we used a mixer to set down the W band to its detection window. The accuracy of the mixer conversion losses can explain the discontinuities after correction.

The RF output power has been measured as a function of Optical power arriving on the grating coupler (figure 9). As expected, the optical input power has a significant impact on the RF output one: 16 dB can be gained on the output for optical input power increasing from 6 to 14 dBm, which is the expected quadratic effect. Nevertheless, the photodiode will saturate if the input power continue to increase. Typically, these photodiodes begin to saturate around 17 dBm input, corresponding approximately 8 to 10 mA of photocurrent.

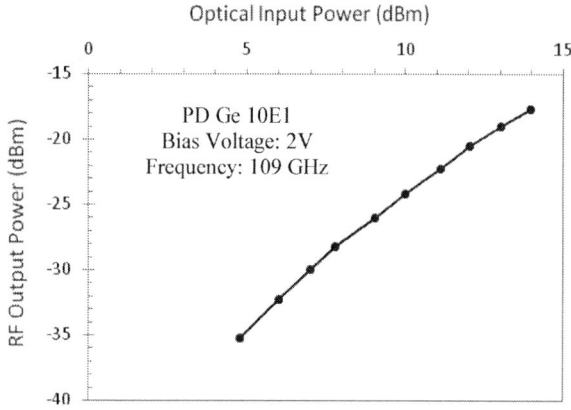

Figure 9. Optical Input Power effect on RF Output Power with -2V bias at 109 GHz

Increasing the optical input power to 14 dBm leads to a maximum of -17 dBm delivered at 109 GHz. These results are very promising in using the in-situ photodiode as millimeter wave noise source, and above. Next steps will analyze the potentiality of these photodiodes for on-wafer white noise generation, including the determination of its excess noise ratio up to 220 GHz.

VI. CONCLUSION

In this paper, the behavior of Silicon integrated Photodiode planned to be used as MMW noise source has been studied through the proposed developed Opto-Electrical test structure and test bench. The bench is compatible with on wafer characterization up to THz frequencies at the electrical access of the photodiode. On the considered technology, output RF power of the photodetector reached -17 dBm @ 109 GHz which is very promising for using it as a noise source. As perspective, this device will be characterized up to 220 GHz in electrical path and will be tested to extract advanced HBT transistor Noise Figure on wafer.

ACKNOWLEDGMENT

The samples have been fabricated under the European Community's Seventh Framework Program (FP7/2007-2013) under grant agreement n°224312 HELIOS. The authors acknowledge the LETI clean room staff for successful fabrication.

REFERENCES

[1] Y. Tagro, D. Gloria, S. Boret, Y. Morandini, G. Dambrine, "In-Situ Silicon Integrated Tuner for Automated On-Wafer mmW Noise Parameters Extraction using Multi-Impedance Method for Transistor Characterization", Microelectronic Test Structures, 2009, ICMTS, pp 184-188

[2] L. Poulain, N. Waldhoff, D. Gloria, F. Danneville, G. Dambrine, "Small signal and HF noise performance of 45 nm CMOS technology in mmW range", Radio Frequency Integrated Circuits Symposium (RFIC), 2011, IEEE, pp 1-4.

[3] H.-J. Song, N. Shimizu, N. Kukutsu, T. Nagatsuma, Y. Kado, "Microwave Photonic Noise Source From Microwave to Sub-Terahertz Wave Bands and Its Applications to Noise Characterization", IEEE Transactions on Microwave Theory and Techniques, Vol. 56, NO. 12, December 2008, pp 2989-2997.

[4] L. Virot, L. Vivien, "40 Gbit/s Germanium Waveguide Photodetector on Silicon", Silicon Photonics and Photonic Integrated Circuits III, conference Volume 8431, Brussels, Belgium, April 16, 2012

[5] Agilent Technologie, Palo Alto, CA, "Noise figure measurement accuracy : The Y-factor method", 1976. Application Note 57-2. 58, 68

[6] A. Ramaswamy, M. Piels, N. Nunoya, T. Yin, J. E. Bowers, "High Power Silicon-Germanium Photodiodes for Microwave Photonic Applications", IEEE Transactions on Microwave Theory and Techniques, Vol. 58, No. 11, Nov 2010, pp 3336-3343.

978-1-4673-4845-4/13 $31.00 © 2013 IEEE

Measurements of SRAM Sensitivity against AC Power Noise with Effects of Device Variation

Takuya Sawada[†], Kumpei Yoshikawa[†], Hidehiro Takata[††], Koji Nii[††], and Makoto Nagata[†, †††]

[†]Graduate School of System Informatics, Kobe University, Kobe, Japan
[††]Renesas Electronics Corporation, Tokyo, Japan
[†††]CREST, JST
Email: {sawada, kumpei, nagata}@cs26.scitec.kobe-u.ac.jp

Abstract— SRAM exhibits the sensitivity of false operation against static and sinusoidal supply voltage variation. A measurement system combines direct radio frequency (RF) power injection, on-chip monitoring of voltage variation on power supply lines, and built-in self test of memory read/write operations. The bit error rate (BER) of an SRAM core exponentially increases when the lowest instantaneous voltage on the power supply line of SRAM cells during RF injection linearly decreases. Test dice on wafers at five different process corners in a 1.5 V 90 nm CMOS technology were tested. The minimum allowable voltage with BER of less than a single bit failure in average becomes smaller, thus more tolerant, when n-channel devices are at the slow corner in a conventional 6-transistor SRAM cell. The measurement technique enables to experimentally evaluate dynamic noise margin of SRAM cores in a given technology.

Keywords— Electromagnetic compatibility; Integrated circuits; Static random access memory; Direct power injection

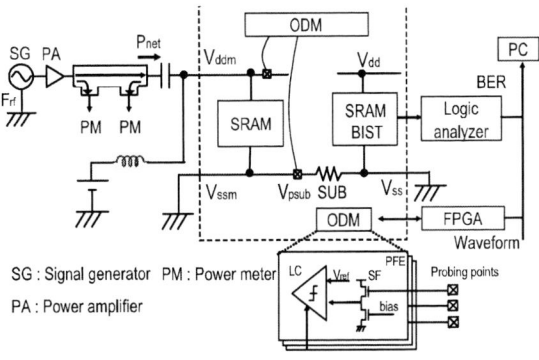

Figure 1. Measurement system incorporating DPI, ODM, and BIST of SRAM core.

I. INTRODUCTION

SRAM is substantially important in semiconductor chips with programmable functionality, for storing codes and data as well as caching for the mitigation of memory wall. Memory capacity of SRAM cores enlarges for the higher computation performance, with consuming a considerable part of silicon area[1, 2]. The variation of SRAM performance on the static as well as dynamic variation of MOS transistors has been margined in the design of SRAM products [3-5].

While static defects in a SRAM cell array will be saved by redundancy, the resiliency of SRAM operation against dynamic environmental variation, such as power supply noise (PSN) and electromagnetic interference (EMI), becomes more essentially important for computing systems used in applications with high reliability and safety. However, it is not feasible to evaluate the sensitivity of an SRAM core against dynamic environmental disturbances by using standard transistor-level circuit simulation, because of the high level of integration.

This paper describes a measurement methodology and interpretation about the failure bits in SRAM cells under static as well as sinusoidal variation of power supply voltage. The impacts of device variation will be also considered. Measurement system will be briefly introduced in Sect. 2. Experimental results are discussed in detail in Sect. 3. Conclusion of this paper will be addressed in Sect. 4.

II. MEASUREMENT SYSTEM

The measurement system of Fig. 1 includes built-in self test mechanism (BIST) of SRAM operation and on-die voltage monitoring (ODM) [6], combined with direct radio frequency (RF) power injection to induce sinusoidal voltage variation on power supply lines of SRAM cells, V_{ddm} [7]. The bit error rate (BER) is evaluated by BIST. Dynamic voltage waveforms on power nodes of the SRAM cell array are captured by ODM and then analyzed to derive the magnitude of voltage variation. The measurement procedures of on-chip waveforms and bit errors under RF power injection are documented in [8].

The ODM senses voltage variation on multiple supply nodes of SRAM cells (V_{ddm}), SRAM peripheral circuits (V_{dd}), and associated ground lines of V_{ssm} and V_{ss}, respectively. The voltage variation of p-type silicon substrate, V_{psub}, can also be measured. The nodes of V_{ssm}, V_{ss}, and V_{psub}, are resistively coupled to each other through ground wiring, substrate taps, and the silicon substrate. We will focus mainly on V_{ddm} within SRAM cell array in the following measurements and BER will be related with static and sinusoidal voltage variation on V_{ddm}.

The ODM has power supply and ground wiring isolated from SRAM cores, by covering the entire ODM circuit with deep Nwell islands. The voltages within SRAM cores are measured in reference to the stable ground voltage of ODM.

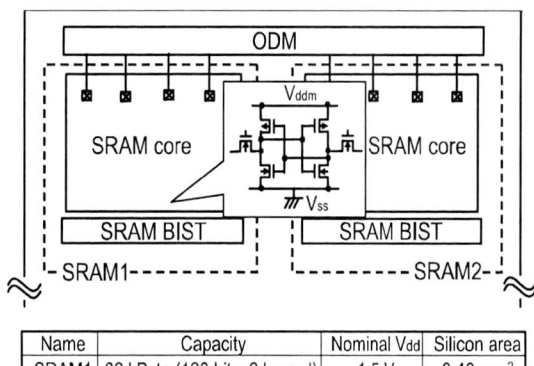

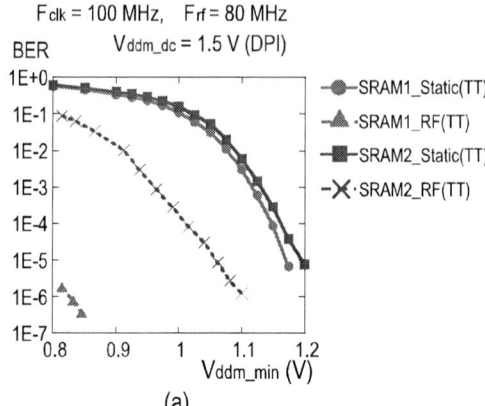

Name	Capacity	Nominal Vdd	Silicon area
SRAM1	32 kByte (128 bit x 2 k word)	1.5 V	0.49 mm^2
SRAM2	16 kByte (64 bit x 2 k word)	1.5 V	0.28 mm^2

Figure 2.   Test chip construction.

The operation sequences are programmed in BIST to measure BER. Binary words with the predefined bit patterns of such as a checker and a line are written into the whole of SRAM cells, and then all the words are read out for comparison with expected bit values. The number of failure bits is evaluated in every single run of BIST. It should be noted that DPI is continuous during the entire BIST operation.

A test chip of Fig. 2 includes 32 and 16 kByte SRAM cores along with ODM and BIST. The SRAM uses a standard 6-transistor memory cells as given in the inset. A 90 nm CMOS technology with the nominal supply voltage of 1.5 V is used for fabrication of test dice. The dice are picked up from post-production wafers with the five different process corners of {TT, SS, FF, SF, FS}, denoting the combination of n- and p-channel MOS transistors with typical (T), slow (S), and fast (F) device characteristics. The process corners are primarily dominated by the variation of threshold voltages of MOS transistors. The ODM uses high-voltage MOS transistors at 3.3 V for covering all supply voltages.

III.   EXPERIMENTAL RESULTS

A.   BER versus voltage variation

The BER of SRAM is normally evaluated under statically reduced supply voltage given on V_{ddm} and exhibits exponential increase as given in Fig. 3(a), with the notation of SRAM_static. The supply voltage is statically maintained at the given value during the whole sequences of BIST operation of the SRAM core under test. The response of SRAM cells against sinusoidal supply voltage variation is measured, as denoted by SRAM_RF in Fig. 3(a).

The voltage varies in time with a sinusoidal shape during RF injection, as captured by ODM in the V_{ddm} wiring in Fig. 3(b). The lowest instantaneous voltage on the power supply during RF injection is measured as V_{ddm_min} from the captured waveforms.

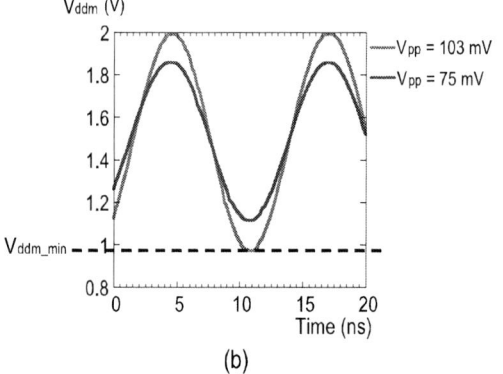

Figure 3.   (a) Comparison of BER versus static and sinusoidal voltage variations. (b) Sinusoidal waveform induced on V_{ddm}.

The BER again increases exponentially with smaller V_{ddm_min}, induced by the larger RF power. It is also observed that the SRAM is much less impacted by the sinusoidal voltage variation than the static voltage reduction. The larger deviation from the nominal standard voltage of V_{ddm} is tolerated during RF injection when BER is assumed to be allowable down to 1E-4 in the 16 kByte SRAM (named "SRAM2"), in comparison with the static voltage reduction. This observation is further enhanced for the case with 32 kByte SRAM (named "SRAM1"). The frequency of RF was chosen as 80 MHz while SRAM operated at 100 MHz.

The measurement with RF injection is combined with the static voltage reduction, as shown in Fig. 4(a) and Fig. 5(a), for 16 kByte and 32 kByte SRAM cores, respectively. The curve is horizontally shifted with the size of static voltage reduction, without significant change in the rate of increase of BER, as denoted as ΔV_{dd} / decade of BER. Therefore, the occurrence of failure bits is linearly superimposed among static and dynamic voltage variation, as consistently observed among the SRAM cores with different size of cell array.

The threshold value of V_{ddm_min} to reach a single failure bit in average among SRAM cells exhibits the clear dependency on the combination of MOS transistors at different process

978-1-4673-4845-4/13 $31.00 © 2013 IEEE

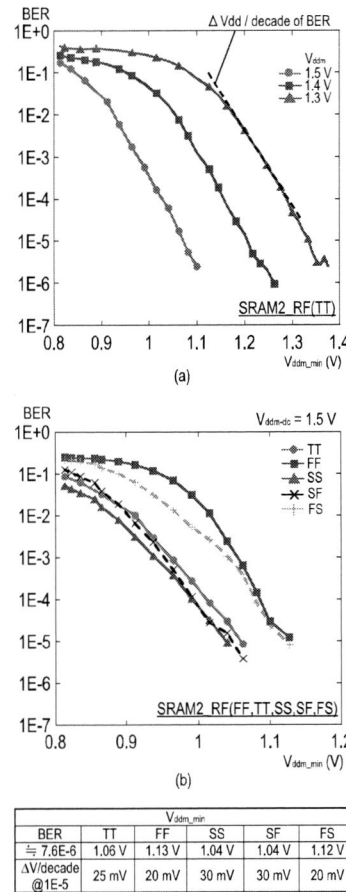

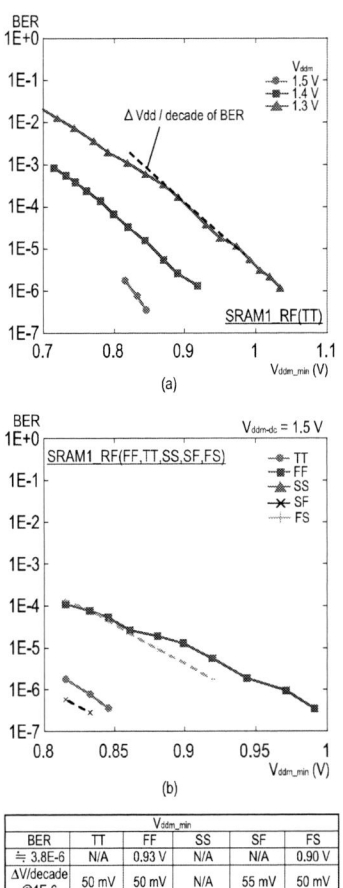

	V_{ddm_min}				
BER	TT	FF	SS	SF	FS
$\doteqdot$ 7.6E-6	1.06 V	1.13 V	1.04 V	1.04 V	1.12 V
ΔV/decade @1E-5	25 mV	20 mV	30 mV	30 mV	20 mV

Figure 4. (a) BER versus V_{dd_min} with different DC level of V_{ddm} in 16 kByte SRAM. (b) BER versus V_{dd_min} for different combinations of process corners. Key values of V_{dd_min} are extracted in table.

	V_{ddm_min}				
BER	TT	FF	SS	SF	FS
$\doteqdot$ 3.8E-6	N/A	0.93 V	N/A	N/A	0.90 V
ΔV/decade @1E-6	50 mV	50 mV	N/A	55 mV	50 mV

Figure 5. (a) BER versus V_{dd_min} with different DC level of V_{ddm} in 32 kByte SRAM. (b) BER versus V_{dd_min} for different combinations of process corners. Key values of V_{dd_min} are extracted in table.

corners, as shown in Fig. 4(b) and Fig. 5(b). This condition corresponds to BER of 7.6E-6 and 3.8E-6 for 16 kByte and 32 kByte SRAM cores, respectively. The inclusion of the n-channel MOS transistors at the slow corner in a 6-transistor SRAM cell, in the cases of "SS" and "SF", results in a larger tolerance, compared with having the n-channel MOS transistors at the fast corner. On the other hand, the rate of increase in BER is approximately calculated as 30 mV (SRAM2) or 50 mV (SRAM1) of V_{ddm_min} per a decade of BER and is constantly applicable among the five different combinations of process corners, as derived from the values listed in the table. This implies two different mechanisms of SRAM sensitivity. The tolerance against the static voltage reduction is affected by the variation of threshold voltages, while the response of BER against the dynamic voltage variation is governed by probabilistic processes of exchanging bi-stable states through the signal driving from SRAM cells to sense amplifiers or vise versa.

B. Distribution of Failure bits

The distribution of failure bits in the SRAM cell array is pictured in a fail bit map (FBM) of Fig. 6(a), under the RF

injection at P_{net} of 30 dBm for the average BER of 1.0 E-1. The ratio of failure bits to the total bits of 512 in a single row is also drawn in Fig. 6(b). It is observed that the failure bits are randomly and evenly located in the cell array. Therefore, the sensitivity of SRAM cells against voltage variation is independent on their positions in the physical layout of the array.

Further analysis is executed among the failure bits. The bit cells that occasionally fail during the iteration of BIST runs under the RF injection at the average BER of interest are named "random fail bits (RFB)." On the other hand, the other bit cells that are always erroneous in every BIST run are named "fixed fail bits (FFB)." The number of BIST runs is chosen as 50 in this experiment.

These identifications are visualized in the FBM of Fig. 6(a), and also the number of RFB in a single row is given in Fig. 6(c) for comparison with FFB. It becomes noticeable by these classifications that the random fail bits dominate the failure bits in the lower physical part of the SRAM core. This suggests the necessity of post-layout extraction of parasitic components and in-depth analysis of impedance distributions in a power-supply

978-1-4673-4845-4/13 $31.00 © 2013 IEEE

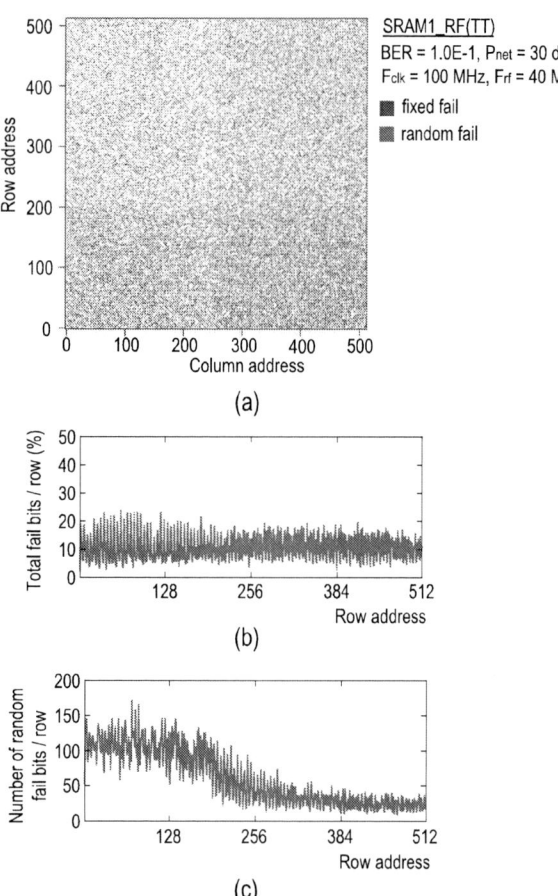

(a)

(b)

(c)

Figure 6. (a) Distribution of random and fixed fail bits in 32 kByte SRAM cell array. (b) Ratio of total fail bits drawn per every single row. (c)Number of random fail bits drawn per every single row.

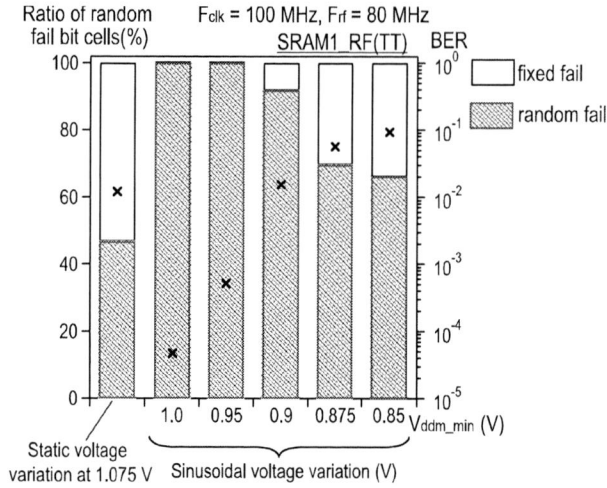

Figure 7. Ratio of random fail bits to fixed fail bits under static and sinusoidal voltage variations in 32 kByte SRAM.

network of the SRAM core, for the better resilient design against dynamic voltage variation.

The ratios of RFB and FFB to the total fail bits are evaluated in Fig. 7, for different level of BER. The comparison between static and sinusoidal voltage variation is also given. It is interestingly shown that both ratios approach to 50% for the higher level of BER under the sinusoidal voltage variation and also for the case of static voltage reduction at 1.075 V.

The size of voltage reduction, V_{ddm_min}, is evaluated by ODM, while BER is measured by BIST.

IV. CONCLUSION

The paper demonstrated the combination of on-die waveform monitoring and direct RF injection on power supply wiring in SRAM cells. The intuitive understanding of the sensitivity of SRAM cells against static and dynamic voltage variation is provided from the measurement results. Further interpretation may be partly helped by post-layout analysis of

SRAM design, however, extremely high-capacity and high-cost computing resources will be demanded. Instead, silicon measurements with test structures incorporating these techniques can be the most efficient and solid solution for deriving design principles for the high resiliency against dynamic variation of operating environments.

ACKNOWLEDGMENT

This work was in part supported by CREST, JST.

REFERENCES

[1] N. S. Kim, K. Flautner, D. Blaauw, and T. Mudge, "Circuit and microarchitectural techniques for reducing cache leakage power,", IEEE Trans. VLSI systems, vol. 12, no. 2, pp. 167-184, Feb. 2004.

[2] S. Rusu, S. Tam, H. Muljono, D. Ayers, J. Chang, B. Cherkauer, J. Stinson, J. Benoit, R. Varada, J. Leung, R. D. Limaye, and S. Vora, "A 65-nm Dual-Core Multithreaded Xeon® Processor With 16-MB L3 Cache,", IEEE J. Solid-State Circuits, vol. 42, no. 1, pp. 17-25, Jan. 2007.

[3] E. Seevinck, F. J. List, J. Lohstroh, "Static-Noise Margin Analysis of MOS SRAM Cells,", IEEE J. Solid-State Circuits, vol. SSC-22, no. 5, pp. 748-754, Oct. 1987.

[4] S. O. Toh, Z. Guo, T.-J.K. Liu, B. Nikolic, "Characterization of Dynamic SRAM Stability in 45 nm CMOS," IEEE J. Solid-State Circuits, vol. 46, no. 11, pp. 2702-2712, Nov. 2011.

[5] D. E. Khalil, M. Khellah, Nam-Sung Kim, Y. Ismail, T. Karnik, V. De, "Accurate Estimation of SRAM Dynamic Stability," IEEE Trans. VLSI Systems, vol. 16, no. 12, pp. 1639-1647, Dec. 2008.

[6] Y. Araga, T. Hashida, and M. Nagata, "An On-Chip Waveform Capturing Technique Pursuing Minimum Cost of Integration," in Proc. IEEE Intl. Symp. Circuits and Systems, pp. 3557 - 3560, May 2010.

[7] " Direct RF power injection to measure the immunity against conducted RF-disturbances of integrated circuits up to 1 GHz," IEC 62132-4, 2003, International Electrotechnical Commission: Geneva, Switzerland.

[8] T. Sawada, T. Toshikawa, K. Yoshikawa, H. Takata, K. Nii, M. Nagata, "Immunity Evaluation of SRAM Core Using DPI with On-Chip Diagnosis Structures," in Proc. IEEE EMC Compo 2011, pp. 65-70, Nov. 2011.

On the length of THRU standard for TRL de-embedding on Si substrate above 110 GHz

A. Orii, M. Suizu, S. Amakawa, K. Katayama, K. Takano, M. Motoyoshi, T. Yoshida, and M. Fujishima

Graduate School of Advanced Sciences of Matter
Hiroshima University
1–3–1 Kagamiyama, Higashihiroshima 739-8530, Japan
Email: fuji@hiroshima-u.ac.jp

Abstract— It is known that the THRU standard (a transmission line) used for thru-reflect-line (TRL) calibration/de-embedding for S-parameter measurement has to be long enough that only a single electromagnetic mode propagates at its center for it to work reliably. But ideally, TRL standards should occupy as little precious silicon real estate as possible. This paper attempts to experimentally find out how long a THRU is long enough above 110 GHz up to 170 GHz through measurements of transmission lines of various lengths. The results indicate that the length of a THRU should be at least 400 micrometers, excluding pads and pad-to-line transitions.

I. INTRODUCTION

The thru-reflect-line (TRL) [1] network analyzer calibration and de-embedding technique requires a set of transmission line standards and offset shorts or opens. The shortest transmission line standard is called THRU [Fig. 1(a)] and the longer ones LINEs [Fig. 1(b)]. While there are other de-embedding methods that use the same or a smaller set of standards as TRL (e.g. [2]–[4]), they make some arbitrary assumption about the embedding 2-ports A and B, shown in Fig. 1, in order to reduce the number of unknowns that must be determined [5]. Such methods work reasonably well at low frequencies, but the results produced by them deviate from the correct results as the frequency becomes higher [6]. On the other hand, TRL does not make such simplifying assumptions and therefore the formulation is reliable even at millimeter-wave frequencies [6]. This explains why we focus on TRL in this study.

It has been known that the THRU used in TRL has to be long enough. This is to ensure that only a single mode of wave propagates at its center [7], [8] and that the so-called leakage terms [9], which are ignored in TRL, are indeed negligibly small. The leakage terms account for the coupling between the left and right ports not via A and B. Ref. [10], for example, recommends launch spacing of two wavelengths. This recommendation, however, is difficult to follow when using advanced Si CMOS platforms. This is because the larger area required for long samples translates to an enormous cost, and also because the long, lossy on-chip transmission lines that intervene between RF probes and the device under test degrade the measurement dynamic range.

Successful use of a very short THRU in a variant of TRL has actually been reported up to 110 GHz [11]. But very few reports are available on de-embedding on Si substrate above 110 GHz [12]. Our experience with TRL with not

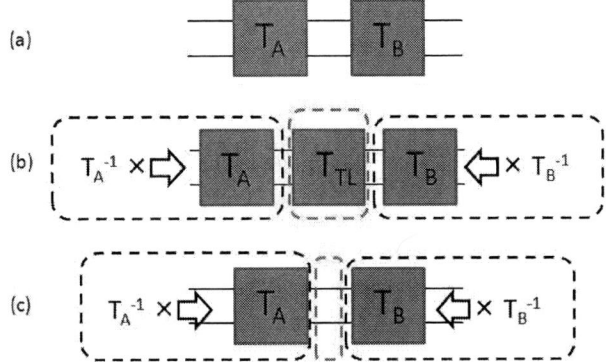

Fig. 1. (a) THRU standard used by multiline TRL. It consists of pads and launchers at both ends and a length of transmission line in between. See also Fig. 3. It is here represented as a cascade of two 2-ports, A and B. T_A and T_B are their wave cascading matrices [1]. (b) In a LINE standard, a length of transmission line, TL, sits between A and B. T_{TL} can be de-embedded by multiplying T_A^{-1} from left and T_B^{-1} from right. (c) When the same de-embedding operation is performed on a THRU itself, a T matrix of a 'zero-length transmission line' results.

so long THRU is that characterization of transmission lines and devices and their modeling with predictive ability [13] become particularly difficult beyond 140 GHz. At such a high frequency, there are so many possible causes of measurement uncertainties and inaccuracies that it makes sense to go back to the basics. In the following, we try to experimentally find out how long a THRU is long enough in the D band (110 GHz–170 GHz).

II. CRITERIA FOR CHECKING THRU LENGTH

We use multiline TRL [14] for determining the propagation constant γ of a transmission line and the S or T matrices of the embedding 2-ports, A and B (Fig. 1). It offers better accuracy and wider frequency coverage than any of the individual line TRL calibrations [14]. Once the calibration is complete, the measurement reference planes are moved to the center of the THRU. The reference impedance for the reference planes equals the frequency-dependent characteristic impedance of the transmission line in the LINE standards.

We prepared and measured four types of transmission lines with different lengths, shown in Fig. 2 and Table I. Fig. 3 shows micrographs of the type-A transmission lines. One of

978-1-4673-4845-4/13 $31.00 © 2013 IEEE

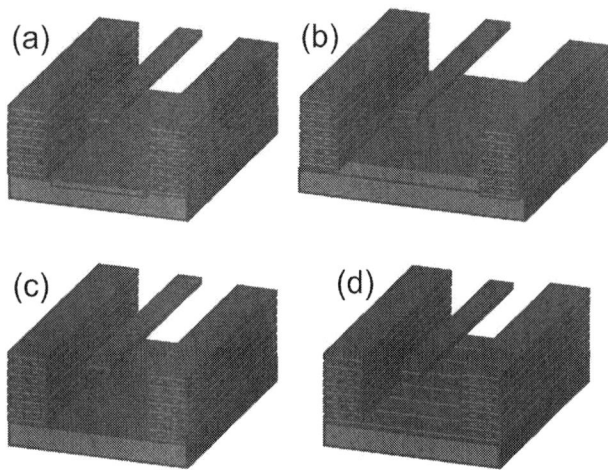

Fig. 2. Structures of the measured transmission lines. The signal line is $4\,\mu$m wide. (a) Type A: the bottom ground plane is actually a mesh. (b) Type B: sidewalls are farther than in type A. (c) Type C: no bottom ground mesh. (d) Type D: bottom ground metal is slotted.

TABLE I

TRANSMISSION LINE TYPES AND LENGTHS AVAILABLE.

Length	A	B	C	D
$0\,\mu$m	x	x	x	x
$100\,\mu$m	x	x	x	x
$200\,\mu$m	x	x		
$300\,\mu$m	x			
$400\,\mu$m	x	x	x	x
$500\,\mu$m	x			
$600\,\mu$m	x			
$1000\,\mu$m	x	x	x	x
$1200\,\mu$m	x	x	x	x

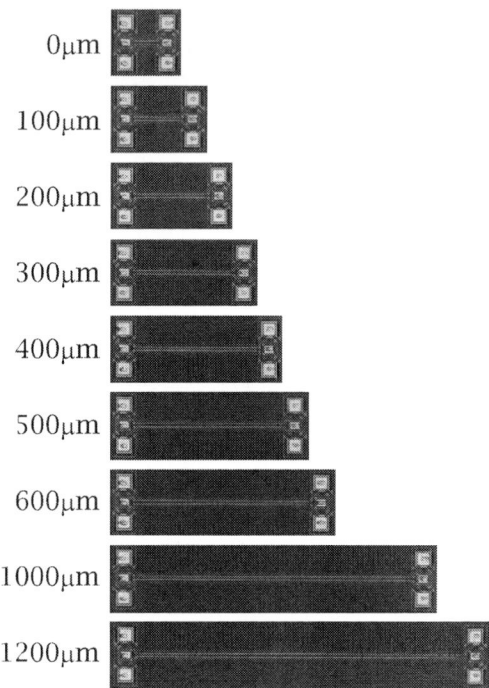

Fig. 3. Micrographs of type-A transmission lines. The shortest one still has a very short section of transmission line but is designated as '$0\,\mu$m,' thereby defining the edges of the launchers at its center.

the short lines ($\leq 600\,\mu$m) was used as a THRU, and 1000-μm and 1200-μm lines were used as LINES. An offset short having a half the length of the THRU was used as a REFLECT standard. Each standard was measured four times. As shown in Table II, when performing a multiline TRL calculation, three of the four measurement data of the two LINES ($1000\,\mu$m and $1200\,\mu$m) and one of the measurement data of a THRU and a REFLECT were used.

Ideally, the values of γ, $\mathsf{T_A}$, and $\mathsf{T_B}$ should be the same regardless of the length of the THRU chosen. But if the trouble of a THRU not being long enough is present, it should somehow manifest itself as the THRU gets shorter.

One way of checking this would be as follows. Suppose that the transmission lines contained in a THRU and a LINE have the lengths ℓ_1 and ℓ_2 ($\ell_1 \leq \ell_2$), respectively. If a transmission

TABLE II

DATA SET USED IN A SINGLE RUN OF MULTILINE TRL.

Standard	Number of data used (out of 4)
THRU of a given length	1
REFLECT of half the length	1
1000-μm LINE	3
1200-μm LINE	3

line is de-embedded from those by the procedure shown in Fig. 1(b), the resultant S matrix of the line will be

$$S = \begin{bmatrix} 0 & e^{-\gamma\ell} \\ e^{-\gamma\ell} & 0 \end{bmatrix}, \qquad (1)$$

where $\ell = \ell_2 - \ell_1$ is the length of the de-embedded line. Suppose now that the LINE considered above actually has the line length of $\ell_2 = \ell_1$. In effect, a 'zero-length transmission line' is de-embedded in this situation, as shown in Fig. 1(c). Since $\ell = 0$ in this case, the zero-length line's S matrix will ideally assume the following form:

$$S_{\text{zero}} = \begin{bmatrix} 0 & 1 \\ 1 & 0 \end{bmatrix}. \qquad (2)$$

Because of Eq. (2), S_{11} and S_{22} of the de-embedded zero-length line should stay at the center of a Smith chart. Also, S_{21} and S_{12} should stay at $(1,0)$ on a polar chart. In reality, the calculated value of S_{zero} for each length of THRU deviates from the ideal value in Eq. (2). The amount of deviation from Eq. (2) could be used as an indicator of nonideality. Note that this check does not require the knowledge of the frequency-dependent reference impedance, and hence is robust. Note also that this check cannot be applied to the plain TRL [1] because it, in effect, computes $\mathsf{T_B}$ from $\mathsf{T_A}$ using the equality in Eq. (2).

To allow one to readily tell if a given deviation is 'substantial' or not, we propose the following as a quick checklist.

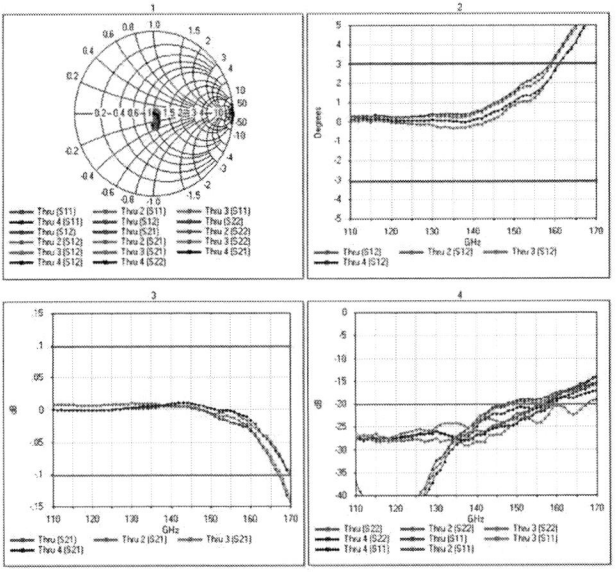

Fig. 4. S parameters of the de-embedded zero-length type-A line when the THRU is $0\,\mu$m. The same THRU was measured four times, and multiline TRL calculations and subsequent de-embedding were performed four times. The four results show a level of repeatability.

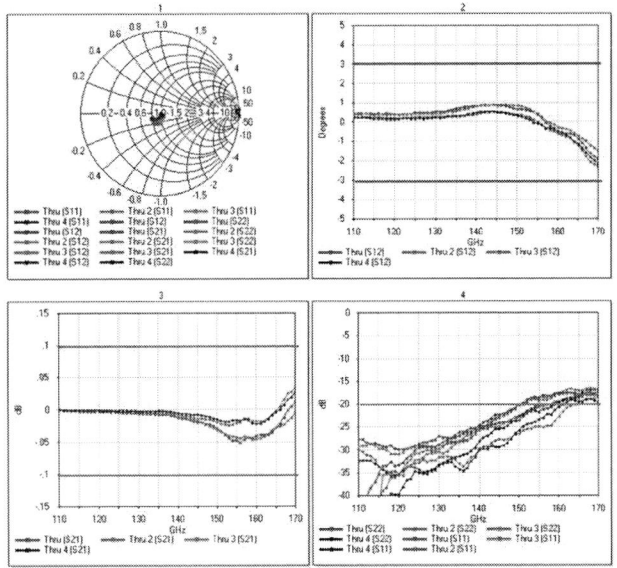

Fig. 5. S parameters of the de-embedded zero-length type-A line when the THRU is $100\,\mu$m. Results from four calculations shown as in Fig. 4.

TABLE III

DISPERSION IN α (MAX MINUS MIN) AT 170 GHz, RESULTING FROM FOUR MULTILINE TRL CALCULATIONS WITH FOUR INDEPENDENT MEASUREMENTS OF EACH TYPE-A THRU.

THRU length	Dispersion in α at 170 GHz
$0\,\mu$m	0.42 dB/mm
$100\,\mu$m	0.24 dB/mm
$200\,\mu$m	0.21 dB/mm
$300\,\mu$m	0.28 dB/mm
$400\,\mu$m	0.12 dB/mm
$500\,\mu$m	0.21 dB/mm
$600\,\mu$m	0.49 dB/mm

1) For the reflection coefficients S_{11} and S_{22}:

$$|S_{11}|,\ |S_{22}| < -20\,\text{dB}. \tag{3}$$

2) For the transmission coefficients S_{21} and S_{12}:

$$|S_{21}|,\ |S_{12}| < 0.3\,\text{dB}, \tag{4}$$

$$|\arg S_{21}|,\ |\arg S_{12}| < 3°. \tag{5}$$

While the choice of the numbers in Eqs. (3) through (5) are somewhat arbitrary, the check does serve its purpose.

Figs. 4 through 10 show the de-embedded type-A S_{zero} when different lengths of THRU are used in multiline TRL. For the THRUs that are $200\,\mu$m or shorter, at least one of the conditions are not met (Figs. 4–6). But THRUs that are $300\,\mu$m or longer meet all the conditions (Figs. 7–10). Note that the repeatability levels seen in Figs. 4–10 are well within Eqs. (3)–(5).

For the other three line types (Table I), the shortest THRUs that meet the conditions are all found to be $400\,\mu$m, and the corresponding S_{zero} are shown in Figs. 11, 12, and 13.

The above results indicate that de-embedding by multiline TRL become unreliable when the transmission line length in a THRU is shorter than $400\,\mu$m in the frequency range of 110 GHz–170 GHz. In terms of the guided wavelength λ_g at 110 GHz and 170 GHz, $400\,\mu$m is roughly $0.3\lambda_g$ and $0.4\lambda_g$, respectively.

III. TRANSMISSION LINE CHARACTERISTICS

Extracted attenuation constants α and the phase constants β, where $\gamma = \alpha + \mathrm{j}\beta$, of the type-A line are shown in Figs. 14 and 15, respectively.

As mentioned earlier, we measured each standard four times, and altogether eight measurement data are used in a single run of a multiline TRL calculation (Table II). To see an effect of measurement repeatability on α, Fig. 16 shows α that resulted from different measurements of the same THRU. The dispersion of α at 170 GHz is tabulated in Table III. Note that in Fig. 16 and Table III, the measurement data for LINEs and REFLECTs are fixed. For all the thru lengths, the maximum dispersion in α at 170 GHz is under 0.5 dB/mm.

Looking back at Fig. 14, which we plotted by taking the most typical data of the four (Fig. 16 and Table III) for each length, the maximum difference in α at 170 GHz due to the difference in THRU lengths ($0\,\mu$m–$600\,\mu$m) is about 1.3 dB/mm. But if we look only at the lengths that meet the conditions presented in Section II, that is $300\,\mu$m–$600\,\mu$m, the maximum difference is only about 0.4 dB/mm. This, together with Fig. 16, could imply that something that cannot simply be ascribed to measurement repeatability problem due to RF probe contacts is present in the measurement data of the short THRUs ($0\,\mu$m–$200\,\mu$m).

Similar THRU-length dependence was also seen for other

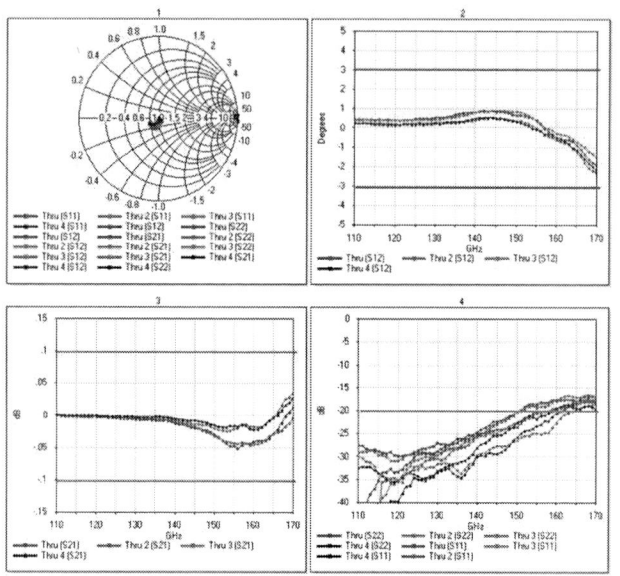
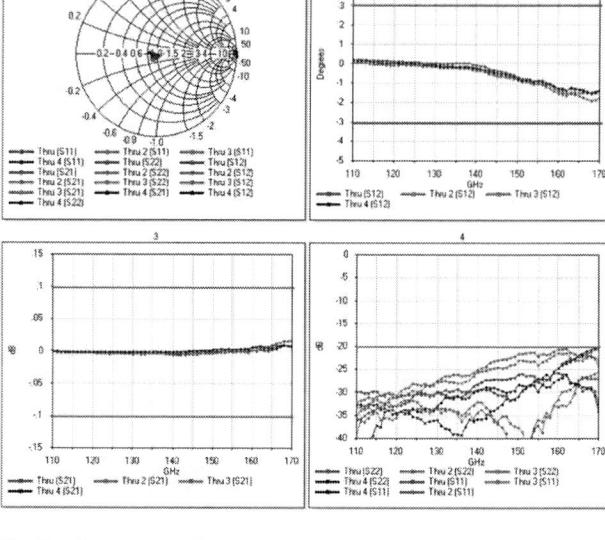

Fig. 6.   S parameters of the de-embedded zero-length type-A line when the THRU is 200 μm. Results from four calculations are shown as in Fig. 4.

Fig. 7.   S parameters of the de-embedded zero-length type-A line when the THRU is 300 μm. Results from four calculations are shown as in Fig. 4.

line types (B, C, and D). Again, it appears safer to have a transmission line section in a THRU that is 400 μm or longer in the frequency range 110 GHz–170 GHz.

IV. CONCLUSIONS

We studied how the length of the THRU standard affects multiline TRL de-embedding above 110 GHz up to 170 GHz. We proposed a simple check that can be used to tell if a result of multiline TRL is self-consistent. We used it to see whether a given length of THRU is long enough. The experimental results indicate that a THRU for that frequency range should be 400 μm or longer, excluding pads and pad-to-line transitions.

ACKNOWLEDGMENTS

This work was partly supported by a STARC joint research program and VDEC in collaboration with Agilent Technologies Japan, Ltd.

REFERENCES

[1] G. F. Engen and C. A. Hoer, " 'Through-reflect-line': an improved technique for calibrating the dual six-port automatic network analyzer," *IEEE Trans. Microw. Theory Tech.*, vol. 27, no. 12, pp. 987–993, Dec. 1979.

[2] Y. Tretiakov, J. Rascoe, K. Vaed, W. Woods, S. Venkatadri, and T. Zwick, "A new on-wafer de-embedding technique for on-chip RF transmission line interconnect characterization," *63rd ARFTG Conf.*, pp. 69–72, Jun. 2004.

[3] M. J. Kobrinsky, S. Chakravarty, D. Jiao, M. C. Harmes, S. List, and M. Mazumder, "Experimental validation of crosstalk simulations for on-chip interconnects using S-parameters," *IEEE Trans. Adv. Packag.*, vol. 28, no. 1, pp. 57–62, Feb. 2005.

[4] A. M. Mangan, S. P. Voinigescu, M.-T. Yang, and M. Tazlauanu, "De-embedding transmission line measurements for accurate modeling of IC designs," *IEEE Trans. Electron Devices*, vol. 53, no. 2, pp. 235–241, Feb. 2006.

[5] T. Sekiguchi, S. Amakawa, N. Ishihara, and K. Masu, "On the validity of bisection-based thru-only de-embedding," *Int. Conf. Microelectronic Test Struct.*, pp. 66–71, Mar. 2010.

[6] S. Amakawa, K. Takano, K. Katayama, M. Motoyoshi, T. Yoshida, and M. Fujishima, "On the choice of cascade de-embedding methods for on-wafer S-parameter measurement," *IEEE Int. Symp. Radio-Frequency Integration Technology*, pp. 137–139, Nov. 2012.

[7] D. F. Williams, R. B. Marks, and A. Davidson, "Comparison of on-wafer calibrations," *38th ARFTG Conf.*, pp. 68–81, Dec. 1991.

[8] D. F. Williams and R. B. Marks, "Calibrating on-wafer probes to the probe tips," *40th ARFTG Conf.*, pp. 136–143, Dec. 1992.

[9] R. A. Speciale, "A generalization of the TSD network-analyzer calibration procedure, covering n-port scattering-parameter measurements, affected by leakage errors," *IEEE Trans. Microw. Theory Tech.*, vol. 25, no. 12, pp. 1100–1115, Dec. 1977.

[10] Agilent Technologies, "Network analysis applying the 8510 TRL calibration for non-coaxial measurements," Product Note 8510-8A, 2006.

[11] A. Rumiantsev, P. L. Corson, S. L. Sweeney, and U. Arz, "Applying the calibration comparison technique for verification of transmission line standards on silicon up to 110 GHz," *73rd ARFTG Conf.*, pp. 132–135, Jun. 2009.

[12] K. H. K. Yau, I. Sarkas, A. Tomkins, P. Chevalier, and S. P. Voinigescu, "On-wafer S-parameter de-embedding of silicon active and passive devices up to 170 GHz," *IEEE MTT-S Int. Microwave Symp.*, pp. 600–603, 2010.

[13] L. O. Chua, "Device modeling via nonlinear circuit elements," *IEEE Trans. Circuits Syst.*, vol. 27, no. 11, pp. 1014–1044, Nov. 1980.

[14] R. B. Marks, "A multiline method of network analyzer calibration," *IEEE Trans. Microw. Theory Tech.*, vol. 39, no. 7, pp. 1205–1215, Jul. 1991.

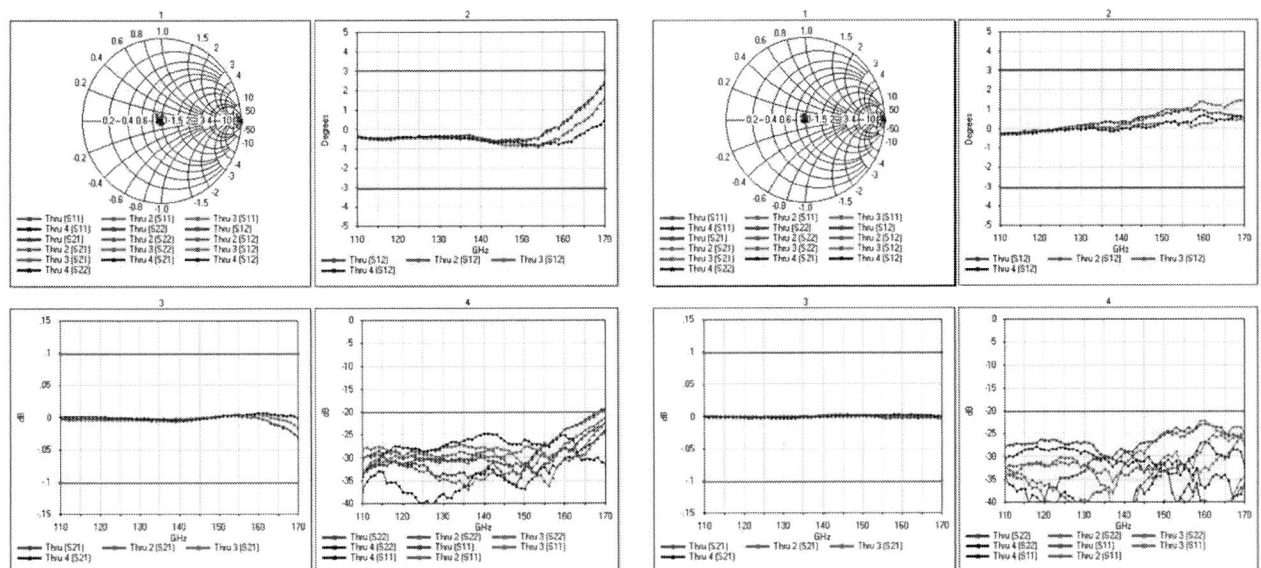

Fig. 8. S parameters of the de-embedded zero-length type-A line when the THRU is 400 μm. Results from four calculations are shown as in Fig. 4.

Fig. 10. S parameters of the de-embedded zero-length type-A line when the THRU is 600 μm. Results from four calculations are shown as in Fig. 4.

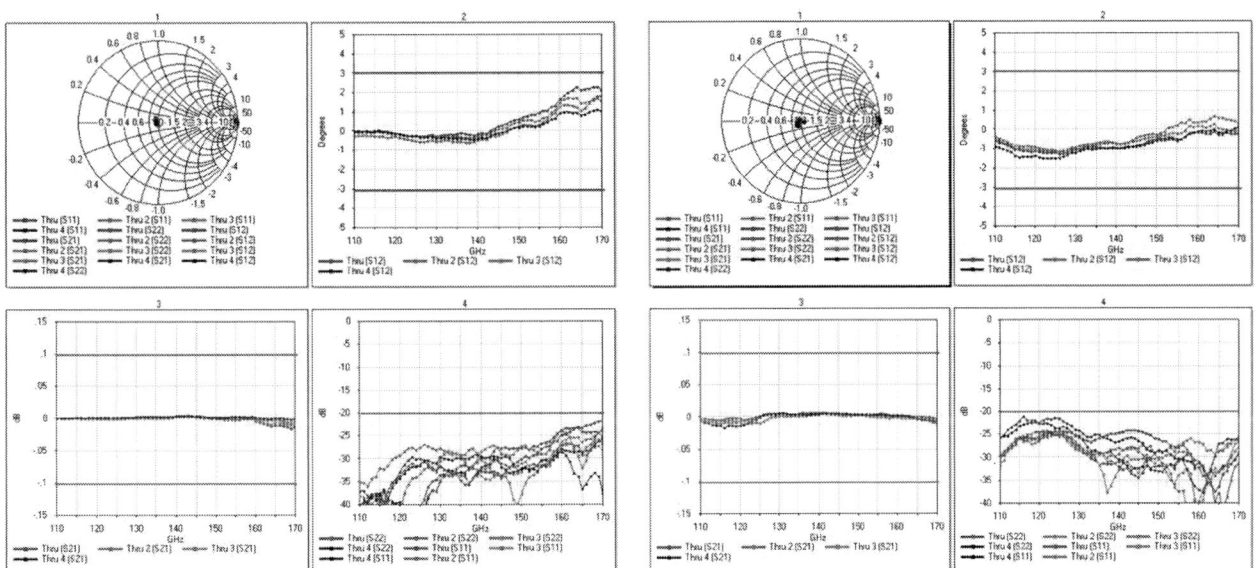

Fig. 9. S parameters of the de-embedded zero-length type-A line when the THRU is 500 μm. Results from four calculations are shown as in Fig. 4.

Fig. 11. S parameters of the de-embedded zero-length type-B line when the THRU is 400 μm. Results from four calculations are shown as in Fig. 4.

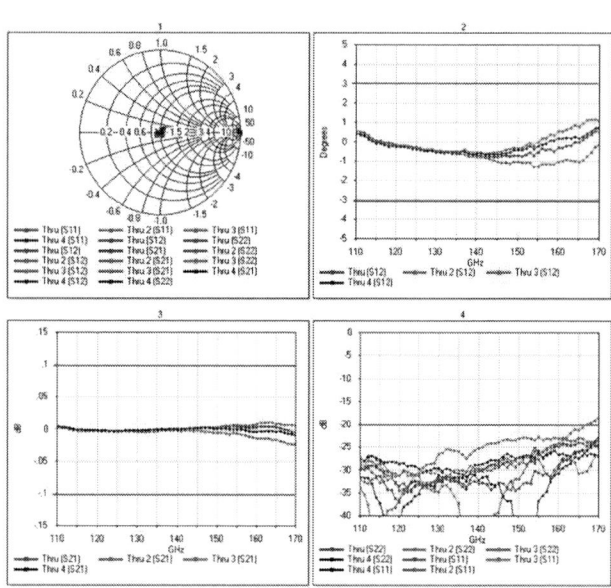

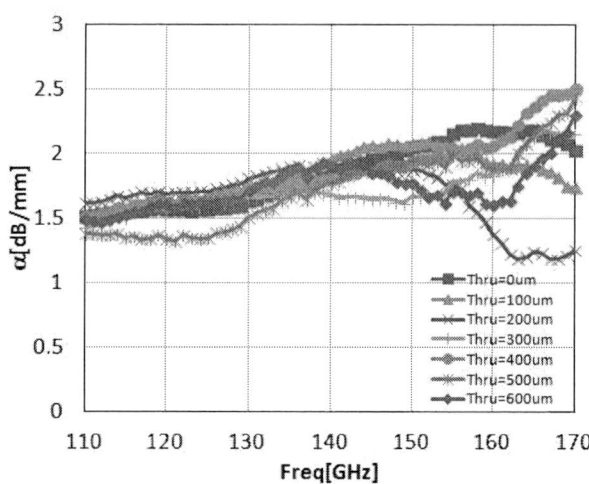

Fig. 14. Attenuation constant α of the type-A transmission line extracted by multiline TRL using different lengths of THRU. LINES are 1000 μm and 1200 μm long.

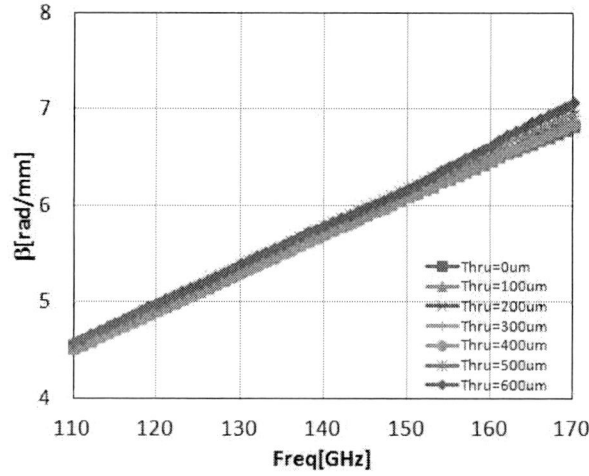

Fig. 15. Phase constant β of the type-A transmission line extracted by multiline TRL using different lengths of THRU. LINES are 1000 μm and 1200 μm.

Fig. 12. S parameters of the de-embedded zero-length type-C line when the THRU is 400 μm. Results from four calculations are shown as in Fig. 4.

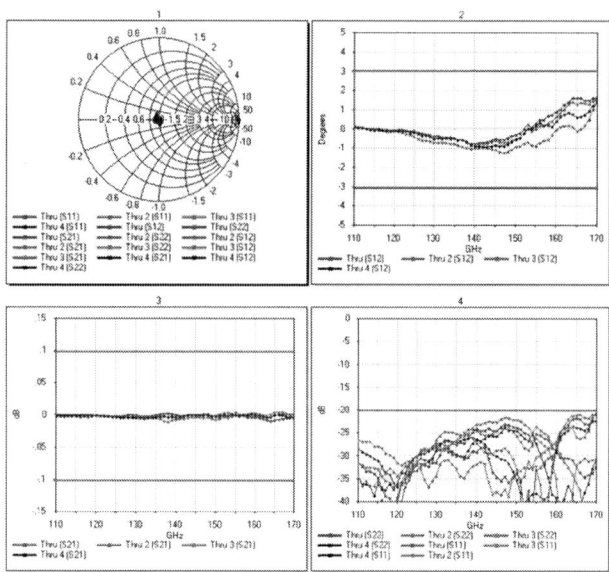

Fig. 13. S parameters of the de-embedded zero-length type-D line when the THRU is 400 μm. Results from four calculations are shown as in Fig. 4.

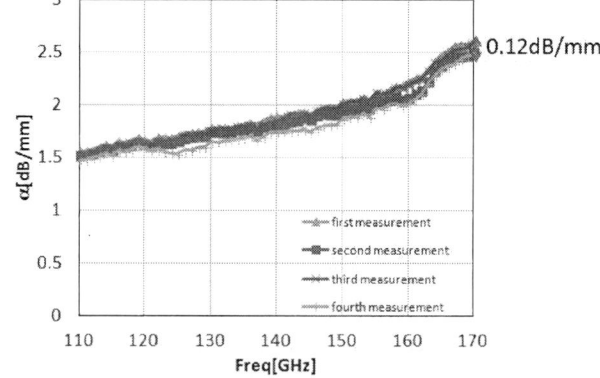

Fig. 16. Effect of measurement repeatability on α. Multiline TRL was performed four times to calculate α with four different measurement data of the same 400-μm type-A THRU.

Evaluation of 1/f noise variability in the subthreshold region of MOSFETs

Hans Tuinhout and Adrie Zegers-van Duijnhoven

NXP Semiconductors – Design Platforms

High Tech Campus Eindhoven, the Netherlands

Abstract — this paper discusses the challenges of characterization of 1/f noise and its variability under weak-inversion operating conditions of MOSFETs. A dedicated test module was designed with a range of MOSFET types with different layout implementations, particularly focusing at the noise behavior of very wide transistors. Through extensive use of a commercial noise characterization system it proved possible to evaluate the variability of 1/f noise in weak-inversion, revealing several interesting and important subtleties of low frequency noise.

Keywords — *1/f noise, low frequency noise measurements, variability, MOSFET, weak inversion, subthreshold*

I. INTRODUCTION

Low frequency noise (LFN) is well known to limit the performance of high performance mixed-signal and RF circuits. A relatively new aspect in this field is that, due to the persistent drive towards lower power consumption for instance for mobile applications, low-noise analogue circuits are not seldom operating in the weak-inversion (subthreshold) regime. This paper summarizes a study that was initiated to investigate whether 1/f noise models require adaptations for such applications.

MOSFET LFN models are usually derived and fitted to measurement data in strong inversion (and generally in saturation), typically using transistor geometries ranging from the technology's node minimum dimension up to something of the order of 2 to 10 µm, both in width and length. Moreover, LF noise measurement systems are generally designed to characterize noise at device current levels of micro-Amps and higher [1,2,3], hence typically well above the threshold voltage of test transistors usually available on test structure modules for device modeling.

Last but not least, LFN power spectral densities (PSD) of small devices tend to show large device-to-device variability that can span several orders of magnitude [4]. This is a trending topic in the current literature on device variability and RTN (Random Telegraph Noise) in emerging deca-nanometer CMOS technology nodes e.g. [5].

One of the novel learning points from the study described in our paper is that in the subthreshold operating region of even fairly large area devices, 1/f noise PSD's can also vary over several orders of magnitude (see for instance Fig. 1).

Large random (or deterministic) variability of any observable quantity implies that substantial populations of devices must be measured for drawing statistically significant conclusions regarding subtle effects. Fig. 1 gives a flavor of the kind of interpretation challenges that are encountered when large numbers of noise PSD spectra are collected, in this case for

GO2 W/L=20/0.5 NMOSFETs operating in the weak-inversion regime. V_T for these devices is typically 0.7 V. Note the (expected) large range of PSD's for different gate biases, the (less expected) large variability within a population for a particular gate bias, and the occurrences of Lorentzians (humps in the spectra) which vary from gate bias to gate bias.

This paper summarizes techniques used to address the issues sketched above. A dedicated LF noise characterization module was designed, processed and measured. A fully automated state-of-art commercial measurement system was used to collect sufficient data to reveal several subtle layout related LFN effects in the subthreshold regime of MOSFETs. Elaborate statistical analysis is used to reveal subtleties of LFN.

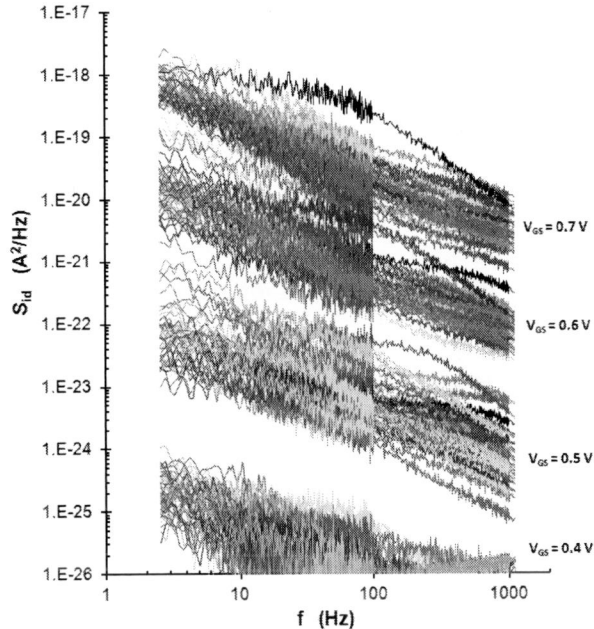

Figure 1. Illustration of challenges regarding interpretation of 1/f noise PSD variability for a population of 44 W/L=20/0.5 GO2 NMOSTs, depicted at four different gate biases (V_{GS} = 0.4, 0.5, 0.6 and 0.7 V) and V_{DS}=0.5 V. For clarity, only the frequency bands from 2.5 Hz to 1.1 kHz are plotted.

The remainder of this paper is organized as follows: after a description of the noise characterization test module, the basic LFN measurement and characterization approach is described. Then, the challenges regarding the appropriate data analysis of noise variability are discussed, followed by a brief discussion of some useful results. Finally, the most important conclusions are summarized.

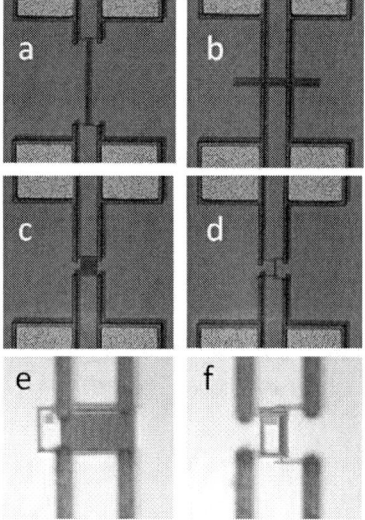

Figure 2. LF noise transistor test module.

Figure 3. Close-up photographs of some of the transistors in the test module. Different layout styles (a, b, c, d) and addition of simple gate protection diodes (e,f)

II. TEST STRUCTURE

Fig. 2 shows an overview of the test module that was designed and manufactured for this study. The module was processed in a 0.14 µm triple gate-oxide CMOS technology. The module consists of thirty different transistors laid out in a conventional 2x12 probe pad configuration. To exclude leakage paths and signal cross-coupling, no shared pads are used, and each transistor has its own well- or substrate-pad. The channel lengths of the transistors are always larger than the minimum allowed for the technology, as is often customary in analog circuits (smaller g_{ds}, higher voltage gain). The main purpose of the test module was to assess the LF noise of MOSFETs at gate biases well below the threshold voltage. For

this purpose a selection of very wide transistors[1] (typically 480 µm wide) were laid out in several different architecture implementations. Fig. 3 depicts examples of layouts for the GO2 NMOSTs that form the basis of this paper. Fig. 3 a, b and c are W/L=480/0.5 (µm/µm) transistors with the total transistor width split-up (folded) in 4, 120 and 24 fingers respectively. Fig 3 d shows a conventional single finger W/L=20/0.5 transistor normally used for modeling purposes (enabling comparison with the 24 finger device of Fig 3 c). Some of the transistors are also available in a version equipped with a simple 5x5 µm² n⁺-p-substrate gate protection diode. Close-up photographs of two of those are depicted in Fig, 3 e and f.

III. LOW FREQUENCY NOISE CHARACTERIZATION

To be able to cope with LFN variability, the measurements must be done on conveniently large populations of each of the different transistors on the test module described above. In this study we measured 48 dies distributed evenly over 200 mm wafers. Devices were selected at least 10 mm away from the edge of the wafers. All LFN and DC measurements were done with the commercial flicker noise measurements system depicted in Fig. 4 [1].

Figure 4. The used EDGE Flicker noise measurement system.

The fully integrated measurement system provides the combination of LF noise measurements and full DC I-V measurements (Fig. 5), and includes automatic stepping from device to device. The major advantage of this combined data collection is that it is always possible to verify afterwards whether an unexpected noise spectrum is related to a malfunctioning transistor, a probing failure, or an EMC disturbance.

After the DC IV-curve, LF noise spectra are measured at multiple V_{GS} bias points. V_{DS} is set to 0.5 V for all measurements presented in this paper. As explained above, bias points were selected predominantly in weak (and moderate) inversion. As depicted in Fig. 5, V_{GS} values of 0.35, 040, 0.45, 0.50, 0.55, 0.60, 0.7 and 1.00 V were used. For the narrower (20 µm wide) transistors the lowest bias did not yield meaningful spectra (drain current too low), while for the 480

---

[1] Noise spectral density S_{id} is proportional to I_D^2, hence W^2.

978-1-4673-4845-4/13 $31.00 © 2013 IEEE

μm wide devices the highest point was usually discarded because the high current level (order 10 mA) frequently showed unrealistic variability, attributed to probe-to-pad contact resistance variation of the single tungsten needle used for contacting.

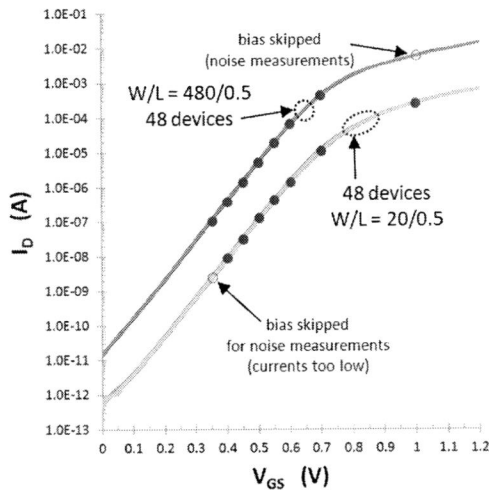

Figure 5. Wafer populations of DC I_D-V_{GS} curves of 2x48 devices. V_{DS}=0.5 V. Solid dots indicate V_{GS}'s used for LF noise measurements.

LF noise measurements were done on all transistors over a frequency range from 1 Hz to 100 kHz. Averaging was set at 20 x for the lowest band (1 - 100 Hz) and 80 x for the higher bands. The system can –in principle- handle frequencies up to 40 MHz, but as in this study we are predominantly looking at the noise at low current levels, we focused on the lower frequency bands.

To reduce possible sources of interpretation confusion, in most graphs in this paper the (horizontal) frequency axes in the spectra are limited somewhat. Below 2.5 Hz, the PSD's drop, probably related to the lower bandwidth edge of the LNA. The parts >> 1 kHz are "cut-off" from the spectra, that is from the figures, as they provide no information on the LF noise variability that we are currently interested in. At the high frequency ends of the spectra, the combination of the LNA's load resistor and load capacitance (cables, probe, and device) can give a bandwidth limitation. By individual adjustment of the load resistors, the high (1/f) end of the spectra can be extended, but that was not done for this study. All load resistors were set automatically by the system based on the biasing and the measured device currents and conductances (g_{ds} and g_m).

One of the consequences of the objective of this study to characterize LF noise in the subthreshold region (with relatively low currents), is that a lot of the data must be analyzed at relatively low frequencies. To first order, noise spectral densities S_{id} scale with I_D^2 and 1/f, so by definition, if one wants to characterize 1/f noise at low current levels, one is naturally limited to low frequencies. In this paper, the actual quantification of noise variability is done by looking at the noise spectra around 20 Hz. Note for instance in Fig. 1 that for V_{GS}=0.4 V, the noise levels for some devices barely rise above the noise floor of the system (which is found to be a little below 0.1 pA/√Hz) for frequencies above 10 Hz.

Statistical analysis and device comparisons of the LF noise performance are done using a figure of merit (FOM) called "AxS_{id}/I_D^2" (μm^2/Hz), in which A (Area) is equal to (the designed) W x L of the transistor, I_D its drain current during the noise measurement, and S_{id} a calculated noise PSD at 1 Hz. The exact extraction algorithm is as follows (Fig. 6).

- First, a power fit is done on S_{id} vs. frequency (yielding the thin straight line in Fig 6.), using 120 frequency bands around 20 Hz (fat line in Fig. 6), hence in this case spanning an extraction window of 5 Hz to 35 Hz.

- The resulting fit equation will be S_{ide} x f$^\alpha$. In the example depicted in Fig 6, this yields an alpha of -0.89 and an S_{ide} of $1.7x10^{-22}$ (A^2/Hz).

- From this fit, the best guess PSD at 20 Hz is calculated (in this case approximately $1.2x10^{-23}$).

- Next, this value at 20 Hz is extrapolated to 1 Hz using a (theoretical) alpha of -1 (hence, simply multiplied by 20), yielding a value of $2.38x10^{-22}$.

- This value is then multiplied with the device area A (in this case 240 μm^2), and divided by the square of drain current (for this transistor bias 3.82 $x10^{-7}$ A), finally yielding an AxS_{id}/I_D^2 of $3.92x10^{-7}$ (μm^2/Hz).

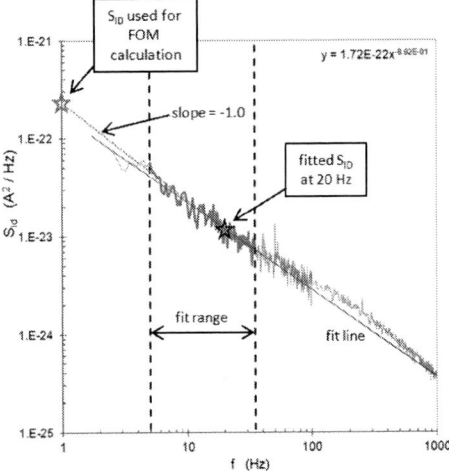

Figure 6. Example of the analysis of (20 x averaged) LF noise spectrum around 20 Hz.

Note that the variance of the (1 Hz) FOM calculated like this represents the variance band at 20 Hz, and is not unrealistically amplified by the fact that alpha varies from device to device. Imagine for instance the extreme case of a Lorentzian with a fall-off point just above the fit range, which could result in an alpha of zero. This could result in a much too low value for the extrapolated S_{ide} at 1 Hz, particularly if the center frequency for the extraction is high. As mentioned before, the choice of 20 Hz is a bit arbitrary. The same algorithm can be applied to other (higher) frequency bands, albeit that this limits the lowest attainable current levels substantially. The multiplication with effective area is done to enable comparison of the FOM to the ones as extracted for smaller devices.

978-1-4673-4845-4/13 $31.00 © 2013 IEEE

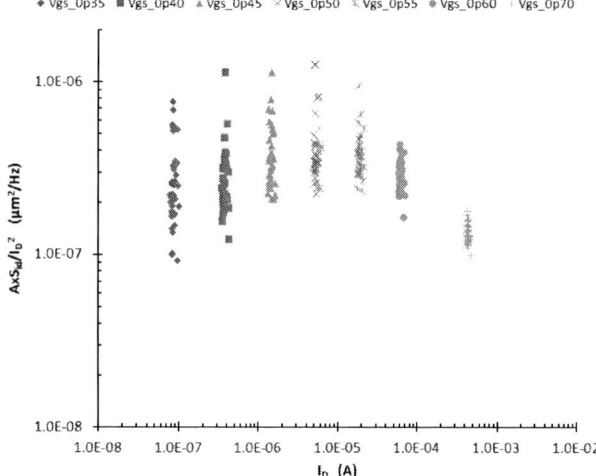

Figure 7.　Typical example of LF-noise FOM AxS_{id}/I_D^2 vs. drain current for 480/0.5 um NMOSTs for 7 different gate biases in weak and moderate inversion. (VGS = 0.35 to 0.70 V) .

Fig.7 shows a typical example of AxS_{id}/I_D^2 as measured on a population of 480/0.5 devices for 7 different gate biases in weak and moderate inversion. Note in this figure that there is a substantial variability, which clearly increases when the devices are biased deeper into the subthreshold region. Furthermore, one is easily tempted to appoint a few outliers, and it seems as if there is a horn-shaped curvature in the observations.

The remainder of this paper discusses some of the challenges and rewards of the detailed analysis of this type of results.

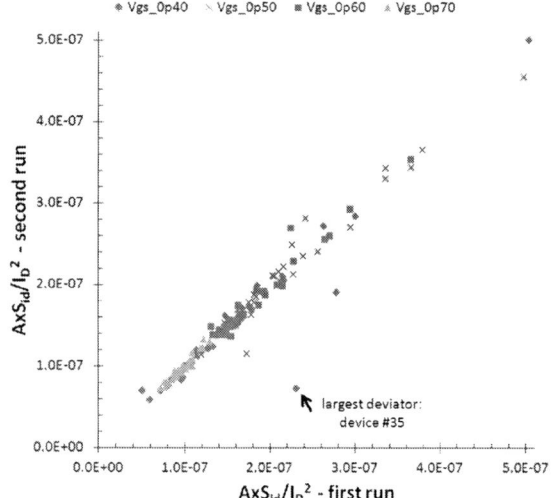

Figure 8.　Repeatability of AxS_{id}/I_D^2 measurements. Excellent agreement for most observations; largest deviations explained in figures.

IV.　REPEATABILITY AND GSM DISTURBANCES

A logical initial concern when trying to interpret effects associated with inherently noisy variables (and a new measurement system) is the question how reliable the determined parameters or estimators are. The first thing to

assess is the repeatability of the measurements. Fig. 8 shows a scatter plot of AxS_{id}/I_D^2 for four different gate biases for a population of 480/0.5 transistors that was measured twice. In a general sense, the overall agreement between the first and second run appears excellent and hence provides very high confidence in the quality (repeatability) of the measurements. In fact, the majority of the data yield a repeatability corresponding with a standard deviation of about 5 %. This should be considered as quite acceptable for this type of noise measurements. Nevertheless, it will be clear when looking at Fig. 8, that there also are a few AxS_{id}/I_D^2 observations (at different V_{GS}'s) that show substantially larger unrepeatability. By looking at the original PSD spectra, it becomes clear what causes this unrepeatability.

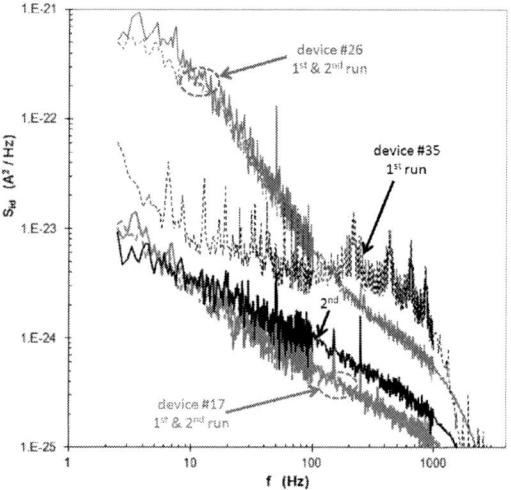

Figure 9.　Repeatability of PSD spectra for three devices measured with V_{GS}=0.4 V. 1st run: dashed lines, 2nd run solid lines. Black dashed spectrum (device #35) severely affected by mobile phone interference.

Fig. 9 shows the first and second run spectra for three transistors from the population used for Fig. 8 (V_{GS}=0.4 V). The two spectra for the transistor corresponding to a high noise level (red) and the one with low noise (green) agree very well. The two spectra (in black) for the device # 35, belonging to the largest deviator in Fig 8, however differ substantially.

The characteristic comb-like shape of the spectrum for the first measurement of transistor #35 is an indication of a GSM mobile phone disturbance. This is the spectrum that everybody recognizes when a mobile phone is held too close to an audio system while it is transmitting its burst-like data packages. Note that applying the AxS_{id}/I_D^2 calculation procedure around 20 Hz as presented in the previous section, will easily result in an almost 4 times larger number for the dashed (1st) spectrum compared to the solid (2nd) one.

A closer look at the total population of (first run) spectra at V_{GS}=0.4 V (Fig. 10), exemplifies the typical dilemmas when analyzing collections of wildly varying data. At first sight it proves difficult to filter-out unreliable measurements. In fact, there is more than one LFN spectrum affected by mobile phone disturbances. Note that although the spectral signature is comparable for the four disturbed measurements (inset Fig 10), the level of the impact certainly is not. And whereas the two largest disturbances in this case indeed correspond to the two largest repeatability deviators in Fig. 8, the lower two will have little or no effect on the FOM calculated in the 5 to 35 Hz frequency band, and hence gave no warning in the repeatability scatter plot.

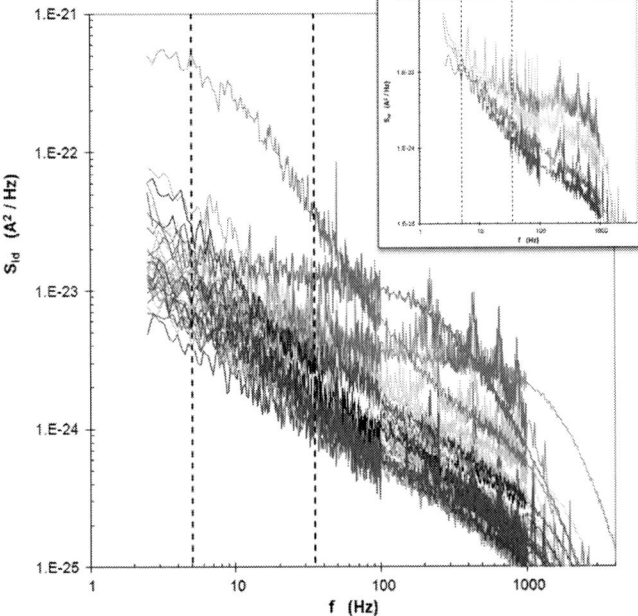

Figure 10. Overview of the PSD spectra of the first run of the repeatability test of Fig. 8 for V_{GS}=0.4 V. Inset depicts the isolated 4 spectra that were affected by mobile phone disturbances.

The occurrence of frequent GSM-related disturbances complicates the interpretation of subtle noise effects. It should be kept in mind that LF noise measurements for large populations of devices over multiple biases run unattended for many hours (>20 hrs is not exceptional). The use of mobile phones in a measurement lab, near the prober can in principle be regulated (forbidden). The example of Fig. 10 shows however, that much lower noise disturbances also show up (even fairly frequently in our lab environment). This probably indicates that some of these disturbance signals are generated at a much larger distance away from the prober, for instance in the corridor outside the measurement lab, or coming from the floor below or above the measurement lab. Although the ELITE 300 wafer prober [1] undisputedly forms the best engineered and cleanest on-wafer measurement solution for high-precision measurements, this example demonstrates that extreme care must be taken when it comes to proper interpretation of LF noise behavior and its variability. Apparently, even the best shielded systems are not immune for the harsh wireless communication tools that form an indispensable element of the contemporary high-tech work environment. Note that it was observed before that (especially) non-linear devices such as bipolar transistors and MOSFETs in subthreshold can be very sensitive to GHz range of EMC disturbances [6].

V. LF NOISE DATA INTERPRETATION

To systematically study both the deterministic behavior, as well as the variability of the LF noise in the weak inversion region, data interpretation is based on the three following steps:

- First all spectra (for all devices and all biases) are collected and visually inspected for suspicious spectra contaminated by GSM disturbances. When a particular spectrum is found to be contaminated, the spectra at the other (7) biases are discarded for this device as well. This assures the indispensible consistency of the population when doing the statistical analysis of the variability at different biases. Typically, this leaves us with a population of 40 to 48 samples out of the original 48 devices that were measured for each type.

- Next, the AxS_{id}/I_D^2 values are calculated and plotted as exemplified in Fig. 7. Subsequently, for each bias, the main statistical estimators (average and standard deviation) are calculated. These provide the basic plot for interpretation of the 1/f noise as a function of the bias.

- Finally, a 5999 draw bootstrap analysis[2] is applied to the population, to assess the statistical uncertainties of the measured (accepted) populations. Uncertainties are visualized through plotting the 5 and 95 percentile values of the bootstrap's 5999 sample population.

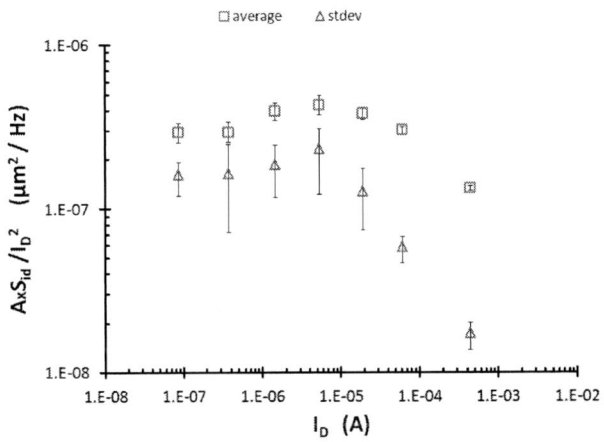

Figure 11. Result of statistical analysis of Fig. 7 LF noise data. Averages per bias are depicted by the red squares; triangles represent the associated standard deviations. Error bars are the 5 and 95 percentile points of the 5999x bootstrap analyses for both estimators.

Fig. 11 shows the result of applying the above procedure to the transistor population that was used to collect the AxS_{id}/I_D^2 data of Fig 7. As was probably obvious from Fig. 7, the standard deviations are not much more than a factor of 2 to 3 smaller than the averages. The statistical uncertainty on the average is however small and confirms the horn-shaped

---

[2] In a bootstrap analysis of a population of N samples, N "new" samples are drawn randomly from the original population (hence some samples are drawn multiple times). This is repeated M times (here M=5999) after which the resulting distributions are subsequently analyzed statistically.

curvature observed in Fig 7. It seems as if for this population of devices, the noise spectral density indeed peaks just below the onset of strong inversion and reduces a bit when the device is pushed deeper into weak inversion. What was also clear to the engineers' eyes in Fig. 7, but now quantified in Fig. 11 is that the LF noise variability (i.e. the standard deviation) of AxS_{id}/I_D^2 is at least an order of magnitude larger in the weak inversion regime compared to the values normally encountered in strong inversion. To the best of our knowledge, this aspect has not been published before in the LF-noise literature.

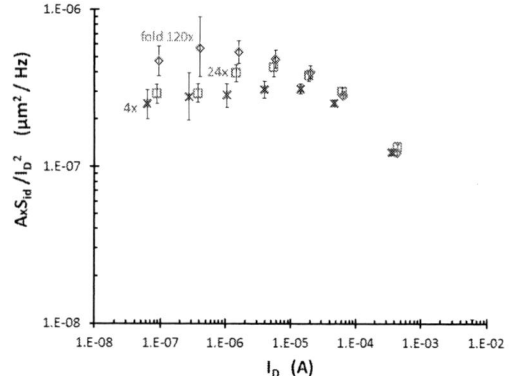

Figure 12. Comparison of the noise FOM averages (incl. 5 and 95 percentile uncertainty estimates) for three different GO2 W/L=480/0.5 NMOS transistor layout realizations (see Fig. 3 a,b and c).

VI. SOME RESULTS AND DISCUSSION

The purpose of the study described in this paper was to assess whether the compact modeling of LF noise of GO2 NMOS transistors in the subthreshold regime would require refinements. Our experiments show that this is by and large not the case. If anything, a small reduction of the average noise FOM is observed when the devices are pushed deeper into subthreshold. This could be attributable to the fact that, as the Fermi-level sweeps through the Si bandgap in the subthreshold region, some of the traps that contribute to the 1/f noise are de-activated. It did become clear however that the variability of LF noise does increase substantially in the subthreshold region. This places new demands on corner and Monte-Carlo models for variability-aware circuit design. The proper way to deal with the observed noise variability distributions in models is a point of further study, and beyond the scope of this paper.

This paper is topped-up with Fig. 12 and 13, summarizing some effects of transistor layout. The downward trend in the average AxS_{id}/I_D^2 values is also seen in the other populations of devices. Fig. 12 demonstrates that folding of the transistor has a statistically significant effect on LF noise. The 120x folded transistor population clearly exhibits higher noise, particularly in the subthreshold regime. This suggests that some of the most dominant LF noise contributors are located along the STI edges of the transistors, of which there obviously are a lot more in a 120x folded transistor. Last but not least, Fig. 13 demonstrates a clear benefit of adding Gate protection diodes to the test transistors. This suggests that an appreciable part of the measured LF-noise can be attributed to process induced

damage, which is (at least partly) mitigated by adding an n^+ to p-substrate diode.

Note that a high-precision, high throughput system such as the EDGE is required to filter such subtleties out of the large PSD variability in subthreshold.

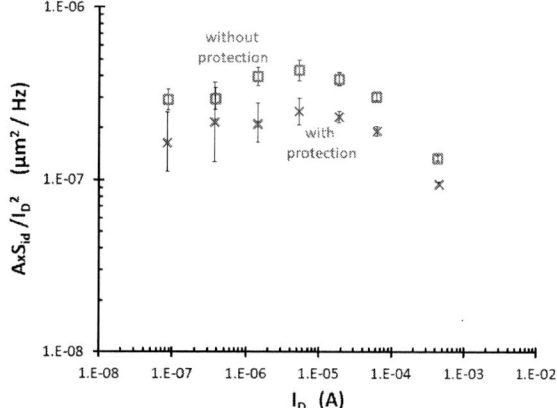

Figure 13. Effect of adding a gate protection diode on the noise FOM averages for two different GO2 W/L=480/0.5 NMOS transistor layout realizations (see Fig. 3 c. and e).

VII. CONCLUSIONS

This paper summarizes a study of low frequency noise levels and their variability for MOSFETs operating in the subthreshold region. We show that the LF noise variability can be assessed and quantified in this challenging operating region using a dedicated test module with very wide MOS transistors, in combination with extensive statistical measurements measured with a state-of-art commercial LF noise measurement system, and a newly developed elaborate data analysis procedure. As a demonstration, we reveal some subtle differences of the noise levels related to the transistors' layout implementation and the use of protection diodes. To the best of our knowledge, comparable results were not published on this scale before.

ACKNOWLEDGMENT

Cascade-Microtech Europe is kindly acknowledged for placing and supporting a demo EDGE system in our lab. The work described in this paper would not have been possible without this generous offer.

REFERENCES

[1] Edge Flicker noise measurement solution, www.cmicro.com

[2] NoisePro / Proplus 9812B system, www.proplussolutions.com.

[3] A. Blaum et al., "A new robust on-wafer 1/f noise measurement and characterization system", *ICMTS-2001*, pp. 125-130, 2001.

[4] G. Ghibaudo and O. Roux-dit-Buisson, "Impact of GHz disturbances on DC parametric matching measurements", *ESSDERC-1994*, pp. 693-700, 1994.

[5] N. Tega, et al., "Increasing Threshold Voltage Variation due to Random Telegraph Noise in FETs as Gate Lengths Scale to 20 nm", *Symposium on VLSI Technology*, pp. 50-51, 2009.

[6] H. P. Tuinhout and P. G.M. Baltus., "Impact of GHz disturbances on DC parametric matching measurements", *ICMTS-2006*, pp. 71-75, 2006.

SESSION 5: Variability and Yield

978-1-4673-4845-4/13 $31.00 © 2013 IEEE

978-1-4673-4845-4/13 $31.00 © 2013 IEEE

Newly developed Test-Element-Group for detecting soft failures of the low-resistance-element using doubly nesting array

Shingo Sato, Hiroki Shinkawata, Atsushi Tsuda, Tomoaki Yoshizawa, Takio Ohno

Renesas Electronics Corporation,

Devices and Analysis Technology Div., Production and Technology Unit.,

4-1-3, Mizuhara, Itami-shi, Hyogo, 664-0005, Japan

{shingo.sato.xa, hiroki.shinkawata.fn}@renesas.com

Abstract—**We report newly developed Test-Element-Group for detecting soft failures of low-resistance-element like interconnect via using doubly nesting array. We detected the soft failure of fine via which resistance had about 10 times larger resistance than normal via using this structure manufactured in 40nm CMOS technology.**

Keywords—soft failure, Back End Of Line, interconnect via, array structure, variability, yield

I. INTRODUCTION

With scaling-down of the LSI devices, variation of elements used in LSI chip like contact, interconnect via and so on is crucial for managing yield. Soft failures are one of the most interesting problems from the viewpoint of product reliability since the chip including soft failures cannot be screened out with electrical test before shipping in different from the chip including hard failures. To detect the soft failure in early stage of process development is needed for improving the product reliability since the soft failure has the potential of degrading to the hard failure after shipping. Various Test-Element-Group (TEG) used for process development has been reported for managing yield and detecting the failure of the element [1-4]. We report newly developed TEG for detecting soft failures of low-resistance-element like interconnect via using doubly nesting array, which is named as High-sensitivity-Screening-and-Detection-decoder TEG (HSD-TEG). This will be applicable to any manufacturing process which requires high reliability.

II. EXPERIMENTAL SET UP

A. TEG Structure

In this section, the architecture of HSD-TEG is described. Figure 1 shows the chip diagram of HSD-TEG. It consists of 5-to-32 decoders at X and Y-direction, block array, and I/O bus for force/sensing current/voltage signals as shown in figure 1(a). Unit block has sub-array of 8 x 8 (= 64) bit-cells selecting with 3-to-8 sub-decoders at x and y-direction as shown in figure 1(b). The sizes of the unit cell and the unit block including the peripheral circuit are about 7.8 um x 7.8um and 80 um x 80 um. The maximum size of HSD-TEG is 3.0 mm x 3.0 mm when we place the maximum block arrays limited by outputs of the decoders.

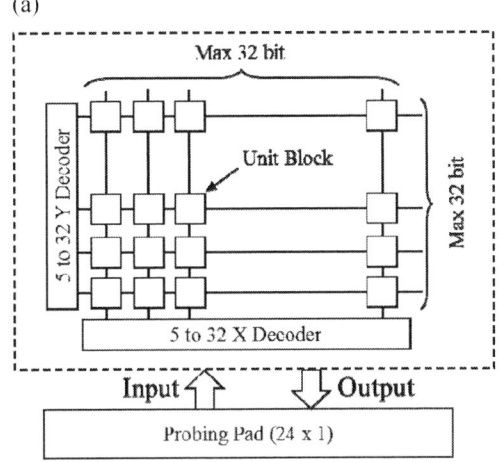

(a)

(b)

Figure 1: Chip diagram of HSD-TEG.
A Block has sub-array of 8 x 8 bit-cells.

Figure 2 shows a schematic circuit of the unit cell, which is Kelvin force/sensing scheme to eliminate parasitic resistance. R_{DUT} is a low-resistance-element like interconnect via to be measured. VDD_DOE and I_FORCE are terminals of supply

978-1-4673-4845-4/13 $31.00 © 2013 IEEE

voltage and forcing current for Device-Under-Test (DUT). SEL_DOE is a terminal for selecting DUT. The voltages at both sides of R_{DUT} are output to VOUT by switching SEL_VH and SEL_VL. The number of pad needed is 24. It will be reduced by utilizing logic circuit like counter or shift register when the arrays are addressed. HSD-TEG needs at least 2Cu-metal standard CMOS technology to construct the peripheral circuit.

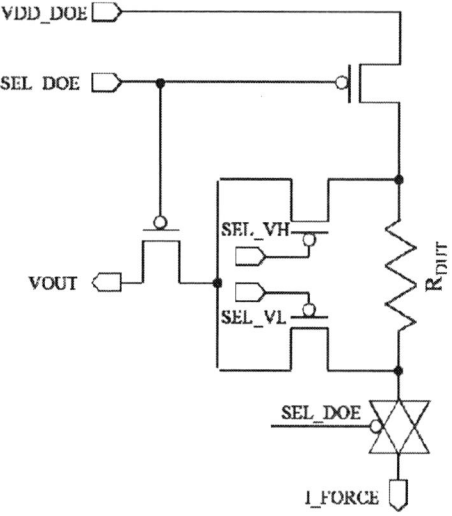

Figure 2: Schematic circuit of the unit cell

B. Measurement mode and how to detect the soft failure

Two modes of the measurements using the peripheral circuit are prepared. One is Chain-mode connecting 32 bit-cells by the peripheral circuit and measuring a couple of the chains per block. The other is Kelvin-mode measuring all of the unit cells. We can easily detect the soft failure by utilizing the measurement program to operate the peripheral circuit as follow. At first, all of the blocks for desired module are measured with Chain-mode and calculate difference of the resistance for each block. Then the blocks are sorted into descending order of absolute value of the differences. The cells inside a block are measured with Kelvin-mode in sorted order at last. The measurement system used is parametric tester like Agilent 4070 series with switching matrix system. The number of retested blocks affects the measurement time. The number of retested block is 1 and integration time, which is defined in Agilent 4070 series, of the measurement is *Short* as default throughout this paper unless it's mentioned. It is strongly recommended that the number of retested block is adjusted with the stage of the process development. It takes approximately 45 seconds to measure the module the maximum number of the arrays is placed, which is effectively equivalent to evaluate 64Kbit cells. This is about 30 times shorter in comparison to measure with Kelvin-mode only. These architectures of TEG and measurement program provide us reduction of the testing time and easiness of the physical analysis for failure bit since we can directly specify the coordinate to be analyzed using TEM without any other analysis like Voltage-Contrast technique.

III. MEASUREMENT RESULT

At first, we evaluate performances of HSD-TEG like accuracy and error degraded by the peripheral circuit. Figure 3 shows correlation characteristic for various resistors. The vertical line shows the median value of the resistance measured with Kelvin mode of HSD-TEG for a fixed DUT repeatedly and the horizontal one shows the resistance of Kelvin devices, which layouts are as same as previous ones, placed in other pad. Both resistances are well correlated for wide range of the resistance from less than 1Ω to more than 100Ω.

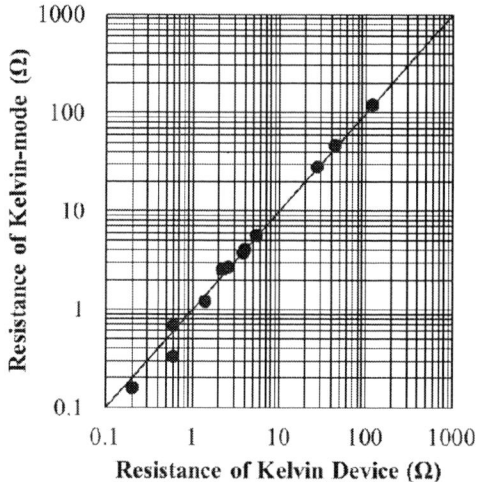

Figure 3: Correlation characteristic between Kelvin device and Kelvin-mode of HSD-TEG for various resistors

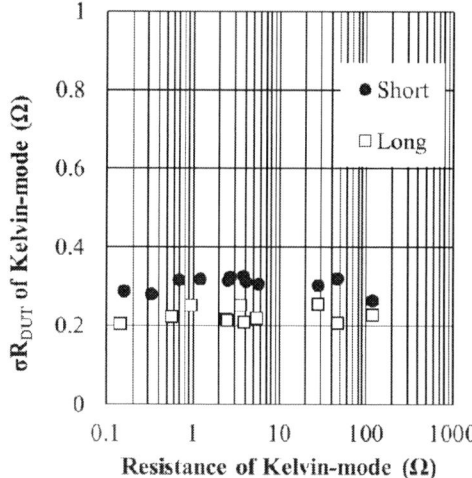

Figure 4: Comparison of standard deviation between integration times for various resistors with Kelvin-mode of HSD-TEG.

Figure 4 shows the characteristic of the standard deviation for measuring a fixed DUT with various integration times. Closed circle and open square symbol indicate the measurement result with *Short* and *Long* integration time. The standard deviation of Kelvin-mode is about 0.3Ω for any resistor with *Short* integration time and is reduced by increasing the integration time. These results indicate that HSD-TEG can evaluate the low-resistance-element like interconnect via, metal resistor and so on with high accuracy. Measurement error will be also reduced by optimizing the length and the width of the pass transistor inducing the leakage current [5].

Various DUT like 1st via chain, via-chain stacking from 1st via to 2nd via, via-chain stacking from 1st via to 4th via and metal resistors are prepared and measured to improve the manufacturing process of Back-End-Of-Line (BEOL). Via-chain stacking from 1st via to 4th via which number of the blocks placed is 384 with $(X, Y) = (32, 12)$ is demonstrated in this paper.

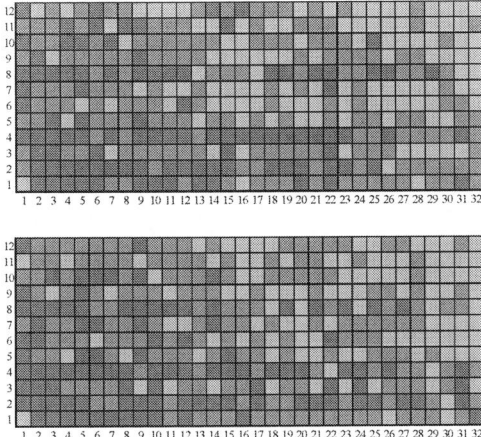

Figure 5: Bit-map-images of resistance measured with Chain-mode. Darker bit has larger resistance.

Figure 5 shows bit-map-images of the resistance measured with Chain-mode. We can obtain two bit-map-images of the resistances since Chain-mode connects 32 bit-cells in series and make two chains per block. Figure 6(a) and (b) show bit-map-image and cumulative distribution of the differences between a pair of the chains inside each block. Bit-map-image in figure 5 and figure 6(a) is consistent with figure 1(a) and a bit corresponds to a block. Darker bit in the bit-map-image has larger resistance in figure 5 and larger difference of the resistance in figure 6(a). The distribution of the difference is uniformly distributed inside the chip as shown in figure 6(a) nevertheless the resistance of the chains slightly becomes larger near the bottom of the chip as shown in figure 5. This result indicates that the effect of the variation inside chip can be eliminated by adopting the difference when we try to detect the soft failure. The characteristic of the local mismatch is powerful and useful for detecting the failures [6].

Since block $(X, Y) = (25, 10)$, a black bit in figure 6(a), has largest difference in figure 6(b), the cells of its block are measured with Kelvin-mode and its results are shown in figure 7. Figure 7(a) is corresponding to the schematic of the unit block shown in figure 1(b). The resistance is normalized by effective number of the vias in Figure 7(b) and its median

value is about 4. We could detect 8 times larger resistance assumed to become one of the vias soft failure.

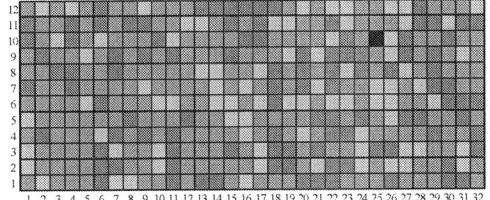

(a) Bit-map-image

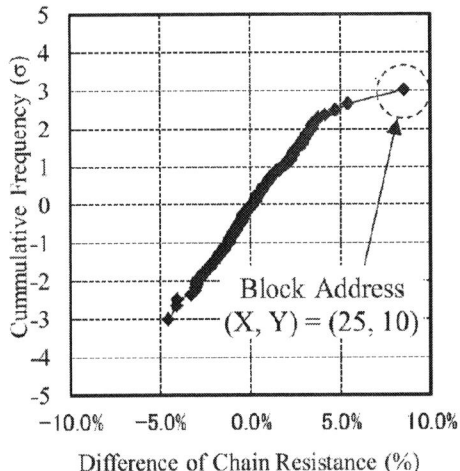

(b) Cumulative distribution

Figure 6: The difference of the resistance measured with Chain-mode

Figure 8 shows the result of the physical failure analysis using TEM. This result indicates that the soft failure is caused by insufficiency of filling Cu metal. Figure 9 show wafer-map of the soft failure detected with HSD-TEG. The shots detecting the soft failure are hatched with gray color. The detected soft failure randomly distributes on the wafer. It will be confirmed whether or not this failure degrades to hard one by accelerated aging test to construct the highly product reliability in future work:

IV. CONCLUSION

Newly developed TEG for detecting soft failures of low-resistance-element like interconnect via using doubly nesting array was reported. We demonstrated the usefulness of this TEG by detecting the soft failure of fine via, which resistance is about 10 times larger than that of normal via. We can construct the manufacturing process with high reliability utilizing developed TEG called as HSD-TEG.

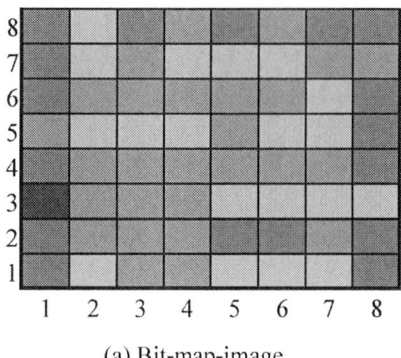

(a) Bit-map-image

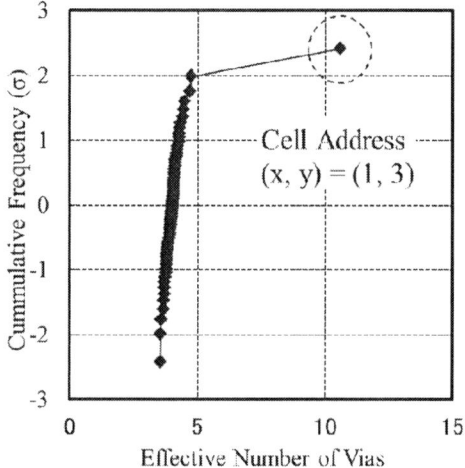

(b) Cumulative distribution

Figure 7: Measurement results with Kelvin-mode.

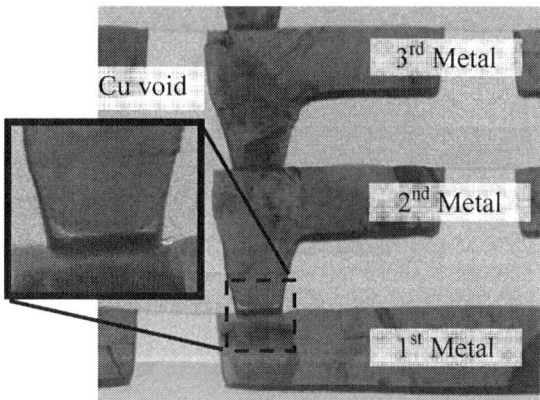

Figure 8: TEM image for fail bit.

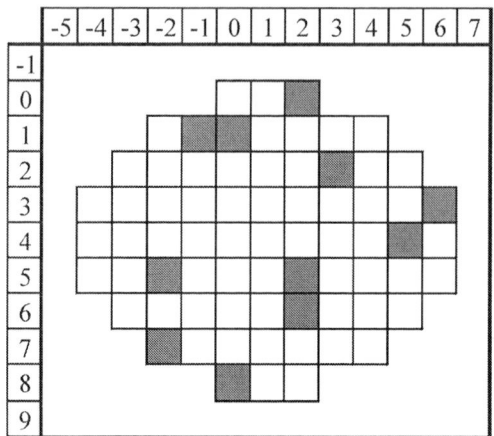

Figure 9: Wafer-map of the soft failure

ACKNOWLEDGMENT

We would like to thank H. Tasaka, S. Honbu, K. Kasai and N. Morimoto of Renesas Electronis Corporation for some fruitful discussion and comments to develop this TEG. We also would like to thank J. Bordelon of Stratosphere Solution Inc. for previous work.

REFERENCES

[1] C. Hess, M. Squcciarini, S. Yu, J. Burrows, J. Cheng, R. Lindley, A. Swimmer and S. Winters, "High Density Test Structure Array for Accurate Detection and Localization of Soft Fails", ICMTS, 2008, pp.131-136.

[2] M. Karthikeyan, S. Fox, W. Cote, G. Yeric, M. Hall, J. Garcia, B. Mitchell, E. Wolf, S. Agarwal, "A 65nm random and systematic yield ramp infrastructure utilizing a specialized addressable array with integrated analysis software", ICMTS, 2006, pp.104-109.

[3] K. Y. Y. Doong, J. Bordelon, K. J. Chang, L. J. Hung, C. C. Liao, S. C. Lin, R. S. Ho, S. Hsieh and K. L. Young, "Field-Configurable Test Structure Array (FC-TSA) : Enabling design for monitor, model and manufacturability", ICMTS, 2006, pp.96-103.

[4] J.R.D. DeBord, T. Grice, R. Garcia, G. Yeric, E. Cohen, A. Sutandi, J. Garcia, G. Green, "Infrastructure for successful BEOL characterization and yield ramp at the 65 nm node and below", IITC, 2005, pp.27-29.

[5] S. Ohkawa, M. Aoki, H. Masuda, "Analysis and characterization of device variations in an LSI chip using an integrated device matrix array", IEEE Trans. Semiconductor Manufacturing, vol. 17, 2004, pp.155-165.

[6] T. Weidong, P. Steinmann, E. Beach, I. Khan, P. Madhani, "Mismatch characterization of a high precision resistor array test structure", ICMTS, 2008, pp.11-16.

New methodology for drain current local variability characterization using Y function method

L. RAHHAL[1,2], A. BAJOLET[1], C. DIOUF[1,2], A. CROS[1], J. ROSA[1], N. PLANES[1], G. GHIBAUDO[2]

1) STMicroelectronics 850, rue Jean Monnet, 38926 Crolles, France
2) IMEP-LAHC, Minatec /INPG, BP 257, 38016 Grenoble, France
lama.rahhal@st.com

Abstract—**Y function is well known to overcome the influence of source/drain series resistance (Rsd) in MOSFETs. In this work we present a new methodology for drain current local variability characterization using Y function method. Thus, we show that the study of Y function statistical variability permits the extraction of threshold voltage (V_{TH}) and current gain factor (β) local variability without the influence of Rsd values. We also demonstrate a simple drain current local variability model taking into account the influence of Rsd and its variability in strong inversion regime. This new V_{TH} and β extraction method, and drain current variability model were applied with success to advanced FDSOI and Bulk devices with different dimensions.**

Keywords—Y function; drain current; source/drain series resistance; MOSFET; threshold voltage; current gain factor, local variability, FDSOI, Bulk.

I. INTRODUCTION

Drain current variability is one of the critical problems in scaled MOSFET's. It has been analyzed since the beginning of variability studies showing that the threshold voltage (V_{TH}) and the current gain factor (β) local fluctuations are the major sources of drain current (I_D) variability [1] [2].

As the channel length is scaled down, the source/drain series resistance (Rsd) is becoming non-negligible in the total device resistance (Rtot). Markov shows in [3] [4] that, for short channel lengths, the Rsd local fluctuations are an additional contributor to the drain current variability. This phenomenon is more pronounced in ultra thin silicon body Fully Depleted Silicon On Insulator (FDSOI) devices [3] [4]. The need for a reliable drain current variability model taking into account Rsd impact and variability is thus mandatory.

In this paper, we propose a new current variability model based on V_{TH}, β and Rsd local variability. We also propose an original V_{TH} and β variability extraction method, based on the well-known Y-function [5].

II. THEORETICAL MODEL

A. Drain current local variability model

It has been demonstrated that the Rsd limits the drain current performance of advanced MOSFET's [6][7]. Fig.1 shows that, in strong inversion regime and linear region, drain current is lowered when Rsd is increased. Note that the curves are obtained using Eqs 1-4 where Qi, Gd_0, Gd, K, T, n, and $\theta1$, $\theta2$ are respectively the channel inversion charge, the intrinsic

channel conductance, the extrinsic channel conductance, the reduced Boltzmann constant, the temperature, the subthreshold slope ideality factor, and the mobility reduction factors, with Rsd contribution for typical parameters (L=0.05μm/W=0.08μm and Rsd=0 and Rsd=2775Ω and drain voltage V_D=50mV).

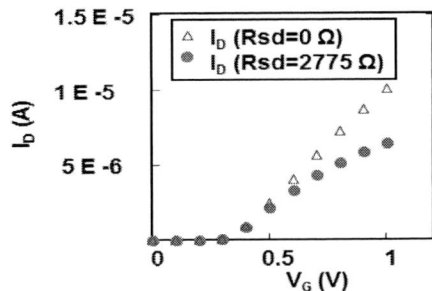

Figure 1: Drain current values as a function of gate voltage, for Rsd=2775Ω and Rsd=0Ω. (FD SOI transistors with W=0.08μm/L=0.05μm and V_D =50mV).

$$Qi = C_{ox}K.T.n.\ln(1 + e^{\frac{V_G - V_{TH}}{n.K.T}}).$$

$$.(1 - \frac{\ln(1 + \ln(1 + e^{\frac{V_G - V_{TH}}{n.K.T}}))}{2 + \ln(1 + e^{\frac{V_G - V_{TH}}{n.K.T}})}) \quad (1)$$

$$Gd_0 = \frac{\beta}{1 + \theta_1.\frac{Q_i}{C_{ox}} + \theta_2.(\frac{Q_i}{C_{ox}})^2} Q_i \quad (2)$$

$$Gd = \frac{Gd_0}{1 + Gd_0 Rsd} \quad (3)$$

$$I_D = Gd.V_D \quad (4)$$

As Rsd has an important impact on I_D values, it is worth taking into account Rsd influence and its variability contribution in drain current variability.

Departing from first order Taylor approximation, and calculating the drain current variance, the drain current local variability can be written as shown in Eq.5. Note that the correlations between V_{TH}, β and Rsd fluctuations are considered as negligible.

$$\sigma_{\Delta I_D / I_D}^2 = (\frac{1}{I_D}\frac{\partial I_D}{\partial V_{TH}})^2.\sigma_{\Delta V_{TH}}^2 + ... \\ + (\frac{1}{I_D}\frac{\partial I_D}{\partial \beta})^2.\sigma_{\Delta \beta}^2 + (\frac{1}{I_D}\frac{\partial I_D}{\partial Rsd})^2.\sigma_{\Delta Rsd}^2 \quad (5)$$

After calculating the derivative of I_D with respect to V_{TH}, β and Rsd, the drain current local variability model is obtained as shown in Eq.6, where Gm is the transconductance.

$$\sigma_{\Delta I_D / I_D}^2 = (\frac{Gm}{I_D})^2.\sigma_{\Delta V_{TH}}^2 + (1 - Gd.Rsd)^2.\sigma_{\Delta \beta / \beta}^2 + ... \\ + (Gd)^2.\sigma_{\Delta Rsd}^2 \quad (6)$$

B. Threshold voltage and current gain factor local variability extraction using Y function method

The Y function given by Eq.7, which has been introduced as a simple method to extract the MOS transistor's parameters [5] & [8], is immune to Rsd values as shown in Fig.2.

$$Y = \frac{I_D}{\sqrt{Gm}} \quad (7)$$

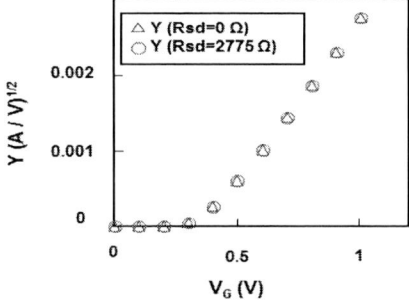

Figure 2: Y function values as a function of gate voltage, for Rsd=2775Ω and Rsd=0Ω. (FD SOI transistors with W=0.08µm/L=0.05µm and V_D =50mV).

The classical $\sigma_{\Delta V_{TH}}$ and $\sigma_{\Delta \beta / \beta}$ extraction consists in extracting V_{TH} and β (using Y method [5] in our case) for N sampling of MOSFET transistor pairs. Thus, after applying a recursive filter to eliminate erroneous data, the standard deviation $\sigma_{\Delta V_{TH}}$ and $\sigma_{\Delta \beta / \beta}$ of the Gaussian distribution are calculated. Instead of passing through V_{TH} and β and then calculate their variances, we propose a direct $\sigma_{\Delta V_{TH}}$ and $\sigma_{\Delta \beta / \beta}$ extracting method based also on the Y function method. Indeed, using the same approach as the current variability model in Eq.5, the Y function local variability can be written as shown in Eq.8. Note that the derivative of Y function versus Rsd is equal to zero.

$$\sigma_{\Delta Y / Y}^2 = (\frac{1}{Y}\frac{\partial Y}{\partial V_{TH}})^2.\sigma_{\Delta V_{TH}}^2 + (\frac{1}{Y}\frac{\partial Y}{\partial \beta})^2.\sigma_{\Delta \beta}^2 \quad (8)$$

For $V_G > V_{TH}$, using Eq.7 & Eq.8, the Y function variability can be written as shown in Eq.9, where $\sigma_{\Delta Y / Y}^2 (V_G - V_{TH})^2$ as a function of $(V_G - V_{TH})^2$ is a linear curve as shown in Fig.3. Its intersection with the ordinate axis and its slope gives respectively $\sigma_{\Delta V_{TH}}^2$ and $\frac{1}{4}\sigma_{\Delta \beta / \beta}^2$ values.

$$\sigma_{\Delta Y / Y}^2 = \frac{\sigma_{\Delta V_{TH}}^2}{(V_G - V_{TH})^2} + \frac{1}{4}\sigma_{\Delta \beta / \beta}^2 \quad (9)$$

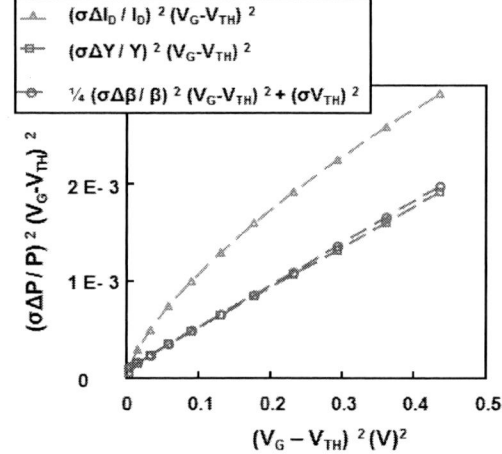

Figure 3: σΔVt and σΔβ/β extraction using $\sigma_{\Delta Y/Y}^2$. $(V_G-V_{TH})^2$ as a function of $(V_G-V_{TH})^2$ and $\sigma_{\Delta Id/Id}^2$. $(V_G-V_{TH})^2$ as a function of $(V_G-V_{TH})^2$ (FD SOI transistors with W=0.08µm/L=0.05µm and V_D =50mV and V_{TH}=0.34V).

C. Drain current local variability model using Y function method.

Different models have been proposed to explain the behavior of I_D local variability: in weak inversion region [9], in weak to strong inversion regime as a function of gate bias and transistor geometry [10], and, in weak to strong inversion regime and in linear to saturation regime for transistors with pockets [11]. A local variability model has been published by [2]. It describes variability in the drain current as a function of threshold voltage and current gain factor local variability, surface roughness scattering and the saturation velocity. It analyzes the mismatch as a function of gate bias and transistor geometries.

In order to validate our model for full drain current local variability, in Fig4: a) theoretical drain current variability represented by Eq.5 (σΔVTH=0.02V, σΔβ/β =0.131 and σΔRsd =0.1Rsd), b) current variability model represented by Eq.6, c) croon's model [2] represented by Eq.10 (where V_{TH}-β correlation, surface roughness scattering and saturation velocity have been neglected) are plotted as a function of V_G.

$$\sigma_{\Delta I_D / I_D}^2 = (\frac{Gm}{I_D})^2.\sigma_{\Delta V_{TH}}^2 + \sigma_{\Delta \beta / \beta}^2 \quad (10)$$

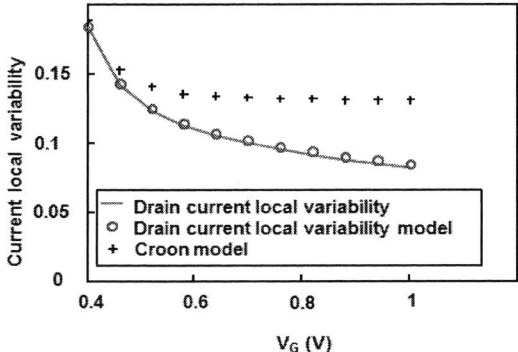

Figure 4: Comparison between simulated full drain current local variability, drain current local variability model and Croon model without correlation, roughness scattering and saturation velocity terms (FD SOI transistors with W=0.08μm/L=0.05μm and V_D=50mV).

This figure shows that, by correcting Croon's model, after multiplying $\sigma^2_{\Delta\beta/\beta}$ by $(1-Gd.Rsd)^2$ (which represent the Rsd contribution and thus the mobility attenuation one) and without V_{TH}-β correlation term, the drain current local variability is reproduced.

III. EXPERIMENTAL DETAILS

Electrical characterizations have been carried out on FD-SOI and Bulk devices integrating High-k gate oxide and metal gate.

A sampling of 70 pairs of identical MOS transistors has been considered. The two MOSFET's of the pair are spaced by the minimum allowed distance, placed in identical environment and electrically independent with symmetric connections.

For local variability study, an electrical parameter P is measured for each of the two paired devices. The difference of P noted ΔP (or ΔP/P) between the pair is calculated. This method is then repeated for N samples. Thus, after applying a recursive filter to eliminate erroneous data, the standard deviation $\sigma_{\Delta P}$ (or $\sigma_{\Delta P/P}$) of the Gaussian distribution is calculated.

All presented results refer to measurements performed in linear regime with drain voltage V_D=50mV, a gate voltage range from 0 to 1V and at 25°C.

IV. RESULTS AND DISCUSSION

A. FD SOI NMOS transistors with moderate gate length (W=1μm/L=0.1μm)

Departing from NMOS transistors with a moderate length (L=0.1μm), the drain current I_D is measured and plotted in Fig.5 as a function of V_G. The Y function is calculated from the drain current using Eq.7 and plotted in Fig.6 as a function of V_G. Fig.5 & Fig.6 show typical behaviors of I_D and Y as a function of V_G.

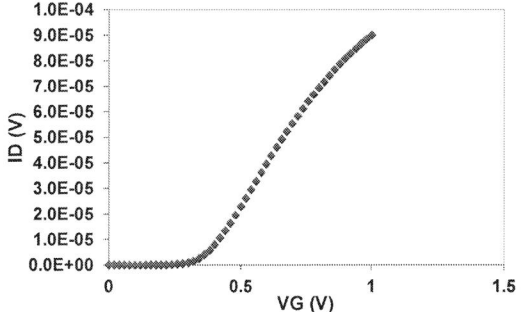

Figure 5: Drain current as a function of gate voltage. (FD SOI transistors with W=1μm/L=0.1μm and V_D=50mV).

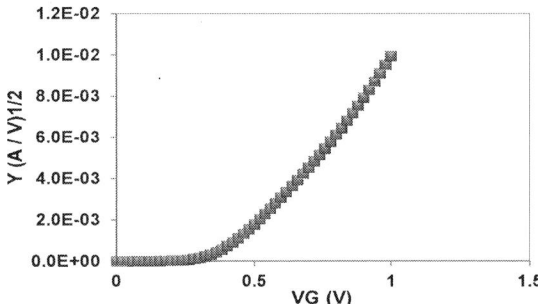

Figure 6: Y function as a function of gate voltage. (FD SOI transistors with W=1μm/L=0.1μm and V_D=50mV).

The differences of I_D noted $\Delta I_D/I_D$ and Y noted $\Delta Y/Y$ between the pair of MOS transistors are calculated. This method is repeated for 70 pairs of identical NMOS transistors. The standard deviations of $\Delta I_D/I_D$ and $\Delta Y/Y$ are then deduced. Thus, $\sigma^2_{\Delta I_D/I_D}$ and $\sigma^2_{\Delta Y/Y}$ multiplied by $(V_G-V_{TH})^2$ are plotted as a function of $(V_G-V_{TH})^2$ in Fig.7. This figure shows that, while $\sigma^2_{\Delta I_D/I_D}(V_G-V_{TH})^2$ as a function of $(V_G-V_{TH})^2$ might be nonlinear due to combined effects of Rsd and mobility degradation. Instead, $\sigma^2_{\Delta Y/Y}(V_G-V_{TH})^2$ presents a better linearity as a function of $(V_G-V_{TH})^2$ with 0.98 correlations. Using Eq.9 $\sigma^2_{\Delta V_{TH}}$ and $\sigma^2_{\Delta\beta/\beta}$ are then deduced. Therefore, this method permits the extraction of $\sigma_{\Delta V_{TH}}$ and $\sigma_{\Delta\beta/\beta}$ without the influence of Rsd values.

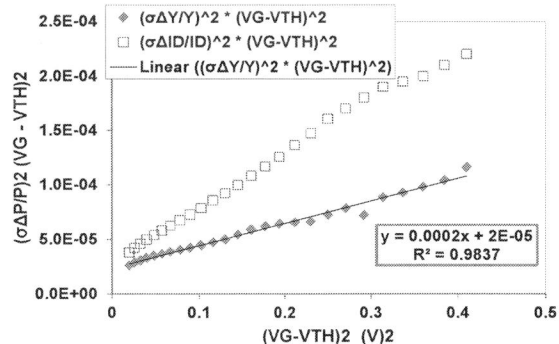

Figure 7: V_G>V_{TH} (V_G range from 0.6->1V with V_{TH}=0.36V), Y function and drain current local variability multiplied by $(V_G-V_{TH})^2$ and plotted as a function of $(V_G-V_{TH})^2$ (FD SOI transistors with W=1μm/L=0.1μm and V_D=50mV).

To verify if the values of V_{TH} and β local variability are correct, $\sigma^2_{\Delta_{V_{TH}}}$ and $\sigma^2_{\Delta\beta/\beta}$ are extracted by the classical Y function method explained in section II.B. TABLE 1 represents a comparison between $\sigma^2_{\Delta_{V_{TH}}}$ and $\sigma^2_{\Delta\beta/\beta}$ extracted by the classical Y function and the new Y Function method. It does not show any significant difference between the two methods.

TABLE 1: Comparison between V_{TH} and β local variability extracted by classical Y function and new Y function extraction methods.

$(\sigma_{\Delta Parameter})^2$	$(\sigma_{\Delta VTH})^2$	$(\sigma_{\Delta\beta/\beta})^2$
Classical Y function	3.10^{-5} V	9.10^{-4}
New Y function	2.10^{-5} V	8.10^{-4}

Using $\sigma^2_{\Delta_{V_{TH}}}$ and $\sigma^2_{\Delta\beta/\beta}$ values, and extracting Rsd value with $R_{tot}=f(1/\beta)$ method shown in [12], the new drain current variability model in Eq.6 is calculated. Note that $(Gd)^2.\sigma^2_{\Delta Rsd}$ term in Eq.6 can be neglected due to the moderate gate length [3] [4]. In order to validate our model experimentally, measured drain current variability, Croon's model [2] represented by Eq.10 (with no V_{TH}-β correlation) and new drain current variability model are plotted as a function of V_G in Fig.8 for strong inversion regime.

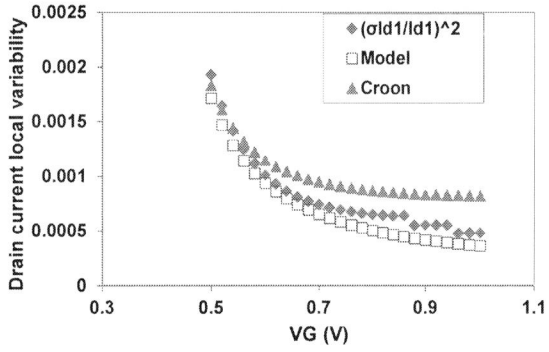

Figure 8: $V_G>V_{TH}$ (V_G range from 0.5->1V with V_{TH}=0.36V), Comparison between drain current local variability model, drain current local variability Croon model and measured drain current local variability as a function of V_G (FD SOI transistors with W=1µm/L=0.1µm and V_D =50mV, Rsd=220Ω).

Fig.8 shows that Croon's model without V_{TH}-β correlation term does not fit the measured data. Thus, after correcting Croon's model, by multiplying $\sigma^2_{\Delta\beta/\beta}$ by $(1-Gd.Rsd)^2$ (which represent the Rsd contribution), the drain current variability is well reproduced without any need for V_{TH}-β correlation, indicating also that Rsd variability is negligible. The interest of this new model is that, while [2] must use V_{TH}-β correlation term to fit the data, the new drain current variability model well accounts for the measured data only by considering the Rsd contribution.

B. FD SOI NMOS transistors with short gate length (W=0.08µm/L=0.05µm)

In this section, the same approach as in section IV.A is considered for short gate lengths. For each pair of MOS transistor I_D and Gm are first measured, Y function is then calculated from I_D using Eq.7. The standard deviations of $\Delta I_D/I_D$ and $\Delta Y/Y$ are then calculated for 70 pairs of identical NMOS transistors. Thus, $\sigma^2_{\Delta I_D/I_D}$ and $\sigma^2_{\Delta Y/Y}$ multiplied by $(V_G-V_{TH})^2$ are also plotted as a function of $(V_G-V_{TH})^2$ in Fig. 9. This figure shows again that, while $\sigma^2_{\Delta I_D/I_D}(V_G-V_{TH})^2$ as a function of $(V_G-V_{TH})^2$ might be nonlinear, $\sigma^2_{\Delta Y/Y}(V_G-V_{TH})^2$ presents a better linear behavior as a function of $(V_G-V_{TH})^2$.

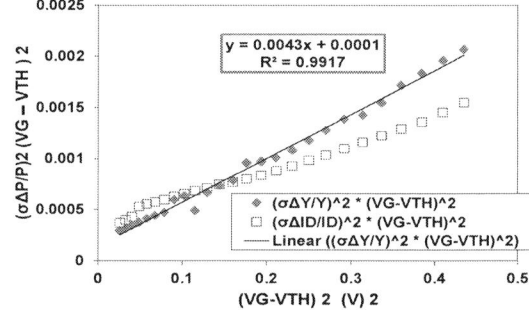

Figure 9: $V_G>V_{TH}$ (V_G range from 0.5->1V with V_{TH}=0.34V), Y function and drain current local variability multiplied by $(V_G-V_{TH})^2$ and plotted as a function of $(V_G-V_{TH})^2$ (FD SOI transistors with W=0.08µm/L=0.05µm and V_D =50mV).

Using Eq.9, $\sigma^2_{\Delta_{V_{TH}}}$ and $\sigma^2_{\Delta\beta/\beta}$ are then deduced. ($\sigma^2_{\Delta_{V_{TH}}} = 10^{-4}V^2$ and $\sigma^2_{\Delta\beta/\beta} = 0.0172$).

Having $\sigma^2_{\Delta_{V_{TH}}}$ and $\sigma^2_{\Delta\beta/\beta}$ values, the new drain current variability model of Eq.6 is calculated (with Rsd extracted by $R_{tot}=f(1/\beta)$ method shown in [12]) . Note that, at first order, $(Gd)^2.\sigma^2_{\Delta Rsd}$ term in Eq.6 is considered as negligible. Thus, measured drain current variability, Croon's model [2] (with no correlation) represented by Eq.10, and, new drain current variability model are plotted as a function of V_G in Fig.10.

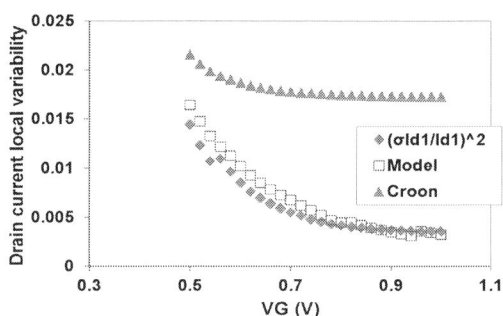

Figure 10: $V_G>V_{TH}$ (V_G range from 0.5->1V with V_{TH}=0.34V), Comparison between drain current local variability model, drain current local variability Croon model and measured drain current local variability as a function of V_G (FD SOI transistors with W=0.08µm/L=0.05µm and V_D =50mV, Rsd=2775Ω).

Fig.10 shows that, for short lengths, Croon's model without V_{TH}-β correlation term also does not fit the measured data. As result, by correcting Croon's model as in Eq.6, the new drain current variability model allows well reproducing the experimental variability data. Note that Rsd variability is also negligible for short lengths. Also that the gap between Croon's model (with no V_{TH}-β correlation) and the new drain current variability model is more important for L=0.05µm than for L=0.1µm. Thus, Rsd value and β variability have more impact on I_D variability for short lengths, emphasizing the interest of our new drain current variability model of Eq. 6.

C. Bulk NMOS transistors with short gate length (W=10µm/L=0.03µm)

The new V_{TH} and β variability extraction method has also been applied with success to bulk NMOS devices (W=10µm/L=0.03µm, V_D=50mV) as illustrated in Fig.11, where $\sigma^2_{\Delta V_{TH}} = 3.10^{-5}V^2$ and $\sigma^2_{\Delta\beta/\beta} = 4.10^{-4}$.

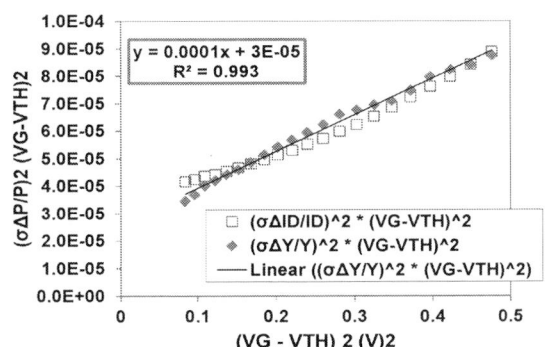

Figure 11: $V_G > V_{TH}$ (V_G range from 0.6->1V with V_{TH}=0.31V), Y function and drain current local variability multiplied by $(V_G-V_{TH})^2$ and plotted as a function of $(V_G-V_{TH})^2$ (BULK transistors with W=10µm/L=0.03µm and V_D=50mV).

Having $\sigma^2_{\Delta V_{TH}}$ and $\sigma^2_{\Delta\beta/\beta}$ values, and extracting Rsd using [12], the new drain current variability model of Eq.6 has also been applied with success to the measurement data. Fig.12 shows that, using the new drain current variability model of Eq.6, the measurement drain current variability data are well reproduced in strong inversion regime.

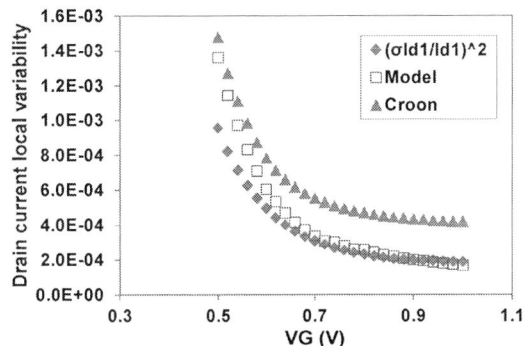

Figure 12: $V_G > V_{TH}$ (V_G range from 0.5->1V with V_{TH}=0.31V), Comparison between drain current local variability model, drain current local variability Croon model and measured drain current local variability as a function of V_G (Bulk transistors with W=1µm/L=0.03µm and V_D=50mV, Rsd=23Ω).

CONCLUSION

We have demonstrated a new method of threshold voltage and current gain factor variability characterization not affected by Rsd values and based on the well-known Y function. We have also demonstrated a new drain current local variability model based on V_{TH}, β, Rsd values and variability. The new extraction method and the new drain current local variability model were applied with success to measured data for SOI and Bulk NMOS transistors and with different lengths and widths. For future technologies this model can also be used to extract the Rsd variability contribution to drain current ones. Note that in our case this variability was demonstrated to be negligible.

REFERENCES

[1] K.R. Lakshmikumar et al, SSC, vol.21, no.6, p.1057 1066, Dec. 1986.

[2] J.Croon et al., IEEE Solid State Circuits, vol.37, no.8, Aug. 2002.

[3] S. Markov et al, SOI Conference (SOI), IEEE 2011.

[4] S. Markov et al, IEEE Electron Device Letters, vol.33, no. 3, mar. 2012.

[5] G.Ghibaudo, IEE electronics Letters, vol. 24, pp.543-545, Apr.1988.

[6] K.K. Ng et al, IEEE Trans. Electron Devices, vol.ED-34, no.3, pp. 503-511, Mar.1987.

[7] S.Thompson et al, VLSI Symp.Tech.Dig., 1998, pp.132-133

[8] D. Fleury et al, in Proc. IEEE ICMTS, Mar2008, pp. 160–165.

[9] F. Forti and M. Wright, , IEEE Journal of Solid-State Circuits, v. 29, n° 2, pp. 138-142, 1994.

[10] T. Serrano-Gotarredona and B. Linares-Barranco, *ESSCIRC*, pp. 627-630, 2003.

[11] C. Mezzomo et al, *ESSDERC*, pp. 122-125, 2010.

[12] D. Fleury et al, VLSI-TSA 2009, p. 109-110.

A Novel BJT Structure for High- Performance Analog Circuit Applications

Seon-Man Hwang, Hyuk-Min Kwon, Jae-Hyung Jang, Ho-Young Kwak,
Sung-Kyu Kwon, Seung-Yong Sung, Jong-Kwan Shin, Jae-Nam Yu, In-Shik Han[1],
Yi-Sun Chung[1], Jung-Hwan Lee[1], Ga-Won Lee and Hi-Deok Lee[*]

Department of Electronics Engineering, Chungnam National Univ., Yuseong, Daejeon 305-764, Korea
[1]MagnaChip Semiconductor, Cheongju, Choongbuk, 361-725, Korea
[*]Tel: +82-42-821-7702, Fax: +82-42-823-9544, E-mail: hdlee@cnu.ac.kr

Abstract—**A novel structure is proposed to improve the matching characteristics of bipolar junction transistor (BJT) based on CMOS technology for high performance analog circuit applications. This paper includes the analysis of electrical and matching characteristics in collector current density (J_C), base current density (J_B) and current gain (β). Although the collector current density J_C of the proposed structure is similar to that of the conventional structure, the base current density J_B is lower than that of conventional structure, which results in higher current gain. The matching characteristics of the collector current density and the current gain of the proposed structure showed improvement of about 12.22% and 36.43%, respectively compared with the conventional structure.**

Keywords— Bipolar junction transistor (BJT); Mismatch; Novel structure; Analog application; Matching coefficient

I. Introduction

A reduction of the differences of device parameters between identically designed devices is one of the key points for analog circuit and/or analog/digital mixed signal applications. As the dimensions of the devices are reduced, impact of matching becomes more important and the high precision circuit design based on matching technique requires accurate device performance [1]-[3]. However, despite the importance of matching characteristics in analog devices, there was little study in this field, especially BJT which is based on CMOS technology [4, 5]. Although MOSFET has become the main device as a high-performance integrated circuits, BJT is still significant for many analog application circuits like Analog/Digital converters, differential amplifier, band gap voltage references, etc [6]-[8] and because of good matching performance, BJT takes advantage of fast and accurate analog signal processing [9].

In this paper, a novel BJT structure is proposed to improve matching characteristics. Then, the electrical characteristics and the analysis of the matching characteristics between conventional and proposed BJT structures in the collector current density J_C, the base current density J_B and the current gain β are performed in depth.

II. DEVICE STRUCTURE

BJTs are fabricated using a standard 0.18-μm CMOS process. Schematic diagrams of the conventional and proposed BJT structures are comparatively shown in Fig. 1

and Fig. 2, respectively. The conventional structure has a symmetrical base and collector structures as in Fig. 1(a). The emitter region is fully surrounded by the base region and the collector region surrounds the emitter and base regions. On the contrary, the proposed structure has an asymmetrical base and collector structures as shown in Fig. 2(a). Both structure show vertical BJT structure as the deep n-well beneath the p-well (base) behaves as the collector region, that is, the emitted electrons from the n+ region will be collected by the deep n-well region as shown in Fig. 1(b) and 2(b).

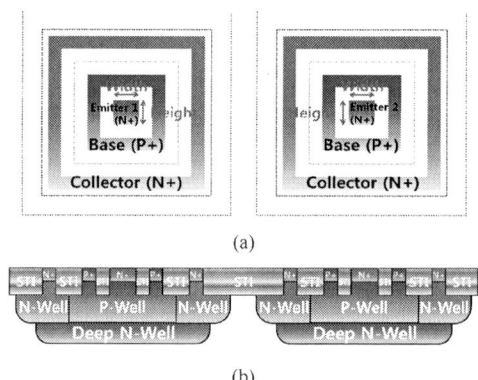

(a)

(b)

Fig. 1. Top view (a) and cross-sectional view (b) of conventional BJT matching structure using CMOS technology.

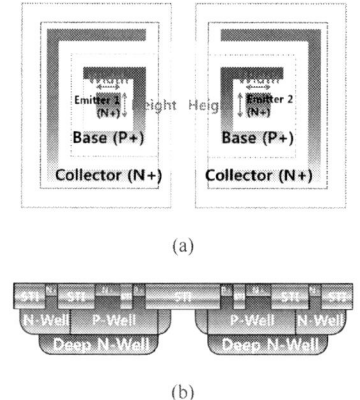

(a)

(b)

Fig. 2. Top view (a) and cross-sectional view (b) of proposed BJT matching structure using CMOS technology.

In previous study, it was shown that the asymmetrical base structure can improve the matching property [4]. The main feature of proposed structure here is the concurrent reduction of the collector region as well as the base region as shown in Fig. 2. The difference of device parameters between the adjacent two BJTs represents the matching performance of the BJTs. In order to evaluate the electrical parameters and matching characteristics, each structure has a split of emitter area with several height/width ratio (2/2, 2/5, 5/5, 5/10, 10/10 μm/μm).

III. ELECTRICAL CHARACTERISTICS

Fig. 3 shows the collector current density, J_C and base current density, J_B as a function of the base-emitter voltage, V_{BE} between the conventional and proposed structures with the emitter area of 10x10 μm². Despite the reduction of base and collector regions, there is little difference in the current density for an overall V_{BE}, which implies that the proposed structure does not affect the current flow from the emitter to collector. Fig. 4 shows the comparison of the current gain as a function of the collector current density between the conventional and proposed structures. The current gain β of proposed structure is greater than both the conventional structure and structure of Ref.[4] up to the current density of 10^{-4} A/μm².

Fig. 5 shows the comparison of the collector and base current density as a function of the emitter area. Although the collector current density of the proposed structure is similar to other structure as shown in Fig. 5(a), the base current density of proposed structure is lower than that of the other structures as in Fig. 5(b). The decrease of the base current density is believed to be due to the decrease of the base and collector contact region as shown in Figs. 2(a) and (b). Then, it can be assumed that the current gain of the proposed structure will be greater than the other structure due to the similar collector current density and smaller base current density. Fig. 6 proves the greater current gain of the proposed structure.

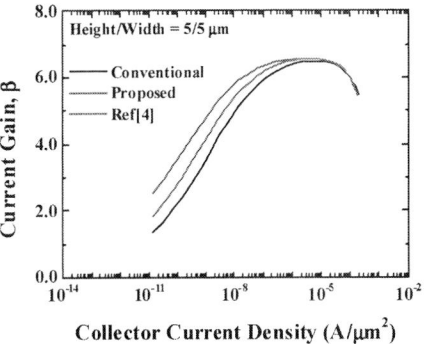

Fig. 4. Comparison of current gain β as a function of collector current density between the conventional and proposed structures. β of the poposed structure is greater than the conventional and previously reported structures [4].

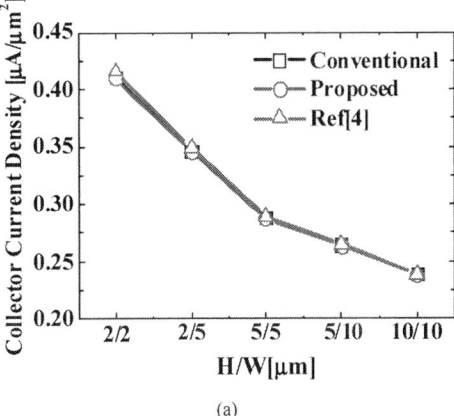

(a)

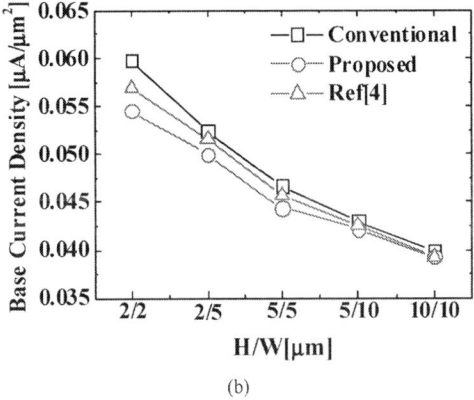

(b)

Fig. 5. Comparison of the collector current density, J_C (a) and base current density, J_B (b) for various emitter areas with a split of emitter height, H and width, W. Applied V_{BE} and V_{CE} are 0.7V and 1.2V, respectively. The proposed structure shows similar collector current density but smaller base current density compared with the conventional structure and the previous structure [4].

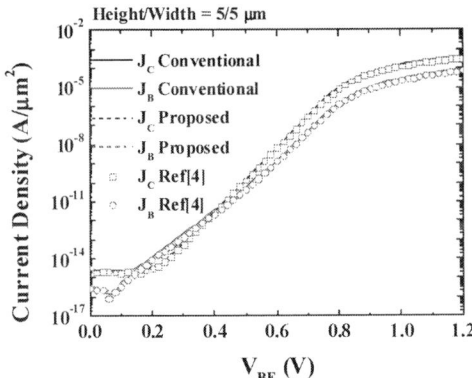

Fig. 3. Comparison of the collector current density J_C and J_B versus V_{BE} between conventional and proposed structures. J_C and J_B of the proposed structure show similar level with the conventional structure as well as previous work [4].

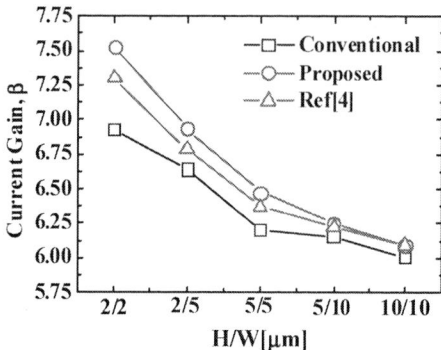

Fig. 6. Comparison of the current gain, β for various emitter areas with splits of emitter height H and width W. Applied base-emitter voltage, V_{BE} and collector-emitter voltage, V_{CE} are 0.7V and 1.2V, respectively. The proposed structure shows greater current gain for all splits.

IV. MATCHING CHARACTERISTICS

In order to analyze the matching characteristics of the BJT structures, the mismatch of the standard deviations of collector current density, J_C and current gain, β are measured for the proposed and the conventional structures. Mismatch in collector current and current gain can be described as a parameter P. Then, the variance of a parameter mismatches ΔP dependence on the emitter area (product of height H and width W) can be described as (1) [10]

$$\sigma^2{}_{\Delta P/P} = \frac{A_P{}^2}{HW} \qquad (1)$$

where, ΔP/P is the relative difference in parameter P between the adjacent two devices and A_P is the area proportionality constant (or matching coefficient) for parameter P. Typically, lower A_P implies the better matching characteristics of two devices.

Fig. 7 and Fig. 8 show the standard deviation of the difference of the collector current density, J_C, the base current density, J_B and current gain, β as a function of $1/(HW)^{1/2}$. The slopes in Figs. 7 and 8 are called as the matching coefficient and represent the degree of matching characteristics. Although the current level of three structures exhibited little difference as shown in Fig. 5(a), the proposed BJT structure show the improvement of matching coefficient about 49.16% and 40.20% compared with conventional structure in the base and collector current, respectively, as shown in Fig. 7. As shown in Fig. 8, the proposed structure shows much greater improvement of the matching characteristics of the current gain about 50.89% than conventional structure. Fig. 7 and Fig. 8 also show that the proposed BJT structure has better matching characteristics than the previously reported BJT structure [4]. The improvement of the matching characteristics of the proposed structure can be explained as the reduced current path both in the base and collector regions. That is, the reduction of the base and the collector current

path for the proposed BJT structure results in the less effect of the variation of the base and collector resistances on the base and collector current. As the base and collector region are formed by the p-well and deep n-well processes which do not use the state-of-the-art lithography, sheet resistance variation of them will be greater than gate and active resistances.

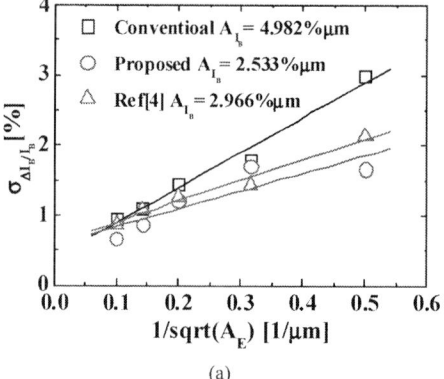

(a)

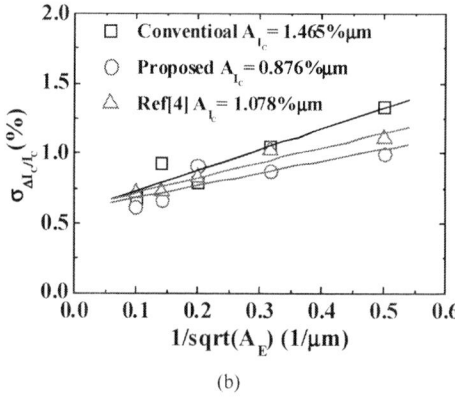

(b)

Fig. 7. Comparison of matching characteristics of the base current density, J_B (a) and collector current density, J_C (b) between three structures with V_{BE} and V_{CE} of 0.7V and 1.2V, respectively.

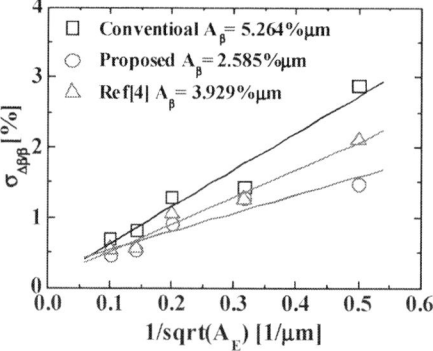

Fig. 8. Comparison of matching characteristics of the current gain, β between three structures with V_{BE} and V_{CE} of 0.7V and 1.2V, respectively.

V. CONCLUSION

The electrical characteristics of the collector current density J_C is similar to those of the conventional structure. However, the base current density J_B of the proposed structure is lower than that of the other structures and the lower base current density results in the higher current gain. The matching characteristics of the base current density, collector current density and current gain for the proposed structure are better than the conventional structure about 49.16%, 40.20% and 50.89, respectively. The improved matching characteristic of the proposed structure is believed to be due to the smaller base and collector current path between the emitter and base and emitter and collector, respectively.

Acknowledgment

This work(research) is financially supported by the Ministry of Knowledge Economy(MKE) and Korea Institute for Advancement in Technology (KIAT) through the Workforce Development Program in Strategic Technology.

References

[1] P. G. Drennan, C. C. McAndrew, "Understanding MOSFET mismatch for analog design." IEEE Journal of Solid-State Circuits, Vol.38, No.3, pp.450-456, Mar. 2003

[2] H. P. Tuinhout, "Design of matching test structures." IEEE International Conference on Microelectronic Test Structures, Vol.7, pp.21-27, Mar. 1994.

[3] H. P. Tuinhout, "Improving BiCMOS technologies using BJT parametric mismatch characterisation." Proceedings of the Bipolar/BiCMOS Circuits and Technology Meeting, pp.163-170, Sept. 2003.

[4] Y. J. Jung, B. S. Park, I. S. Han, H. M. Kwon, S. U. Park, J. D. Bok, Y. S. Chung, M. G. Lim, J. H. Lee, H. D. Lee, "Novel BJT test structure for high-performance matching characteristics in CMOS-based analog applications." IEEE International Conference on Microelectronic Test Structures, pp.198-200, April. 2011.

[5] Y. J. Jung, B. S. Park, H. M. Kwon, S. K. Kwon, J. H. Jang, H. Y. Kwak, Y. S. Chung, J. H. Lee, H. D. Lee, "A Novel BJT Structure Implemented Using CMOS Processes for High-Performance Analog Circuit Applications." IEEE Transactions on Semiconductor Manufacturing, vol.25, No.4, pp.549-554, Nov. 2012.

[6] N. Wils, H. P. Tuinhout, T. Ewert, J. van Berkum, M. Kaiser, R. Weemaes, "Identification and analysis of a new BJT parametric mismatch phenomenon." Proceedings of the Bipolar/BiCMOS Circuits and Technology Meeting, pp. 224-227, Oct. 2005.

[7] G. Lau, W. Einbrodt, W. Sieber, "Improvement of poly emitter n-p-n transistor matching in a 0.6 micron mixed signal technology." IEEE International Conference on Microelectronic Test Structures, pp. 232-237, Mar. 2003.

[8] Tuinhout, H. P., W. C. M. Peters, "Measurement of lithographical proximity effects on matching of bipolar transistors." IEEE International Conference on Microelectronic Test Structures, Vol.11, pp. 7-12, Mar. 1998.

[9] P. G. Drennan, C. C. McAndrew, J. Bates, D. Schroder, "Rapid evaluation of the root causes of BJT mismatch." IEEE International Conference on Microelectronic Test Structures, pp. 122-127. 2000.

[10] M. J. M. Pelgrom, A. C. J. Duinmaijer, A. P. G. Welbers, "Matching properties of MOS transistors." IEEE Journal of Solid-State Circuits, Vol.24, No.5, pp. 1433-1439. Oct. 1989.

978-1-4673-4845-4/13 $31.00 © 2013 IEEE

Reconsideration of the Threshold Voltage Variability Estimated with Pair Transistor Cell Array

Kazuo Terada, Naoya Higuchi and Katsuhiro Tsuji

Faculty of Information Sciences, Hiroshima City University
3-4-1, Ozuka-Higashi, Asa-Minami-Ku, Hiroshima, 731-3194, JAPAN
TEL: +81-82-830-1557, FAX: +81-82-830-1792
E-mail: terada@info.hiroshima-cu.ac.jp

ABSTRACT

The standard deviation of threshold voltage, σ_{VTH}, which is estimated with Pair Transistor cell Array (PTA), is examined using the test chip fabricated by 65-nm technology. It is found that the errors are caused by two problems: 1) the problem in the approximation and 2) leak current in the isolation region. Taking them into account, the application of PTA to the test structure in scribe line is studied.

INTRODUCTION

Pair Transistor cell Array (PTA), with which the standard deviation (σ_{VTH}) of MOSFET threshold-voltage (V_{TH}) can easily be evaluated, has been presented in the previous ICMTS [1-3]. It is shown that the σ_{VTH}-value derived from two DC currents flowing through PTA (σ_{VTH_PTA}) behaves reasonably. This time, we have tried to compare them with σ_{VTH}-value derived from the current data obtained with Device Matrix Array (DMA) [4, 5] (σ_{VTH_DMA}), using the test chip fabricated with 65-nm technology. It is found that the errors are caused in σ_{VTH_PTA} by two problems: 1) the problem in the approximation and 2) leak current in the isolation region. This paper discusses and tries to solve those problems.

PTA STRUCTURE AND MEASUREMENT PRINCIPLE

Fig. 1 shows PTA structure. It consists of the many Pair Transistor (PT) cells, in which paired MOSFETs "1" and "2" are serially-connected each other and the node between them is connected to common wiring through a switch "SW". These cells are parallel-connected each other. Applying suitable voltages to nodes V_{G1} (V_{G1}) and V_{G2} (V_{G2}) such that MOSFETs "1" and "2" operate in subthreshold region, currents flowing from node V_{DD} to node GND are measured for both the cases when "SW" being ON and OFF. Denoting them I_{ON} and I_{OFF}, ratio I_{ON}/I_{OFF} as a function of V_{G1} at a fixed V_{G2} becomes a curve having a peak at $V_{G1}=V_{G2}$ and the peak value is approximated by [3].

$$(1) \qquad I_{ON}/I_{OFF} \cong e^{\frac{a^2\sigma^2_{VTH}}{2}},$$

where $a=q/nkT$, and q, n, k, and T are electronic charge, capacitance ratio of depletion layer to gate oxide film, Boltzmann constant and temperature, respectively.

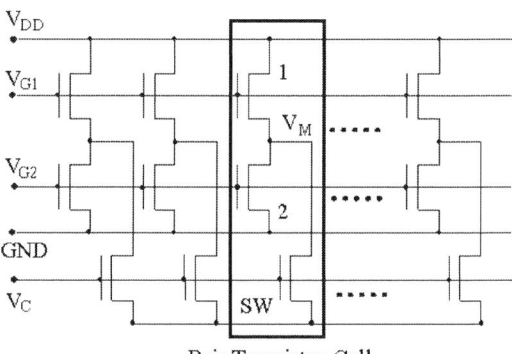

Pair Transistor Cell

Fig.1 PTA structure

When deriving Eq. (1), it is assumed that {1} the MOSFET drain current I_D is approximated by

$$(2) \qquad I_D \cong Ke^{a(V_G - V_{TH})},$$

where K and V_G are gain constant and gate voltage, respectively, {2} the threshold voltages for MOSFETs "1" and "2" in a certain unit cell "i" in Fig. 1, V_{TH1i} and V_{TH2i}, are random variables obeying normal distribution with mean $<V_{TH}>$ and variance σ^2_{VTH}, {3} the current flowing in PT cell "i", I_{OFFi}, is determined by the drain current of the MOSFET whose gate-over-drive $V_{GTki} = V_{Gk} - V_{THki}$ ($k=1,2$) is less than the other's.

TEST STRUCTURES AND MEASUREMENT RESULTS

Test chip including both PTA and DMA is fabricated using 65-nm technology. DMA provides the decoder circuit for selecting device under test (DUT), the transfer gate for connecting it to the common probing pads and 16K DUTs. Drain current (I_D) versus gate voltage (V_G) relations are measured using Kelvin measurement method at 50-mV drain voltage (V_D) for DMA, and then threshold voltages (V_{TH_DMA}) are extracted by linear extrapolation method. σ_{VTH_DMA} is derived from them [6]. PTA has 1000 pair transistor (PT) cells. I_{ON} and I_{OFF} are measured at $V_{DD}=0.1$ V and σ_{VTH_PTA} is calculated by Eq. (1).

Fig. 2 shows I_{ON}, I_{OFF} and I_{ON}/I_{OFF} as a function of V_{G1} when V_{G2}=0.3 V. Since V_{TH} for shorter channel MOSFET is higher due to "reverse short channel effect", I_{ON} and I_{OFF} for shorter channel MOSFET are lower. The I_{ON}/I_{OFF} maximums are given at $V_{G1}=V_{G2}$, and σ_{VTH_PTA} is derived from it.

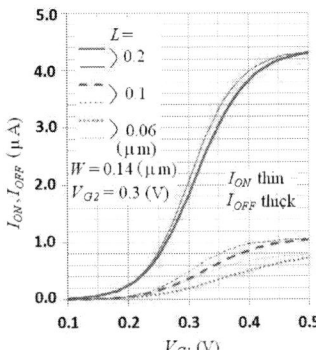

Fig. 2(a) I_{ON}, I_{OFF} vs. V_{G1} relations.

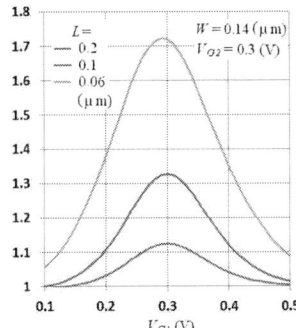

Fig. 2 (b) I_{ON}/I_{OFF} vs. V_{G1} relations.

Fig. 3 compares σ_{VTH_PTA} ($\bigcirc$ indicated by PTA) with σ_{VTH_DMA} ($\times$ indicated by DMA). It is found that σ_{VTH_DMA} is on a straight line and σ_{VTH_PTA} is larger than this line, when the channel is wide and the variability is large. Since σ_{VTH_DMA} obeys Pelgrom's law [7] and is derived by the reasonable procedure, it is considered to be accurate. It, therefore, is a question why σ_{VTH_PTA} is larger than σ_{VTH_DMA}.

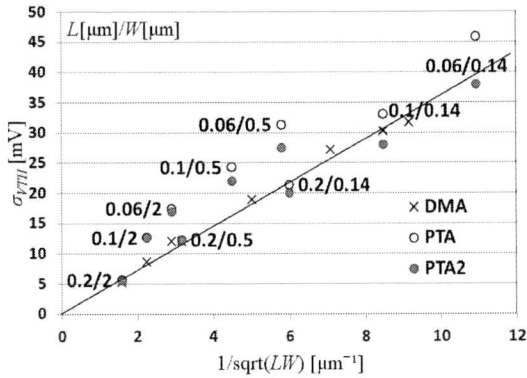

Fig.3 Pelgrom plots for σ_{VTH_PTA} and σ_{VTH_DMA}.

RECONSIDERATION OF APPROXIMATION

To answer the above-mentioned question, let us reconsider the above-mentioned assumptions {1} and {3}. Although Eq. (2) well approximates the MOSFET drain current in subthreshold region (assumption {1}) when $kT/q<<V_D$, it does not for small V_D. It is necessary to use the following equation instead [8].

$$(3) \qquad I_{DS} \cong K e^{\frac{q}{nkT}(V_G - V_{TH})}(1 - e^{-\frac{qV_D}{kT}})$$

If $V_{GT1i} - V_{Mi} \cong V_{GT2i}$, where V_{Mi} is the potential of the node between both the MOSFETs as shown in Fig.1, I_{OFFi} becomes about a half of the drain current for one of the pair MOSFETs. This situation contradicts to the assumption {3}. Since the number of those contradicted PT cell increases as σ_{VTH} increases, the approximate value obtained by Eq. (1) becomes bad. Let us estimate the threshold σ_{VTH}-value at which this bad approximation becomes significant.

Using Eq. (3), currents flowing MOSFET "1", I_{1i} and MOSFET "2", I_{2i}, in unit cell "i" are expressed by

$$(4) \qquad I_{1i} \cong K e^{a(V_{G1} - V_{TH1i} - V_{Mi})}(1 - e^{-\frac{q(V_{DD} - V_{Mi})}{kT}}),$$

$$(5) \qquad I_{2i} \cong K e^{a(V_{G2} - V_{TH2i})}(1 - e^{-\frac{qV_{Mi}}{kT}}),$$

respectively. V_{Mi} is obtained by putting $I_{1i} = I_{2i}$ and solving it. Assuming that $V_{G1}=V_{G2}=V_G$, $V_{TH1i}=V_{TH2i}+\alpha$, $I_{1i} = I_{2i}$ is expressed by:

$$(6) \qquad e^{-aV_{Mi}}(1 - e^{-\frac{q(V_{DD} - V_{Mi})}{kT}}) \cong e^{a\alpha}(1 - e^{-\frac{qV_{Mi}}{kT}}).$$

Let us suppose $kT/q<<V_{DD}$. It is reasonable, because V_{DD}=0.1V is actually used in this measurement and is sufficiently larger than $kT/q\sim 26$mV. When $kT/q<<\alpha$, the gate-over-drive for MOSFET "1" is much less than that for MOSFET "2", and V_{Mi} becomes sufficiently less than kT/q. I_{OFFi} is, therefore, mainly determined by I_{1i}, which is determined by $V_G - V_{TH1} - V_{Mi}$, as shown in Eq. (4). If V_{Mi} is almost independent of α, assumption {3} is right. But, if V_{Mi} depends on α, it shows that I_{OFFi} is affected by MOSFET "2", and consequently assumption {3} becomes wrong. On the other hand, when $\alpha << -kT/q$, the gate-over-drive for MOSFET "2" is much less than that for MOSFET "1", and V_{Mi} becomes more than $|\alpha|$. I_{OFFi} is, therefore, mainly determined by I_{2i}, which is determined by $V_G - V_{TH2}$, as shown in Eq. (5). However, if $|\alpha|$ is not sufficiently lager than kT/q and $e^{-\frac{qV_{Mi}}{kT}}$ term in Eq. (5) is not negligible, I_{2i} depends on V_{Mi}. It shows that I_{OFFi} is affected by MOSFET "1", and consequently assumption {3} becomes wrong.

When $|\alpha|$ is not sufficiently lager than kT/q, $|a\alpha|$ and qV_{Mi}/kT can become less than 1. In that case, the following equation is obtained from Eq. (6) using

978-1-4673-4845-4/13 $31.00 © 2013 IEEE

Taylor expansion and neglecting more than second order terms:

$$(7) \qquad V_{Mi} \cong \frac{kT/q}{1+(\alpha + kT/q)a}.$$

When $\alpha > 0$ and is not negligible compared with kT/q, V_{Mi} depends on α, as shown in Eq. (7). It means that I_{OFFi} is not only determined by $V_G - V_{TH1}$. It is also affected by $V_G - V_{TH2}$. When $\alpha < 0$ and is comparable to $-kT/q$, V_{Mi} becomes nearly equal to kT/q, as shown in Eq. (7), and I_{2i} becomes dependent on the drain voltage V_{Mi}, as shown in Eq. (5). It means that I_{OFFi} is not only determined by $V_G - V_{TH2}$. It is also affected by $V_G - V_{TH1}$. These things means that assumption {3} becomes wrong, when $|\alpha|$ becomes comparable to kT/q.

CIRCUIT SIMULATIONS

To improve the approximation, we calculate I_{ON}/I_{OFF} as a function of σ_{VTH} using a circuit simulator. MOSFET model for 0.18-μm technologies is used. PTA having 200 PT cells is constructed, in which the MOSFET threshold voltages are fluctuated by random numbers. Fig. 4 shows an example of those paired MOSFET threshold voltages. Correlation coefficient between V_{TH1} and V_{TH2} is about -0.00035, and it is considered that they are not correlated.

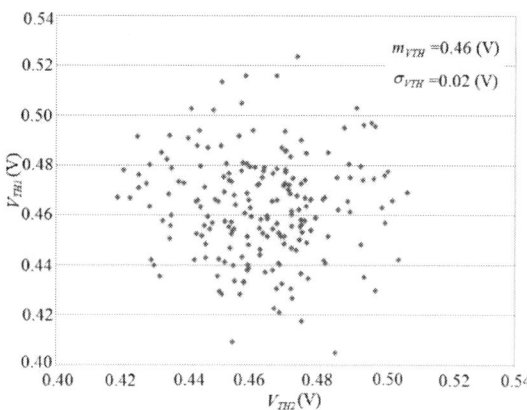

Fig.4 Example of paired MOSFET threshold voltages used in the circuit simulation. Horizontal and vertical axes respectively show V_{TH1} and V_{TH2} for a certain PT cell.

Fig. 5 shows an example of the simulated I_{ON}/I_{OFF} vs. σ_{VTH} relation and confirms the above-mentioned situation. The data (x) obtained by Eq. (1) deviate from the simulated data ($\square$) at $\sigma_{VTH} \sim kT/q$=26mV. This fact coincides with the conclusion derived from Eq. (7). This figure shows four other simulated data, too. They are obtained by intentionally changing subthreshold slope S in the MOSFET model parameters. It is found that I_{ON}/I_{OFF} depends on S. This fact is also explained by Eq. (7), which shows how assumption {3} becomes wrong depends on

a=log10/S. The S–value for the measured samples is $80 \sim 90$mV/dec and almost coincides to the present simulation model. It, therefore, is reasonable that we estimate σ_{VTH} using the simulated I_{ON}/I_{OFF} vs. σ_{VTH} relation of Fig. 5, instead of Eq. (1). σ_{VTH_PTA} ($\bullet$ indicated by PTA2) in Fig. 2 is obtained by this procedure and compared with σ_{VTH_PTA} ($\bigcirc$). It is confirmed that the large σ_{VTH} data are well improved.

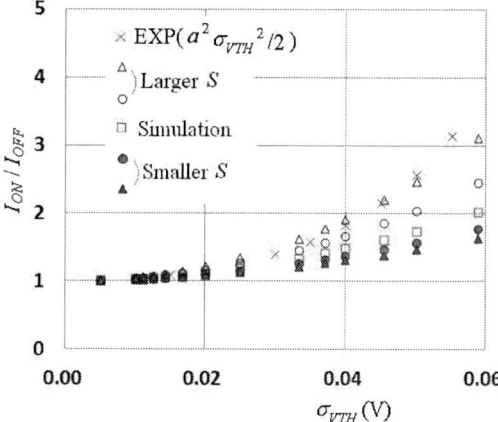

Fig.5 Simulated I_{ON}/I_{OFF} vs. σ_{VTH} relation

EFFECT OF LEAK CURRENT

On the other hand, σ_{VTH_PTA} for short and wide MOSFETs is not improved even if the simulated I_{ON}/I_{OFF} vs. σ_{VTH} relation is used. It is considered that the reason for this error is caused by the leak current flowing through the parasitic MOSFET at the STI (Shallow Trench Isolation) edge [9]. Fig. 6 shows I_D-V_G relations for the short MOSFETs. It is found that a part of lines for wide MOSFETs show a hump. It is considered that the hump is caused by the V_{TH} variability for the parasitic MOSFET at the STI edge and is not avoidable. On the other hand, since σ_{VTH_DMA} is derived from the ON-current by the extrapolation method, it is not affected by this hump. It shows the V_{TH} variability for the intrinsic MOSFET channel region.

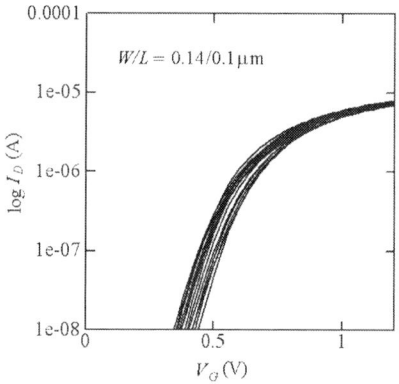

Fig. 6(a) Subthreshold currents for narrow channel MOSFETs

978-1-4673-4845-4/13 $31.00 © 2013 IEEE

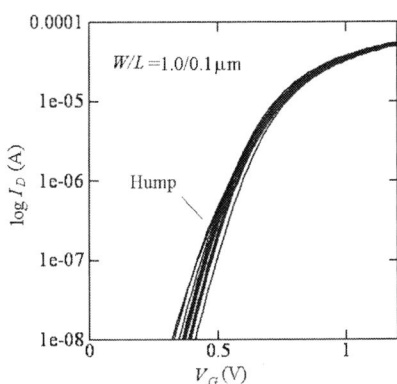

Fig. 6(b) Subthreshold currents for wide channel MOSFETs

SUMMARY

σ_{VTH_PTA} is compared with σ_{VTH_DMA}, using the test chip fabricated with 65-nm technology. It is found that the errors are caused in σ_{VTH_PTA} by two problems: 1) the problem in the approximation and 2) leak current in the isolation region. The assumptions made in the previous study are reconsidered and the more accurate approximate equations are used to calculate I_{ON}/I_{OFF}. It is found that the accuracy of the previous simple approximation becomes bad when σ_{VTH} is comparable to kT/q. More accurate approximate equation is not simple, but numerical simulation data using the similar S-value is useful to calculate I_{ON}/I_{OFF}. Even if this approximation is improved, σ_{VTH_PTA} for short and wide MOSFETs is not improved. It is found that the reason for this error is caused by the leak current flowing through the parasitic MOSFET at the STI (Shallow Trench Isolation) edge. In conclusion, although σ_{VTH_PTA} cannot show the V_{TH} variability for the intrinsic MOSFET, it can show the V_{TH} variability for the overall MOSFET including the STI edge and reasonable test structure in scribe line.

REFERENCES

[1] K. Terada and M. Eimitsu, "A Test Circuit for Measuring MOSFET Threshold Voltage Mismatch", Proc. ICMTS, p.227-231, (2003)

[2] K. Terada and K. Fukeda, "Further Study of V_{TH}-Mismatch Evaluation Circuit", Proc. ICMTS, p.155-159, (2004),

[3] K. Terada, M. Eimitsu and K. Fukeda, "A Test Circuit for Measuring MOSFET Threshold Voltage Mismatch", Solid-State Electronics, Vol.49, p.818-824, (2005)

[4] T. Chagawa, K. Terada, J. Xiang, K. Tsuji, T. Tsunomura and A. Nishida, "Measurement of MOSFET Drain Current Variation Under High Gate Voltage", Proc. ICMTS, p.86-89, (2008)

[5] K. Terada, T. Chagawa, J. Xiang, K. Tsuji, T. Tsunomura and A. Nishida, "Measurement of MOSFET Drain Current Variation Under High Gate Voltage", Solid State Electronics, Vol. 53, p.314-319, (2009)

[6] T. Tsunomura, A. Nishida, F. Yano, A. T. Putra, K. Takeuchi, S. Inaba, S. Kamohara, K. Terada, T. Hiramoto and T. Mogami, "Analyses of 5-Sigma Vth Fluctuation in 65-nm MOSFETs Using Takeuchi Plot", 2008 Symp. on VLSI Technology, Dig. Tech Papers, p.156-157, (2008)

[7] M. Pelgrom, A. Duinmaijer and A. Welbers, "Matching properties of MOS transistors", IEEE J. Solid-State Circuits, Vol. 24, p.1433-1440, (1989)

[8] Y. Taur and T. Ning, "Fundamentals of Modern VLSI Devices", Cambridge University Press (1998)

[9] Y. Joly, L. Lopez, J. Portal, H. Aziza, Y. Bert, F. Julien and P. Formara, "Active 'multi-fingers': Test structure to improve MOSFET matching in sub-threshold area", Proc. ICMTS, p.225-228, (2012).

978-1-4673-4845-4/13 $31.00 © 2013 IEEE

SESSION 6: Thermal and Power

978-1-4673-4845-4/13 $31.00 © 2013 IEEE   114

Comparison of Electrical Techniques for Temperature Evaluation in Power MOS Transistors

A. Ferrara, P. G. Steeneken[1], K. Reimann[1], A. Heringa[2], L. Yan[2], B. K. Boksteen, M. Swanenberg[3], G. E. J. Koops[2], A. J. Scholten[1], R. Surdeanu[2], J. Schmitz, R. J. E. Hueting

MESA+ Institute for Nanotechnology
University of Twente
7500AE, Enschede, The Netherlands
Tel./Fax: +31 53 489 2645/6920. Email: a.ferrara@utwente.nl
NXP Semiconductors: [1]Eindhoven, The Netherlands   [2]Leuven, Belgium   [3]Nijmegen, The Netherlands

Abstract—**Three electrical techniques (pulsed-gate, AC-conductance and sense-diode) for temperature evaluation in power MOS transistors have been experimentally compared on the same device. The device under test is a silicon-on-insulator (SOI) laterally-diffused MOSFET (LDMOS) design with embedded sense-diodes in the center and at the edge of the device for providing local temperature information. On-wafer measurements have been performed on a thermal chuck in the temperature range $25 - 200\,^\circ$C to extract self-heating information and predict the junction temperature for different biasing conditions. Good agreement (within 10%) between the different techniques is achieved, evidencing that reliable temperature estimations can be made using each of the proposed electrical techniques. As a result, factors other than experimental accuracy will play a role in the choice of the most adequate technique for the application of interest. Guidelines for this choice are provided in a benchmarking analysis accounting for ease of application, temperature calibration and accuracy of the results.**

Index Terms—**Temperature, self-heating, thermal resistance, power MOS, pulsed-gate, AC-conductance, sense-diode.**

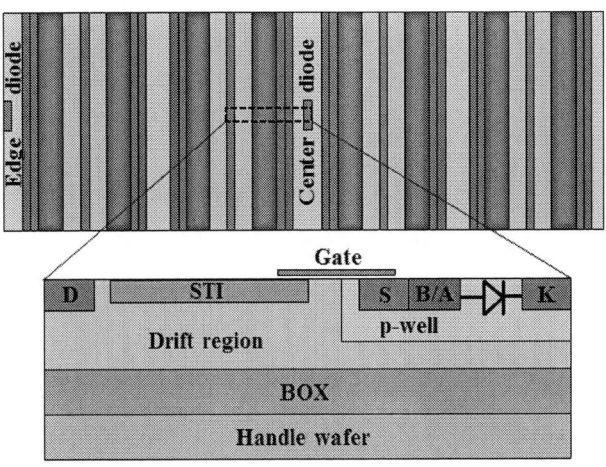

Figure 1.   *Top.* Schematic top view of the LDMOS with embedded center and edge sense-diodes. *Bottom.* Cross-section of the center LDMOS cell showing that the diode is connected between the body contact B (shorted to the source S, and representing the anode A) and the cathode K.

I. INTRODUCTION

Silicon-based power MOS transistors, such as laterally-diffused MOSFETs (LDMOS), find wide application in automotive, smart power, and audio areas [1]. Since optimization and shrinking of the geometry has pushed the current densities and operating temperatures of power devices close to the physical limits [2], temperature reliability issues have become more and more relevant. In order to guarantee high-performance and reliable transistor operation, it is essential to maximize the current density without exceeding a temperature that could lead to device failure. Therefore, good techniques for transistor temperature measurements are of critical importance.

Several techniques for temperature measurements based on different physical principles (electrical, optical or physical contacting) have been reported [3]. Among these, electrical techniques allow a quick, inexpensive and noninvasive estimation of the device thermal properties by exploiting standard electrical equipment. The advantage is that electro-thermal device characterizations can be performed on a single setup allowing the simultaneous extraction of electrical and thermal device parameters. This information can then be used for extracting thermal networks to be implemented in the electro-thermal simulators needed for circuit design.

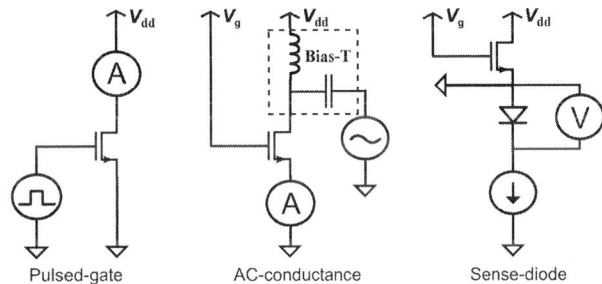

Figure 2.   Schematic electrical configurations for: *a)* pulsed-gate technique, *b)* AC-conductance technique, *c)* sense-diode technique.

This paper aims at benchmarking three electrical techniques reported earlier [4], [5], [6] for accurate temperature estimation by reporting a direct comparison on the same device. The device under test (Fig. 1) is an SOI-LDMOS transistor with embedded sense diodes in the center and at the edge for providing local temperature information.

The operating principle of the selected techniques (*pulsed-gate, AC-conductance* and *sense-diode,* Fig. 2) is presented in Section II, while the experimental results are compared in

978-1-4673-4845-4/13 $31.00 © 2013 IEEE

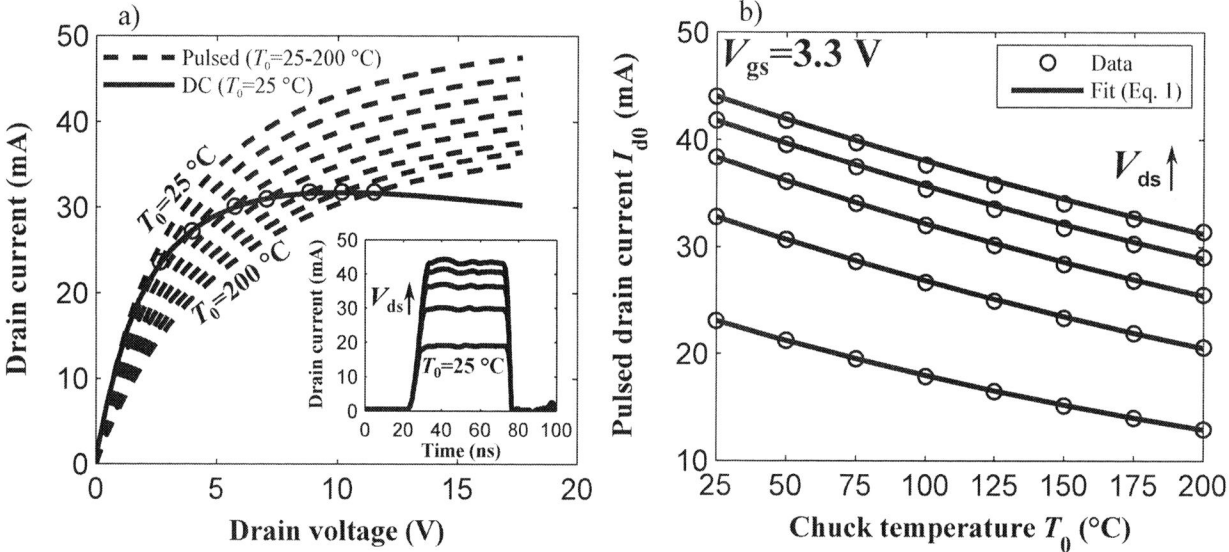

Figure 3. Operating principle of the pulsed-gate technique. *a)* Determination of the junction temperature T_j by comparison between the isothermal $I_{d0} - V_{ds}$ calibration curves at different chuck temperatures (dashed lines) and $I_d - V_{ds}$ DC characteristics at $T_0 = 25\,°C$ (solid line). The circles represent the intersection points where calibration data are available. Inset shows the 50-ns pulsed measurements at different V_{ds}-values ($T_0 = 25\,°C$), from which the isothermal $I_{d0} - V_{ds}$ characteristics were extracted. *b)* Current-temperature ($I_{d0} - T_0$) calibration curves for different drain voltages V_{ds} fitted according to Eq. 1.

Section III. Based on these results, in Section IV a benchmarking analysis is provided, which aims at establishing guidelines for the appropriate choice of the temperature measurement technique according to specified criteria. Finally, the work is summarized in Section V.

II. THE THREE ELECTRICAL TECHNIQUES

Three electrical techniques for temperature evaluation are compared: *a)* the pulsed-gate technique (Fig. 2a), which determines the junction temperature T_j based on the comparison of pulsed (self-heating free) $I_{d0} - V_{ds}$ curves and DC $I_d - V_{ds}$ curves [4]; *b)* the AC-conductance (g_{ac}) technique (Fig. 2b), which determines the thermal impedance Z_{th}, by AC-modulation of the dissipated power P_d, from the frequency-dependent output conductance g_{ac} [5]; and *c)* the sense-diode technique (Fig. 2c), exploiting the linear voltage-temperature relationship in diode sensors embedded in the LDMOS [6].

For each technique, on-wafer measurements were performed in the chuck temperature range $T_0 = 25 - 200\,°C$ and the extracted junction temperatures T_j in the steady-state (DC) limit are compared in Section III.

A. Pulsed-gate technique

The pulsed-gate technique is based on the comparison of the time-domain response of the LDMOS when driven by pulses having either a much shorter or a much longer duration than the device thermal time constant (τ_{th}, in the range of $\sim 10\,\mu s$ for the device under test). In principle, either the gate or the drain terminal could be pulsed while the other terminal is DC-biased. However, in power devices pulsing the gate requires

a much smaller switching energy and represents the standard driving option.

The experimental setup for the pulsed-gate technique is shown in Fig. 2a. A signal generator is used to switch on and off the gate while a DC bias V_{dd} is applied to the drain. For each drain bias, the drain current is monitored by measuring the potential drop across a 50 Ω resistor (the ammeter in Fig. 2a) as a function of time. The output characteristics are extracted for two different pulse widths: *a)* 50 ns (self-heating free characteristics, $I_{d0} - V_{ds}$) and *b)* 2 ms (DC characteristics, $I_d - V_{ds}$). A pulse width of 50 ns is shorter than the thermal time constant of the LDMOS, and self-heating is negligible in this time frame. As a result, the isothermal $I_{d0} - V_{ds}$ characteristics (for $T_0 = 25 - 200\,°C$) are obtained, from which current-temperature ($I_{d0} - T_0$) calibration curves (Fig. 3b) are extracted (for each V_{ds}) using the following exponential fit (see Section IV on temperature calibration):

$$I_{d0}(T_0) = I_{d0}(T_{ref}) \cdot e^{-\frac{T_0 - T_{ref}}{\theta}}, \qquad (1)$$

where T_{ref} is an arbitrary reference temperature ($T_{ref} = 25\,°C$ in this work) and θ is a fitting temperature. I_{d0} represents the self-heating free current, while the DC current is denoted by I_d. The DC $I_d - V_{ds}$ characteristics, affected by self-heating, are extracted using 2 ms pulses and then compared to the $I_{d0} - T_0$ calibration curves in order to extract the junction temperature T_j (for each V_{ds}), as follows:

$$I_d(T_j) = I_{d0}(T_0) \cdot e^{-\frac{T_j - T_0}{\theta}}, \qquad (2)$$

$$T_j = T_0 + \theta \cdot \ln\left(\frac{I_{d0}}{I_d}\right). \qquad (3)$$

978-1-4673-4845-4/13 $31.00 © 2013 IEEE

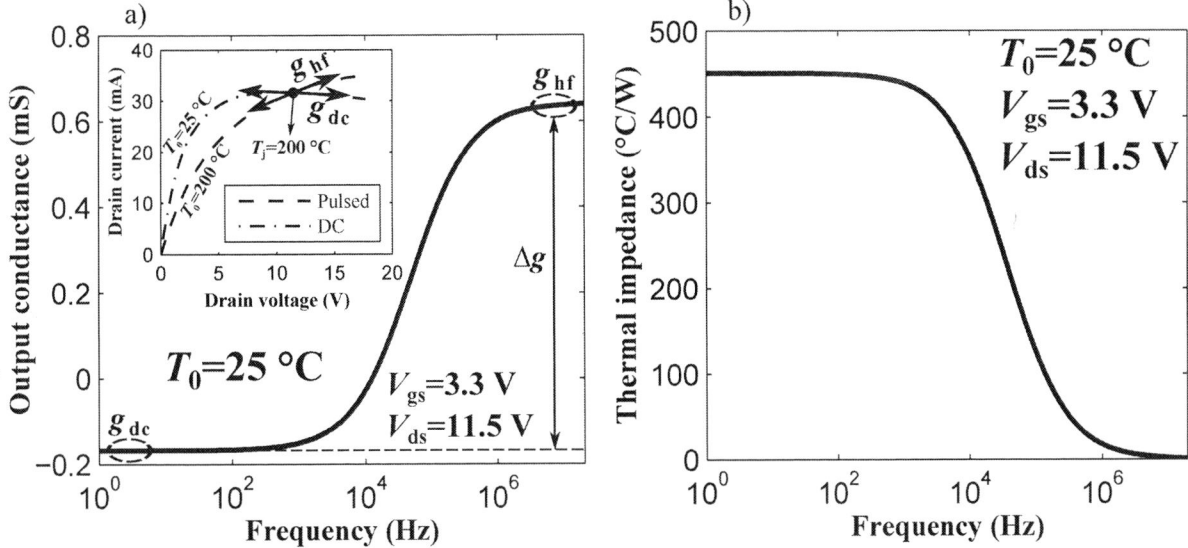

Figure 4. Operating principle of the AC-conductance technique. *a)* Differential output conductance g_{ac} as a function of the frequency. The difference between the high-frequency value (g_{hf}) and the DC value (g_{dc}) of g_{ac} ($\Delta g = g_{hf} - g_{dc}$, see also inset) is used to extract information about the device self-heating [7]. *b)* Frequency-dependent thermal impedance Z_{th} according to Eq. 4.

A graphical representation of the method is shown in Fig. 3a. The exponential fit also allows to extrapolate the junction temperature T_j outside the $I_{d0} - T_0$ calibration range, making it especially useful for devices in strong self-heating conditions. Although the accuracy of the extrapolated temperature cannot be easily predicted, the reliability of the approach is verified by the good agreement between the three different techniques (see Section III) also outside the calibration range.

B. AC-conductance technique

The AC-conductance technique relies on a frequency-domain characterization of the differential (small-signal or AC) output (or drain) conductance g_{ac} to extract self-heating information. The differential output conductance g_{ac} represents the frequency-dependent slope of the $I_d - V_{ds}$ characteristics at a given bias condition (i.e., fixed DC gate and drain potentials).

The experimental setup for the AC-conductance technique is shown in Fig. 2b. While the gate and drain are DC-biased, a small AC signal is coupled to the drain via a bias-tee, and g_{ac} is measured while sweeping the frequency of the AC source. An impedance (or vectorial) analyzer (Agilent 4294A in this work) can be used for this purpose.

An intuitive explanation of the technique (Fig. 4b) can be given as follows [7]. The drain conductance g_{ac} is a measure of the small-signal temperature fluctuations (around a DC junction temperature $T_j \neq T_0$) induced when AC power is dissipated inside the LDMOS. If the frequency f of the AC source is much higher than $1/\tau_{th}$ ($f \gg 1/\tau_{th}$), the device does not respond thermally to the dissipated AC power and its temperature does not vary during the small-signal operation. Under these conditions, there is no dynamic self-heating, and the high-frequency AC-conductance (g_{hf} in Fig. 4a) is simply determined by the current fluctuations induced by the AC

variation of the drain potential V_{ds}. On the other hand, if $f \ll 1/\tau_{th}$, the junction temperature responds to the dynamic drain voltage variation and fluctuates around a DC value. A temperature fluctuation will induce a current fluctuation (since the current is temperature dependent) leading to a variation of the low-frequency AC-conductance (g_{dc}) with respect to the high-frequency value (g_{hf}). Under strong self-heating conditions, the DC drain conductance can become negative, meaning that T_j-fluctuations induce a 180° phase-shift between the AC current and voltage on the drain. This behavior is responsible for the negative slope of the DC output characteristics observed in saturation.

The frequency-dependent thermal impedance Z_{th} (Fig. 4) is extracted according to [8]:

$$Z_{th}(f) = \frac{(g_{ac}(f) - g_{hf}) \cdot (I_d + g_{dc}V_{ds})}{S_I \cdot (I_d + g_{ac}(f)V_{ds}) \cdot (I_d + g_{hf}V_{ds})}. \quad (4)$$

From Eq. 4, the DC thermal resistance R_{th} and junction temperature T_j are given by:

$$R_{th} = Z_{th}(f = 0) = \frac{g_{dc} - g_{hf}}{S_I \cdot (I_d + g_{hf}V_{ds})}, \quad (5)$$

$$T_j = T_0 + R_{th}I_dV_{ds}, \quad (6)$$

where:

$$g_{dc} = g_{ac}(f = 0) = \tfrac{dI_d}{dV_{ds}}, \ S_I = \tfrac{dI_d}{dT_0} \text{ and}$$
$$g_{hf} = g_{ac}(f \gg 1/\tau_{th}). \quad (7)$$

In order to apply Eqs. 4-5, a DC calibration of the derivative of the DC-current I_d with respect to the chuck temperature T_0 ($S_I - T_0$ calibration) is needed.

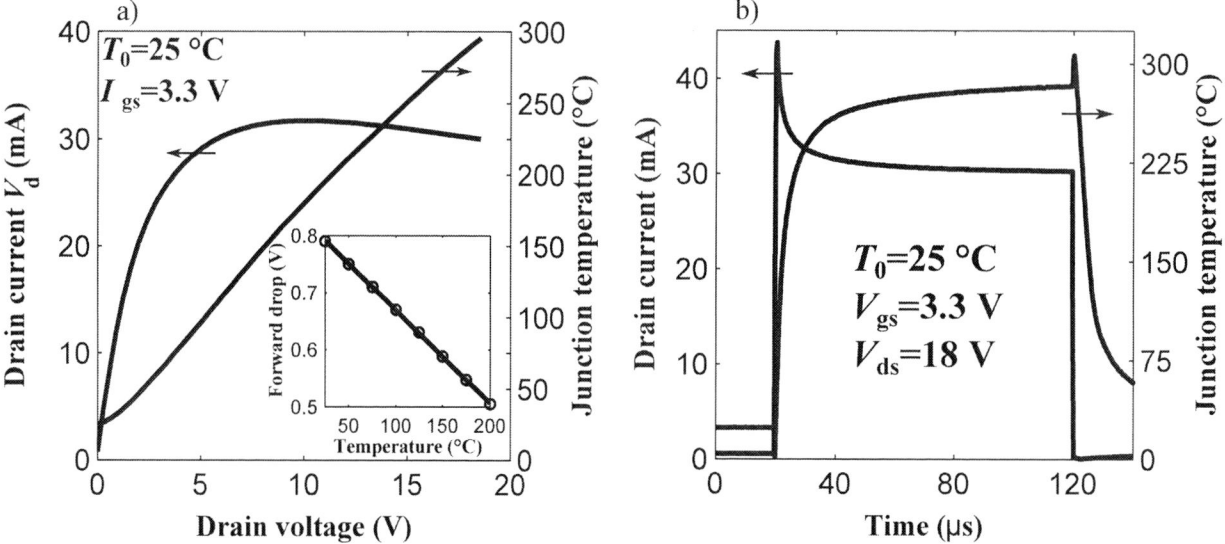

Figure 5. Operating principle of the sense-diode technique. *a)* The forward voltage drop V_f for a 50 μA injected current is measured while sweeping the drain voltage during the $I_d - V_{ds}$ characterization (left axis), and the junction temperature (right axis) is obtained from the $V_f - T_0$ calibration scale (see inset). *b)* Transient behavior of the drain current (left axis) and junction temperature (right axis) during pulsed-gate operation with an on-time of 100 μs. The temperature spikes occuring at turn-on ($t = 20\,\mu$s) and turn-off ($t = 120\,\mu$s) are artifacts inherent to the measurement setup.

C. Sense-diode technique

The sense-diode technique exploits the voltage drop across a forward-biased *pn* junction as a temperature sensitive parameter. A temperature sensor can be integrated in the LD-MOS design with minimal area occupation by introducing an additional n^+-diffusion (the cathode K) in the *p*-well of the active cell where the temperature T_j has to be detected (see Fig. 1). In this way, a *pn* junction is formed between the body/anode (B/A) and the cathode (K) contacts. This principle can be applied to vertical (VDMOS) and horizontal (LDMOS) devices [6].

When a constant current is injected through a *pn* junction, the forward voltage drop V_f is temperature dependent and linearly decays with a rate of approximately 2 mV/K, representing the diode sensitivity. The experimental setup for exploiting the integrated diode-sensor during the LDMOS operation is depicted in Fig. 2c. While the device is biased in the desired operating condition, a small current is injected through the sense-diode and the forward voltage drop is measured. Then, the junction temperature T_j is inferred from the $V_f - T_0$ calibration curve (Fig. 5a).

In this work, the sense-diode technique has been applied while driving the gate with long (2 ms) pulses in the DC limit. Such a configuration has the following advantages: *a)* it allows to analyze thermal transients by monitoring V_f with an oscilloscope (Fig. 5b); *b)* it does not require a prior diode calibration, since the calibration curves are obtained from the V_f waveforms when the device is switched off ($V_{gs} = 0$) and cooled down ($T_j = T_0$).

In order to obtain accurate experimental results, the following assumptions should be verified in the test-structure design and experimental setup:

1) The sense-diode should be located as close as possible

to the active area of the LDMOS cell.

2) The sensing current across the diode should be low enough to prevent self-heating of the diode itself. A more detailed analysis [9] shows that the optimal choice of the diode current derives from a trade-off between negligible diode self-heating and sensitivity and linearity of the $V_f - T_0$ calibration curve. Based on this trade-off, a value of 50 μA has been selected for the measurements.

3) The LDMOS and the diode operation should not mutually interfere. Since the diode current is about three orders of magnitudes smaller than the on-state LDMOS current, its influence on the on-state LDMOS behavior is not relevant. On the other hand, the diode is not affected by the LDMOS electron current since the cathode K is junction isolated from the channel region.

III. EXPERIMENTAL RESULTS

The measurement results obtained with the three techniques are shown in Fig. 6. The pulsed-gate and AC-conductance techniques exhibit good agreement (Fig. 6c) with a minor mismatch for $T_0 = 200\,°$C in a T_j-range for which calibration data are not available. Since both techniques extract a weighted-average temperature across the LDMOS (see Section IV), they underestimate the temperature with respect to the center sense-diode, which is positioned very close to the hottest region of the device. The edge sense-diode data show that the temperature at the LDMOS edge is the lowest and demonstrate that multiple diodes can be useful for examining temperature gradients.

IV. BENCHMARKING ANALYSIS

The advantages and disadvantages of the reported electrical techniques are summarized in Table I. A detailed bench-

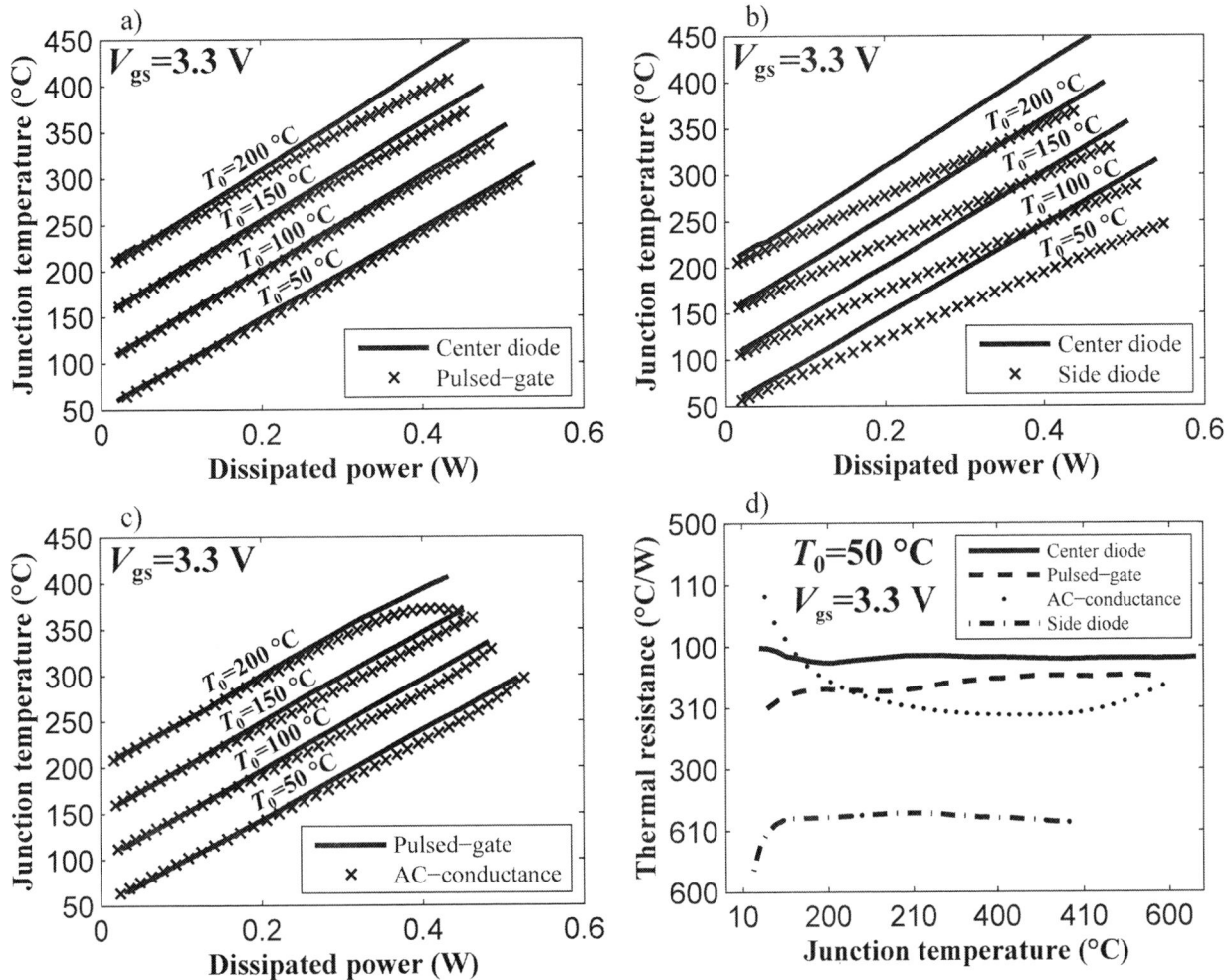

Figure 6. Device temperature (T_j) *vs.* dissipated power (P_d) for different chuck temperatures (T_0) for: *a)* center sense-diode *vs.* pulsed-gate; *b)* center sense-diode *vs.* edge sense-diode; *c)* pulsed-gate *vs.* AC-conductance. *d)* Temperature (T_j)-dependent thermal resistance R_{th} for the different techniques at $T_0 = 50\,°C$.

marking analysis aimed at understanding which technique is best-suited for the application of interest should consider the following aspects: *1)* ease of application, *2)* temperature calibration, *3)* accuracy of the results.

1) Ease of application. From the setup point of view, the easiest technique to apply is the sense-diode technique, as it can be carried out with simple DC measurements when transient thermal information is not needed. The main drawback is the need for a special test structure (MOSFET with integrated diode sensor) which requires extra area and additional measurement terminals. On the other hand, the pulsed-gate and AC-conductance technique can be applied to standard devices, but they require time-domain and frequency-domain measurements, respectively. The pulsed-gate technique allows direct extraction of the average temperature T_j, but it can be difficult to apply to devices with short thermal constants, since very short pulses are needed to acquire the isothermal $I_{d0} - V_{ds}$ characteristics. The AC-conductance technique, on the other hand, only requires an impedance (or network)

analyzer and can be applied to devices with various sizes and thermal constants without particular limitations, but provides an indirect temperature measurement.

2) Temperature calibration. While the pulsed-gate technique requires a pulsed temperature calibration of the transistor isothermal $I_{d0} - V_{ds}$ curves ($I_{d0} - T_0$ calibration), a simpler DC calibration is needed for the AC-conductance and sense-diode techniques. In particular, the AC-conductance technique requires calibrating the derivative of the DC current (S_I) as a function of the temperature T_0, whereas for the sense-diode technique the $V_f - T_0$ calibration curve is needed. If the sense-diode technique is applied in DC conditions, the diode has to be calibrated before (or after) the temperature measurements are performed. As suggested in Section II-C, a possible alternative consists in switching the transistor on and off and get the calibration data from the off-state waveforms.

An important calibration issue is related to the possibility to make temperature predictions outside the calibration range. As shown in Fig. 6, power SOI devices can withstand junction

Table I
OVERVIEW OF THE ADVANTAGES AND DISADVANTAGES OF THE ELECTRICAL TECHNIQUES FOR TEMPERATURE DETERMINATION.

	Pros	Cons
Pulsed-gate	Direct temperature measurement Simple data processing	Weighted-average temperature across transistor Pulsed calibration Difficult on short-τ_{th} devices
AC-conductance	Simple setup DC calibration General purpose	Weighted-average temperature across transistor Indirect temperature measurement
Sense-diode	Local probe DC calibration	Requires extra diode Requires extra measurement terminals

temperatures up to $\sim 400\,°C$, and getting calibration data in this temperature range can be rather difficult, especially for on-wafer measurements. In fact, standard thermal chucks cannot be heated up above $300\,°C$, and the maximum junction temperature transistors can withstand is typically higher than this value. Therefore, a physics-based fitting routine for extrapolating data outside the calibration range is very useful. In the pulsed-gate technique, it is possible to exploit the exponential dependence of the saturated carrier velocity in the drift region, from which the exponential decay of the saturation current with temperature originates [10]. By fitting the parameter θ (in Eq. 1), the exponential fit can also be extended to the linear operating region, where self-heating is less relevant. A similar fitting routine can be applied also to the $S_1 - T_0$ calibration in the AC-conductance technique. In fact, if the current exponentially depends on the temperature, also its temperature derivative will exhibit the same behavior. Calibration issues are less relevant in the sense-diode technique, since the linearity range of the forward voltage drop with temperature extends well beyond $200\,°C$ and can cover the full T_j-measurement range of interest if the diode current is properly selected [9].

3) Accuracy of the results. It is generally recognized that temperature measurements in the electrical domain can only provide a weighted-average temperature across the transistor. The pulsed-gate and AC-conductance technique respectively rely on measuring the current and the output conductance, which are global parameters of the transistor and therefore do not contain any local information. The sense-diode technique also performs a sort of averaging, since the temperature could be non-uniformly distributed across the diode area. Moreover, the diode is not exactly located in the hottest spot (the drift region) of the active cell where the temperature has to be sensed. As a result, temperature gradients could also affect the accuracy of the diode measurements. However, since the diode area is much smaller than the LDMOS area and the diode is positioned very close to the drift region (where most of the power generation occurs), averaging effects are less pronounced than in the pulsed-gate and AC-conductance techniques. As a result, multiple diodes can be used to extract temperature gradients between different active cells.

V. CONCLUSIONS

Three electrical techniques (pulsed-gate, AC-conductance and sense-diode) for temperature measurements in power

MOS transistors have been compared showing a smaller than 10% difference over a large junction temperature range (from room temperature up to $\sim 400\,°C$). For each technique, the measurement setup and the operating principle have been presented, and the experimental results have been discussed. The advantages and the disadvantages of the investigated techniques have been compared in a benchmarking analysis accounting for setup simplicity, calibration issues and experimental accuracy. The pulsed-gate and sense-diode techniques allow direct temperature extraction but require short-pulse calibration measurements and a dedicated test structure, respectively. The AC-conductance benefits from a simpler setup and a broader application range, but relies on an indirect temperature extraction. The added value of using embedded diode-sensors lies in the possibility to measure temperature gradients between adjacent active cells, which might play an important role in device reliability.

ACKNOWLEDGMENTS

This work is a part of the Dutch Point-One program and is supported financially by Agentschap NL, an agency of the Dutch Ministry of Economic Affairs.

REFERENCES

[1] P. Wessels *et al.*, "Advanced BCD technology for automotive, audio and power applications", *Solid-State Electronics* **51**, pp. 195−211 (2007)

[2] B. J. Baliga, "Trends in Power Semiconductor Devices", *IEEE TED* **43**, 10, pp. 1717−1731 (1996)

[3] D. L. Blackburn, "Temperature Measurements of Semiconductor Devices− A Review", *20th SEMI-THERM Symposium* (2004)

[4] C. Anghel, R. Gillon and A. M. Ionescu, "Self-Heating Characterization and Extraction Method for Thermal Resistance and Capacitance in HV MOSFETs", *IEEE EDL* **25**, 3, pp. 141−143 (2004)

[5] W. Redman-White *et al.*, "Direct Extraction of MOSFET Dynamic Thermal Characteristics from Standard Transistor Structures Using Small Signal Measurements", *Electron. Lett.* **29**, 13, pp. 1180−1181 (1993)

[6] M. Pfost *et al.*, "Small Embedded Sensors for Accurate Temperature Measurements in DMOS Power Transistors", *Proc. of ICMTS '10*, pp. 3−7 (2010)

[7] A. J. Scholten *et al.*, "Experimental Assessment of Self-Heating in SOI FinFETs", *Proc. of IEDM '09*, pp. 305−308 (2009)

[8] A. J. Scholten, A. Ferrara *et al.*, to be published

[9] S. Santra *et al.*, "Silicon on Insulator Diode Temperature Sensor− A Detailed Analysis for Ultra-High Temperature Operation", *IEEE Sensors J.* **10**, 5, pp. 997−1003 (2010)

[10] E. Arnold, H. Pein and S. P. Herko, "Comparison of Self-Heating Effects In Bulk-Silicon and SOI High-Voltage Devices", *Proc. of IEDM '94*, pp. 813−816 (1994)

Measurement and Investigation of Thermal Properties of the On-Chip Metallization for Integrated Power Technologies

Martin Pfost,[*] Cristian Boianceanu,[†] Ioana Lascau,[†] Dan-Ionuţ Simon,[†] and Sebastian Sosin[†]

[*]Robert Bosch Center for Power Electronics, Reutlingen University
Alteburgstr. 150, 72762 Reutlingen, Germany
martin.pfost@reutlingen-university.de
[†]Infineon Technologies Romania, ATV PTP TM, 020035 Bucharest, Romania

Abstract—**DMOS transistors in integrated power technologies are often subject to significant self-heating and thus high temperatures. This can lead to device failure and reduced lifetime. Hence, numerical electro-thermal simulations already during circuit design are used to ensure that the device temperature stays within the accepted range. In such simulations, the influence of the on-chip metallization must be considered correctly. Therefore, accurate temperature measurements for different on-chip metallization configurations are required for simulator calibration.**

In this paper, we present test structures with different metal layers and via configurations suitable for that purpose. We will discuss how accurate results can be obtained that show even very small differences between structures with a similar thermal behavior. The measurement results, combined with numerical simulations, give also valuable insights into the heat removal capability of the on-chip metallization.

I. Introduction

Integrated bipolar-CMOS-DMOS (BCD) power technologies such as presented in [1]–[3] offer DMOS transistors with low area-specific on-state resistances. Small devices are therefore possible, but this comes at the expense of more pronounced self-heating. Thus, to prevent failures it is crucial to correctly assess the thermal behavior of the DMOS [1], [4], which is greatly influenced by the thermal conductivity and heat capacitance of the on-chip metallization. This holds especially true if thick power metal layers are used, but also applies to VLSI technologies [5].

Therefore, the metallization influence on the thermal behavior must be considered in the design phase by numerical temperature simulations, e.g. as described in [6]. Since the on-chip metallization of integrated technologies tends to be complex, having several layers and many vias, it is advantageous to use a thermally equivalent layer as a replacement in the numerical temperature simulation to reduce computing time and memory requirements, cf. also [5].

The parameters of such an equivalent layer can easily be extracted from measurements, yielding good results as shown in [6]. However, this implies that the configuration of the on-chip metallization is the same throughout the device, which is not always the case. Transistors with current sensing or with a very large active area usually ask for consideration of a more complex metallization layout in the temperature simulation.

In this paper we present test structures that are suitable to extract the influence of different on-chip metallization configurations on self-heating. Further, we discuss how very accurate results can be obtained as required to distinguish small differences in the thermal behavior. Finally, we will examine our observations with the aid of numerical simulations.

II. Test Structures

The test structures consist of a vertical DMOS with a heat-dissipating active area of $200 \times 162\ \mu m^2$, fabricated in a $0.6\ \mu m$ BCD technology with two thin metal layers and a thick power copper layer based on [2]. Several test structure variants were created with different metallization layouts, see Fig. 1 for two examples. They range from minimum to maximum metal coverage including several via variants as described in Tab. I and Fig. 2.

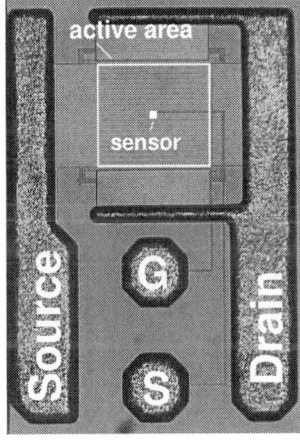
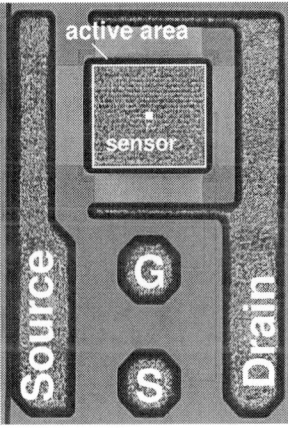

(a) Only metal 1

(b) Metal 1, metal 2, and power metal

Fig. 1. Chip photographs of two test structures with different metallization configurations. Note that only the metal covering the active area of the DMOS was modified. The surrounding wiring and pad structures are the same for all test structure variants. The contact areas for the source and drain needles are shown as well as the pads used for the gate (G) and for the emitter of the temperature sensor (S).

978-1-4673-4845-4/13 $31.00 © 2013 IEEE

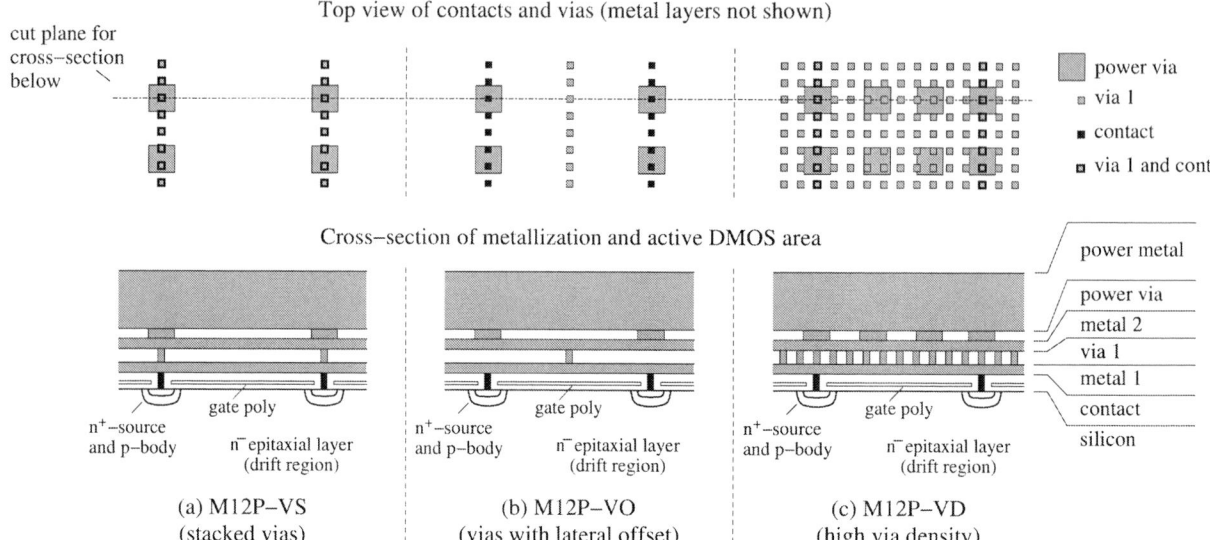

Top view of contacts and vias (metal layers not shown)

Cross−section of metallization and active DMOS area

(a) M12P−VS
(stacked vias)

(b) M12P−VO
(vias with lateral offset)

(c) M12P−VD
(high via density)

Fig. 2. Top view and cross-section (not to scale) of the on-chip metallization for the test structures with (a) stacked, (b) vias with lateral offset, and (c) dense vias, corresponding to M12P-VS, M12P-VO, and M12P-VD in Tab. I, respectively. The drain contact of the (vertical) DMOS is formed by a buried layer below (shown in Fig. 3 below). The heat dissipation takes place mostly in the drift region for the operation conditions under investigation.

label	layers	vias
M1	metal 1	none
M12	metal 1, 2	none
M12P	metal 1, 2, power met.	none
M12PI	metal 1, 2, power met., polyimide	none
M12P-VS	metal 1, 2, power met.	vias only over contacts
M12P-VO	metal 1, 2, power met.	vias with offset
M12P-VD	metal 1, 2, power met.	high via density

TABLE I

Metal, passivation (polyimide), and via configuration of the different test structure variants considered. A cross-section of the variants with vias (M12P-VS, M12-VO, and M12P-VD) is shown in Fig. 2, the material and thicknesses of the layers can be found in Tab. II.

layer	material	thickness
contact	W	1.5 µm
metal 1	Al	1.1 µm
via 1	W	1 µm
metal 2	Al	1.1 µm
power via	Cu	1.3 µm
power metal	Cu	11 µm
passivation	polyimide	3 µm

TABLE II

Metal, passivation, and via materials and thicknesses for the technology under investigation. The inter-layer dielectric is SiO_2.

The metallization on top of the active area of the DMOS is uniform, apart from a small separation in the lower metal layer required to contact the sensor as can be seen from Fig. 1(a). To minimize the influence of the metallization adjacent to the DMOS on its thermal behavior, we used only the lower metal layer for the electrical connection. For the same reason the thick power copper required for routing the pads to the DMOS is separated from the active area of the device even if the latter is covered with power copper as well, see Fig. 1(b). We also aimed at a layout which is as much symmetrical as possible by

adding metal areas even where not required from an electrical point of view.

Sufficiently large contact areas for both force and sense needles are provided for the DMOS source and drain so that the contact resistances can be canceled out. Only one needle each is used for the gate and for the emitter of the temperature sensor where no or only very small currents flow.

The temperature in the middle of the transistor is measured with an embedded sensor as presented in [7], [8], formed by the parasitic bipolar device inherent to the DMOS as shown in Fig. 3. These sensors are very small and close to the heat-generating regions, thus allowing measurements of the junction temperature. (This is not possible with IR thermography because the on-chip metal obscures the silicon below.) By forward-biasing the transistor with a constant emitter current I_E, the device temperature can be determined from its base-emitter voltage V_{BE} over a very wide range. The calibration curves were extracted from the same wafer containing our test structure for highest accuracy.

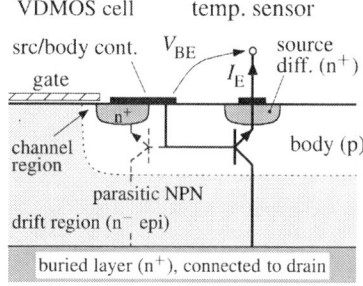

Fig. 3. Schematic cross-section of a temperature sensor embedded in the vertical DMOS cell array, from [8].

III. MEASUREMENT PROCEDURE

To determine the influence of different metallization variants on self-heating we measured the temperature increase in the center of the device caused by a power dissipation pulse of 10 W for 1 ms, achieved by applying a drain-source voltage of 40 V and turning on the gate for 1 ms to obtain a drain current of 250 mA. This power dissipation level will cause a substantial temperature increase, for some test structure variants exceeding 300 °C. A temperature swing that large helps to achieve a high accuracy which is required to distinguish some of the test structure variants. As an example, the thin polyimide layer, being the only difference between the variants M12P and M12PI (cf. Tab. I), is expected to have a very small influence on self-heating.

A. Measurement Setup and Data Acquisition

All measurements presented in this paper are performed on-wafer at a well-controlled chuck temperature of 25 °C. Since the thermal contact of the wafer backside might be affected by dust particles and by imperfections of the chuck, a thermally insulating material is placed between the wafer and the chuck. By this means we ensure a well-defined thermal boundary conditions.

An arbitrary waveform generator is used to control the DMOS gate. We apply a 1 ms long pulse with a carefully chosen rise and fall time of a few μs to minimize overshoot and to avoid ringing caused by parasitic inductances of the wires connecting the measurement instruments with the needles. The pulse length of 1 ms is a good compromise between achieving significant self-heating with moderate power dissipation while at the same time ensuring that the heat capacitance of the wafer is large enough to avoid excessive temperatures. Moreover, the heat wave has not yet fully propagated through the substrate to the wafer backside as was also verified by numerical temperature simulations. In conjunction with the thermally insulating material on the chuck we can thus neglect uncertainties regarding the thermal contact of the wafer backside.

Further, we carefully monitor the DMOS source-drain voltage via the sense needles to ensure that it is the same for all measurements. With this we avoid the influence of varying needle contact resistances, which could cause significant deviations in our measurement results. For the same reason we control the gate-source voltage to obtain a drain current of 250 mA. Further, small variations of the DMOS threshold voltage are canceled out by this approach as well. Similar care has been taken for the temperature sensor. Its base-emitter voltage is measured with a differential probe for high accuracy. The emitter current of 20 μA is supplied by a current source built from discrete components since most standard source-measurement units are too slow to keep the current constant during the 1 ms short pulse.

For each test structure, the DMOS drain-source voltage and the drain current were recorded by a digital storage oscilloscope as well as the base-emitter voltage of the sensor. As described in [8], from the latter the temperature can be obtained. Typical measurement results are shown in Fig. 4.

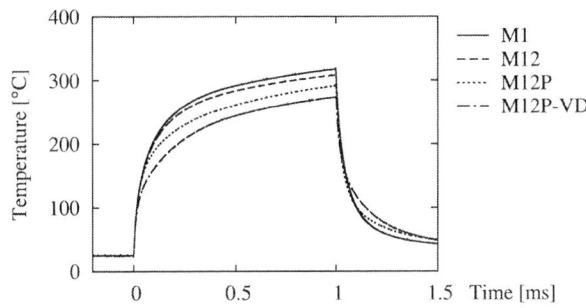

Fig. 4. Measurement results for the center temperature of the DMOS, cf. Fig. 1, for the variants M1, M12, M12P without vias and for M12P-VD with dense vias as described in Tab. I. In all cases we applied 10 W for 1 ms.

The operating point of the DMOS (40 V, 250 mA) is chosen so that the device operates at the temperature compensation point where the drain current does not depend on the temperature, cf. [4]. Thus, electro-thermal coupling can be neglected, and we obtain a uniform power dissipation density. At the same time, the drain current of 250 mA is comparatively low for such a device. This keeps the voltage drop in the source metal small even if only metal 1 is used. Because of this the gate-source voltage can be considered as mostly constant throughout the device. Nevertheless, a perfectly uniform electrical behavior cannot be obtained, also due to the voltage drop in the buried layer which causes a small voltage difference in the drain potential between periphery and center of the DMOS.

B. Post-Processing

For each test structure ten consecutive acquisitions are performed, with a delay large enough to ensure that the structure is at ambient temperature before the next measurement. After that, the ten acquired data sets are averaged to minimize the impact of noise and spurious signals.

To further increase the accuracy of our temperature measurements we looked at the values obtained before turn-on. Since our temperature sensors behave very uniform (as was verified during calibration, see also [8]) we get measurements results very close to the well-controlled chuck temperature with a standard deviation of less than 0.6 °C. Nevertheless, the observed deviations are compensated in a first post-processing step by subtracting the offset from the temperature measurements.

As already mentioned, we checked during the measurements that the drain-source voltage and the drain current and thus the dissipated power meet the desired value. Nevertheless, small deviations (below 1%) from the target could not be avoided. They are compensated in a second post-processing step by scaling the temperature increase accordingly. Here, a linear dependence is assumed, even though the thermal properties of silicon depend on its temperature. This simplification is applicable because the required correction is very small.

With on-wafer measurements we could easily investigate 42 dies for each test structure variant, allowing us to apply statistical checks. We verified that the data follows a normal distribution as shown for the variant M12P in Fig. 5. Because of this, we are confident that averaging over the results of 42 dies further increases the validity of our results.

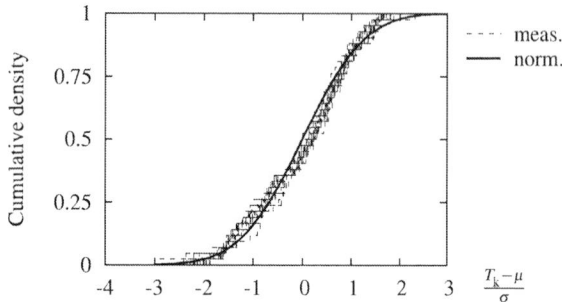

Fig. 5. Cumulative densities for the measured temperatures T_k of the $k = 1 \ldots 42$ examined M12P variants, plotted over the temperature normalized to the mean μ and standard deviation σ. Shown are 20 curves, corresponding to 20 different time steps ranging from 0.04 ms to 0.99 ms. The cumulative density function of the normal distribution is also shown (solid line).

With this approach we obtain a sufficiently high measurement accuracy. This is demonstrated by Fig. 6. Here, the cumulative densities for the 42 dies are plotted for each variant, now only for $t=1$ ms, i.e. the end of the pulse. We can clearly distinguish all investigated variants, even though their differences can be as low as only a few °C.

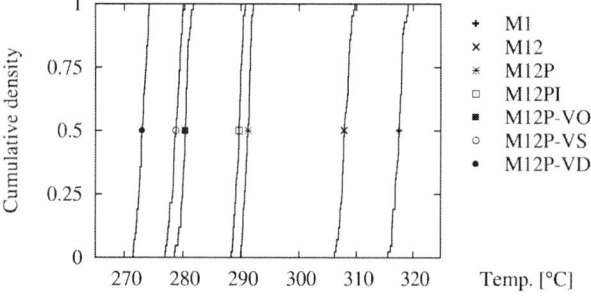

Fig. 6. Cumulative densities for the temperatures obtained at the end of the pulse at 1 ms. Shown are results determined from 42 dies. The mean values are marked by symbols.

IV. DISCUSSION OF MEASUREMENT RESULTS

As expected, the temperature is highest in the M1 variant, see Figs. 4 and 6. The sensor temperature after 1 ms is slightly lowered by 9 °C when adding metal 2 (M12). A more significant decrease by 26 °C is obtained with a power metal layer (M12P), emphasizing the influence of its high specific thermal capacitance. Note that the cooling effect of the thick metal is significant even in the absence of vias between this layer and the heat source. The temperature decreases further as vias are added to the metal stack, having the lowest value for the high via density variant (M12P-VD), being 44 °C smaller than the one measured for M1.

We also calculated the heat removal effect of the additional layers in M12, M12P, and M12PI using the temperature

obtained for M1 as reference. For this, we looked at the temperature decrease versus M1 divided by the temperature of M1, i.e. $(T_{M1} - T)/T_{M1}$. Here, T_{M1} and T correspond to the sensor temperatures in °C in the variants M1 and in the one under investigation, respectively. The results are plotted over time in Fig. 7.

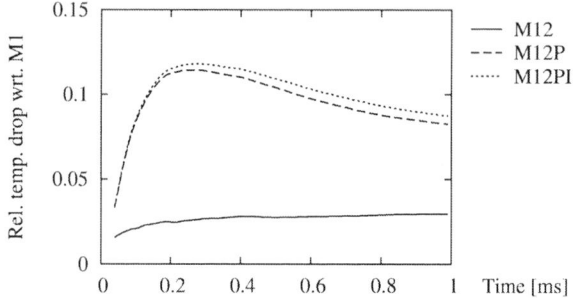

Fig. 7. Measured relative temperature drop for the variants without vias with respect to the structure M1 used as a reference. Shown is the temperature decrease with respect to M1 normalized to the temperature of M1.

We observe that the temperature difference between M1 and M12 is mostly constant throughout the measurement. This is because metal 2 has a low thermal capacitance and is thermally tightly coupled to metal 1. In the case of M12P and M12PI, however, the temperature difference to M1 shows a peak and then settles at a lower value. This can be explained by heat flowing into the power copper, charging its thermal capacitance. This is more pronounced at the beginning of the pulse, abating towards its end. (We will further investigate this in Sec. V.) We also observe that the additional polyimide layer in M12PI does not enhance cooling much, which is due to its lower thermal capacitance.

Similarly, we investigated different via configurations. From the results of Fig. 8 we see that vias significantly improve heat transport to the power metal. As expected, heat removal is best for M12P-VD where the via density is highest. We believe that to be caused mostly by the much better thermal connection between the metal layers, to a lesser extent also by the thermal capacitances of the additional vias. Reducing the via density to M12P-VS lowers the cooling effect significantly. However, by comparing M12P-VS and M12P-VO we observe that lengthening the heat flow path through the metal by offsetting the vias, cf. Fig. 2(b), does not greatly impede heat removal.

V. FURTHER INVESTIGATION OF THICK POWER COPPER

The behavior of the structures has also been investigated with numerical temperature simulations. They allow further insights into the heat-removal effect of the metallization that measurements cannot provide. For this, we applied the numerical simulator presented in [4], [6] and verified its validity for the test structures examined here. Results for M1 and M12P are shown in Fig. 9. An excellent agreement is obtained, the measurements and simulations being hardly distinguishable.

978-1-4673-4845-4/13 $31.00 © 2013 IEEE

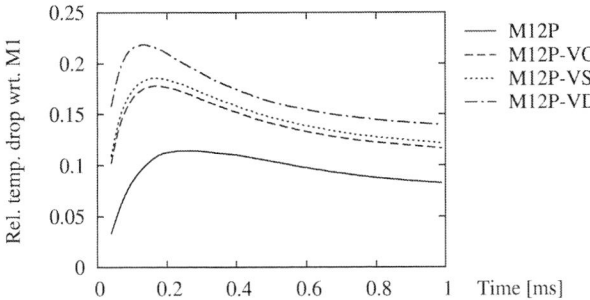

Fig. 8. Measured relative temperature drop defined as in Fig. 7 for variants with same metal layers but different via configurations, cf. Tab. I and Fig. 2.

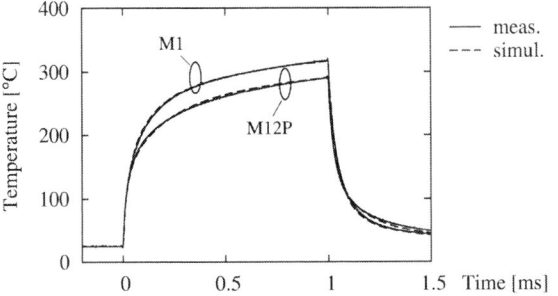

Fig. 9. Measured and numerically simulated temperatures for the test structure variants M1 and M12P.

Using our numerical temperature simulator, we assessed the temperature spread in the DMOS and the power metal, respectively, for the metallization on top of the source for variant M12P. The results are shown in Fig. 10. It can be observed that at the beginning of the pulse the power metal is significantly colder than the active area. In the first 80 μs the minimum temperature of the DMOS even exceeds the maximum temperature in the metal. However, eventually the metal reaches values much closer to the average DMOS temperature. The time required for the power copper to heat up shows its large thermal capacitance. Its good thermal conductivity can be seen from the small spread between maximum and minimum temperature in the metal.

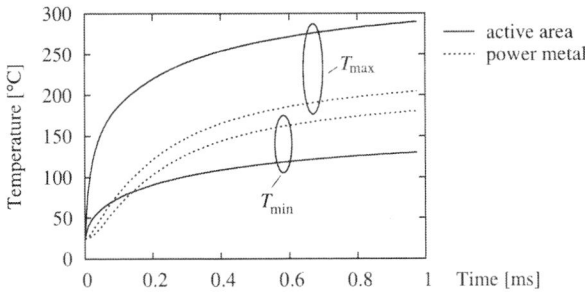

Fig. 10. Numerically simulated minimum and maximum temperatures for the active area (the heat-dissipating region) and the thick power metal for test structure variant M12P.

This is further illustrated by the calculation of the heat flow into the power metal and the energy it stores. The results for M12P are shown in Fig. 11. We see that at the beginning of the pulse the heat flow quickly reaches a peak due to a large temperature difference between DMOS and power copper. After that, the heat flow from the active area decreases as steady-state is being approached, reaching very low values at the end of the pulse. This also explains the reduction of the relative temperature drop with respect to M1 shown in Fig. 7 after 0.2 ms. Moreover, we observe from the energy stored in the power metal that it alone accounts for more than 3% of the dissipated energy. This corresponds well to the reduced peak temperature of M12P with respect to M1, cf. Fig. 4.

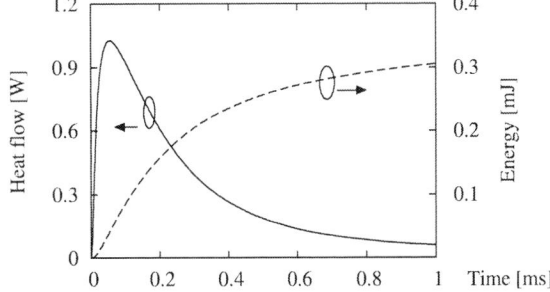

Fig. 11. Heat flow into the power metal and energy stored in it for the test structure M12P determined from a numerical simulation. Total dissipated energy is 10 mJ (10 W for 1 ms).

As we just demonstrated, the thermal capacitance is only effective within the first part of the pulse. Nevertheless, after that thick power copper also helps to reduce the peak temperature, see Fig. 7. This is due to the heat flow within the well-conducting power copper layer. To investigate this we extracted from the numerical temperature simulation the lateral heat flow between the center and the peripheral part of the power copper. The results are plotted in Fig. 12.

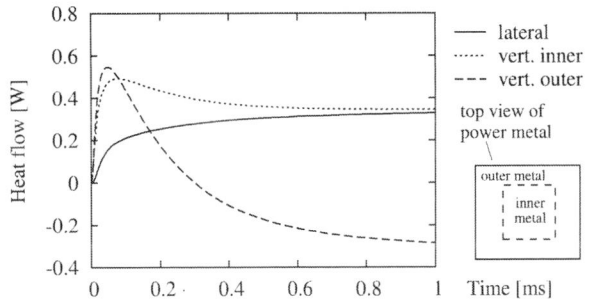

Fig. 12. Numerically simulated lateral heat flow between center and peripheral part of the power metal and vertical heat flow entering these parts from the active area. The boundary between center and periphery of the power metal is indicated by dashed lines in the top view of the power metal.

We observe that the lateral heat flow increases over time, being negligible at turn-on, eventually reaching much higher values than the (total) heat flow charging the thermal capacitance of power copper which becomes very small towards the end of the pulse, cf. Fig. 11. Nevertheless, Fig. 12 shows that even then significant power is entering the thick copper metal in its center, flowing laterally to the periphery, eventually returning to the peripheral part of the heat source. Therefore, the good thermal conductivity of power copper helps to level out temperatures and thus to lower their peak values even if the thick metal covers only the heat-generating area.

A further proof of this is given by Fig. 13, showing the temperature profiles for M12P and M1 at the end of the pulse where the thermal capacitances of the metallization on the DMOS are almost fully charged. The curve for M12P is flatter, having a significantly reduced peak temperature in the transistor center.

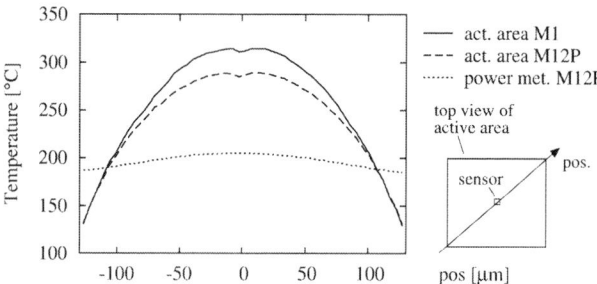

Fig. 13. Simulated temperature plotted over position in the active DMOS area for M1 and M12P and for the power metal in M12P at $t = 1$ ms, i.e. at the end of the pulse. At the center at $0\,\mu$m the temperature sensor is located. It causes a small temperature drop because its power dissipation is lower than that of the active DMOS area.

VI. SUMMARY

Several test structures were introduced to assess the influence of different integrated BCD power-technology metallization variants on the temperature during self-heating of DMOS transistors. We presented measurement procedures and post-processing steps suitable to determine very accurately the temperature in the active device area.

The obtained measurement results are key to more detailed insights into the heat removal mechanisms, which is important for technology optimization. This has been illustrated using thick power copper as an example. By additional numerical simulations we could detail how its high thermal capacitance and its good thermal conductivity reduce the device temperature.

Moreover, we believe that the obtained measurement results will be very valuable to develop and verify methods for the determination of equivalent thermal layers. For this, we recommend to perform measurements at several ambient temperatures to obtain different ratios between the strongly temperature-dependent thermal properties of silicon and the metal layers.

REFERENCES

[1] R. Rudolf, C. Wagner, L. O'Riain, K. Gebhardt, B. Kuhn-Heinrich, B. von Ehrenwall, A. von Ehrenwall, M. Strasser, M. Stecher, U. Glaser, S. Aresu, P. Kuepper, and A. Mayerhofer, "Automotive 130 nm smart-power-technology including embedded flash functionality," in *Proc. ISPSD 2011*, May 2011, pp. 20–23.

[2] M. Stecher, N. Jensen, M. Denison, R. Rudolf, B. Strzalkoswi, M. Muenzer, and L. Lorenz, "Key technologies for system-integration in the automotive and industrial applications," *IEEE Trans. Power Electronics*, vol. 20, no. 3, pp. 537–549, May 2005.

[3] C. Contiero, A. Andreini, and P. Galbiati, "Roadmap differentiation and emerging trends in BCD technology," in *Dig. ESSDERC 2002*, Sep. 2002, pp. 275–282.

[4] M. Pfost, J. Joos, and M. Stecher, "Measurement and simulation of self-heating in DMOS transistors up to very high temperatures," in *Proc. ISPSD 2008*, Orlando, FL, May 2008, pp. 209–212.

[5] L. Jiang, C. Xu, B. Rubin, A. Weger, A. Deutsch, H. Smith, A. Caron, and K. Banerjee, "A thermal simulation process based on electrical modeling for complex interconnect, packaging, and 3DI structures," *IEEE Trans. Advanced Packaging*, vol. 33, no. 4, pp. 777–786, Nov. 2010.

[6] M. Pfost, C. Boianceanu, H. Lohmeyer, and M. Stecher, "Electro-thermal simulation of self-heating in DMOS transistors up to thermal runaway," *IEEE Trans. Electron Devices*, 2013, in press.

[7] M. Pfost, D. Costachescu, A. Podgaynaya, M. Stecher, S. Bychikhin, D. Pogany, and E. Gornik, "Small embedded sensors for accurate temperature measurements in DMOS power transistors," in *Proc. 2010 ICMTS*, Hiroshima, Japan, Mar. 2010, pp. 3–7.

[8] M. Pfost, D. Costachescu, A. Mayerhofer, M. Stecher, S. Bychikhin, D. Pogany, and E. Gornik, "Accurate temperature measurements of DMOS power transistors up to thermal runaway by small embedded sensors," *IEEE Trans. Semiconductor Manufacturing*, vol. 25, no. 3, pp. 294–302, Aug. 2012.

Investigation on Safe Operating Area and ESD Robustness in a 60-V BCD Process with Different Deep P-Well Test Structures

Chia-Tsen Dai and Ming-Dou Ker

Institute of Electronics, National Chiao-Tung University, Taiwan
cttai.ee99g@nctu.edu.tw

Abstract—Safe operating area (SOA) is one of the noticeable reliability concerns for power MOSFETs during the normal circuit operating conditions. Besides, electrostatic discharge (ESD) reliability is another important reliability issue for the power IC products. To save the silicon area of power IC with high-voltage (HV) devices, it is preferable for HV MOSFET to be self-protected without any additional ESD protection device, and to behave wide SOA region. In this work, the impact of deep P-Well (DPW) structure to the electrical SOA (eSOA) and ESD robustness of HV MOSFET has been investigated in a 0.25-μm 60-V BCD process. DPW structure is used to implement the RESURF (reduced surface field) in MOSFET, which make it be able to sustain the high operating voltage. From the experimental results in silicon chip, the ESD robustness and eSOA of HV MOSFET can be improved by the modified DPW structure.

I. INTRODUCTION

Nowadays, the smart power technology with HV MOSFET devices has been developed and used to fabricate the display driver circuits, power switch, motor control systems, and so on [1]. Among the various reliability specifications, safe operating area (SOA) is a noticeable reliability concern during normal circuit operating conditions for power IC with the HV MOSFET [2]. The SOA region of HV MOSFET must be well characterized for using in circuit design to meet the specification of applications, which defines the operating limitation without damaging the IC products. In a HV n-type MOSFET, there is a parasitic n-p-n BJT inherent in the device structure. Once the parasitic BJT was triggered on to initiate a snapback, the gate control over the HV n-type MOSFET would be lost, and the current crowding effect would damage the HV device violently [3]. Thus, the SOA boundary has been defined without triggering on the parasitic BJT. In Fig. 1, SOA region is depicted as the shadow region, and the dotted line represented the SOA boundary. To minimize the self-heating effect under SOA measurement, device under test (DUT) is usually stressed by the pulses with a short pulse width. A transmission line pulse (TLP) system that delivers square pulses with a 100-ns pulse width is usually applied for the measurement of electrical SOA (eSOA) [4], [5]. The eSOA boundary is acquired when thermal effect is not strongly involved during operation.

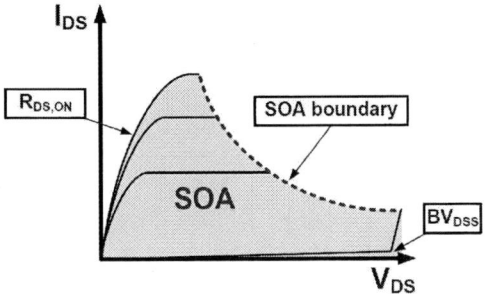

Fig. 1. A diagram showing on-resistance ($R_{DS, ON}$), SOA region, and breakdown voltage (BV_{DSS}) of a HV MOSFET.

Besides, on-chip electrostatic discharge (ESD) protection has been known as one of the important issues in high-voltage (HV) integrated circuits [6], [7]. ESD is an accidental event during fabrication, packaging, and testing processes of integrated circuits. In order to protect the internal circuits from ESD damage, on-chip ESD protection devices are applied to all input/output (I/O), power (V_{DD}/V_{SS}) and switch (SW) pads. For example, the circuit diagram of ESD protection scheme for LED driver and DC-DC buck converter are shown in Fig. 2(a) and 2(b).

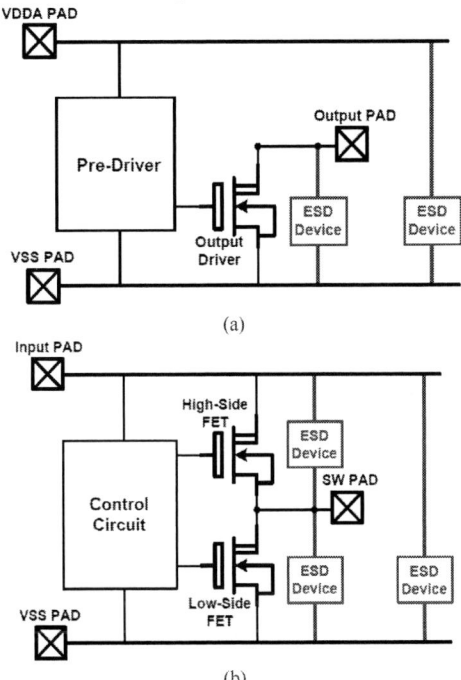

Fig. 2. ESD protection for the applications of (a) output pad of LED driver and (b) switch (SW) pad of DC-DC buck converter.

To save the silicon area, it is preferable for HV MOSFETs to have high ESD robustness and wide SOA region simultaneously without any ESD protection device added into the HV integrated circuits. In this work, the HV MOSFET in a 0.25-μm 60-V BCD process is investigated with different Deep-P-Well (DPW) structures to improve both of ESD robustness and SOA region.

II. TEST DEVICE STRUCTURES IN 60-V BCD PROCESS

Fig. 3(a) and Fig. 4(a) show the device cross-sectional view and layout top view of the standard 60-V n-channel lateral diffused MOSFET (nLDMOS), respectively. The normal operating voltage of the HV MOSFET is $V_{DS} = 0 \sim 60$ V and $V_{GS} = 0 \sim 5V$. Such HV device is surrounded by N-buried layer (NBL), which is connected to drain. The deep P-well (DPW) is used for RESURF (reduced surface field) technique to increase the breakdown voltage (BV_{DSS}) without paying wider layout distance for low on-resistance ($R_{DS,ON}$)

978-1-4673-4845-4/13 $31.00 © 2013 IEEE   127

consideration [8]. In this work, the DPW structure in nLDMOS was modified to investigate the impact of DPW structure to eSOA and ESD robustness. Three test devices investigated in the silicon chip include the standard HV device (nLDMOS_S), the modified HV device (nLDMOS_A), and the 2nd modified HV device (nLDMOS_B).

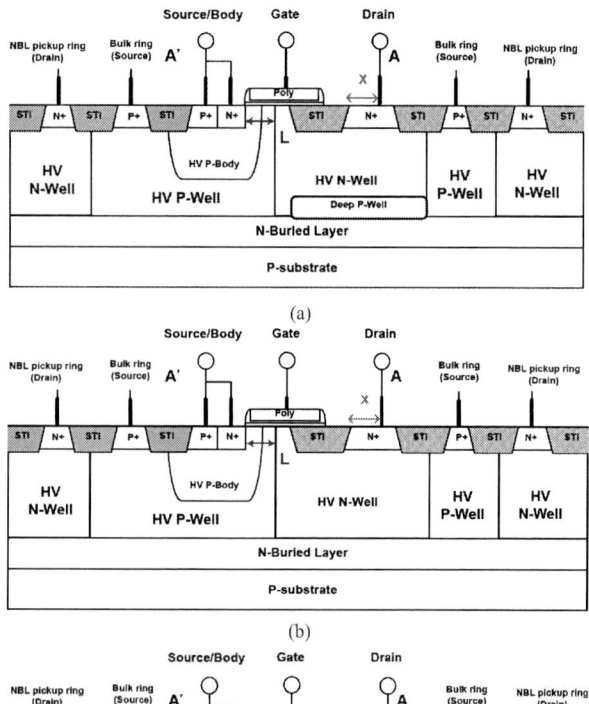

(a)

(b)

(c)

Fig. 3. The cross-sectional view of (a) the standard nLDMOS_S, (b) the modified nLDMOS_A, and (c) the 2nd modified nLDMOS_B, high-voltage devices in a 0.25-μm 60-V BCD process.

As shown in Fig. 3(b) and Fig. 4(b), the DPW structure under drain side is totally erased in the nLDMOS_A. Based on the prior study [9], the vertical BJT path can be induced where ESD current could flow into NBL region without DPW structure. The vertical ESD current path was spread almost along the entire region underneath the device, resulting in a significant reduction of the power density in the drain side, therefore to improve ESD robustness. The 2nd modified test device structure of nLDMOS_B is shown in Fig. 3(c) and Fig. 4(c), whose DPW structure is slotted under the drain N+ region. It is expected that such nLDMOS_B device can sustain high ESD robustness without lowering breakdown voltage. All the test devices are drawn with the same device dimension of W/L = 320 μm/1 μm in the silicon chip. With a large drain N+ region, it is expected that power density under ESD stress can be further reduced. Therefore, the range of drain N+ region, represented by the distance X marked in the figures, is split with 0.14, 5, and 10 μm, respectively (0.14 μm is the minimum distance in this HV

process). Fig. 5 shows the test chip of total test devices fabricated in a 0.25-μm 60-V BCD process.

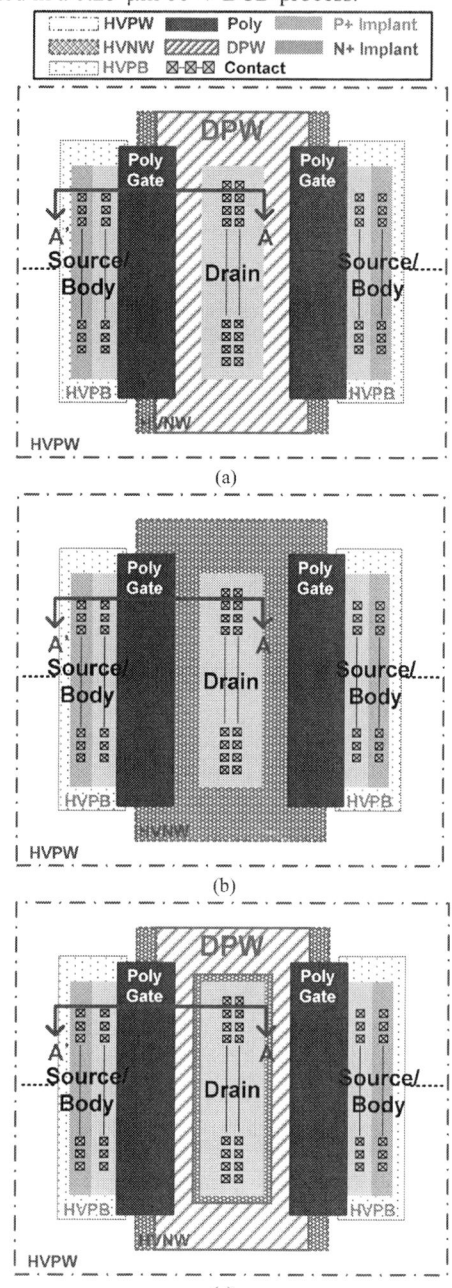

(a)

(b)

(c)

Fig. 4. The corresponding layout top view of (a) the standard nLDMOS_S, (b) the modified nLDMOS_A, and (c) the 2nd modified nLDMOS_B, HV devices in a 0.25-μm 60-V BCD process.

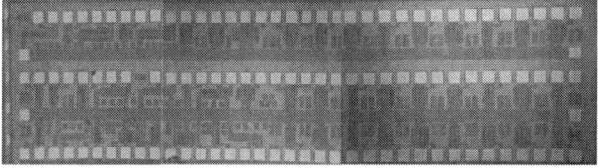

Fig. 5. The test chip of total test devices fabricated in a 0.25-μm 60-V BCD process.

III. EXPERIMENTAL RESULTS

A. Measurement Results of Electrical SOA

To investigate the impact of DPW structure to device ruggedness under normal circuit operating conditions, the

978-1-4673-4845-4/13 $31.00 © 2013 IEEE

characterization of eSOA is measured by 100-ns TLP pulses when giving a DC voltage for gate bias. The measurement setup was shown in Fig. 6. The gate bias is varied from 0 V to 5 V. TLP-measured I-V characteristics under different gate biases and the eSOA boundary of nLDMOS_S with distance X of 0.14μm are shown in Fig. 7, which is acquired by connecting the last *I–V* points under different DC gate biases before snapback.

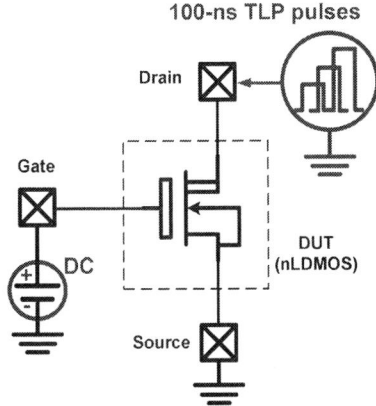

100-ns TLP pulses

Fig. 6. Test setup for eSOA measurement by 100-ns TLP pulses.

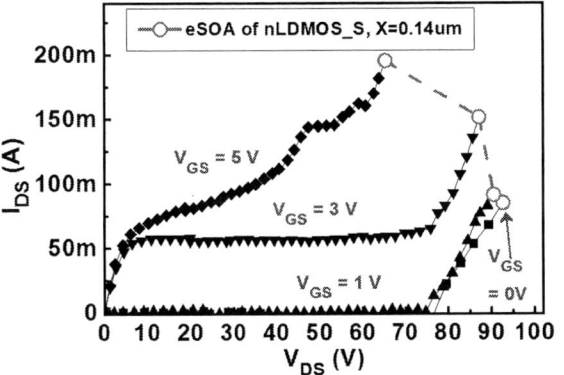

Fig. 7. The measured eSOA boundary of nLDMOS_S with distance X of 0.14 μm by 100-ns TLP pulses.

TABLE I
THE BREAKDOWN VOLTAGE (BV$_{DSS}$) AMONG DIFFERENT TEST DEVICES

Device (V$_{GS}$=0V)	BV$_{DSS}$ (V) (X=0.14μm)	BV$_{DSS}$ (V) (X=5μm)	BV$_{DSS}$ (V) (X=10μm)
nLDMOS_S	77	78	78
nLDMOS_A	53	56	56
nLDMOS_B	74	75	75

The breakdown voltage of the test devices are summarized in Table I. Fig. 8 shows the measured eSOA boundary of nLDMOS_S, nLDMOS_A, and nLDMOS_B with different distances X of 0.14, 5, and 10 μm, respectively. The test devices are measured under gate biases of 0, 1, 3, and 5 V, respectively, where the last *I-V* points before snapback are acquired for eSOA boundary. According to the comparison of different eSOA in Fig. 8, the eSOA of nLDMOS_A is the widest one, while that of the nLDMOS_S is the worst. The eSOA of all test devices are slightly extended with a wide drain region when a large distance X is used. Although the eSOA of nLDMOS_A is the widest one, its breakdown voltage is lower than 60 V, which cannot be used in 60-V applications. However, the test results of nLDMOS_B can

have a better eSOA than nLDMOS_S, as well as its breakdown voltage is about 75 V which is almost not degraded by the slotted DPW structure.

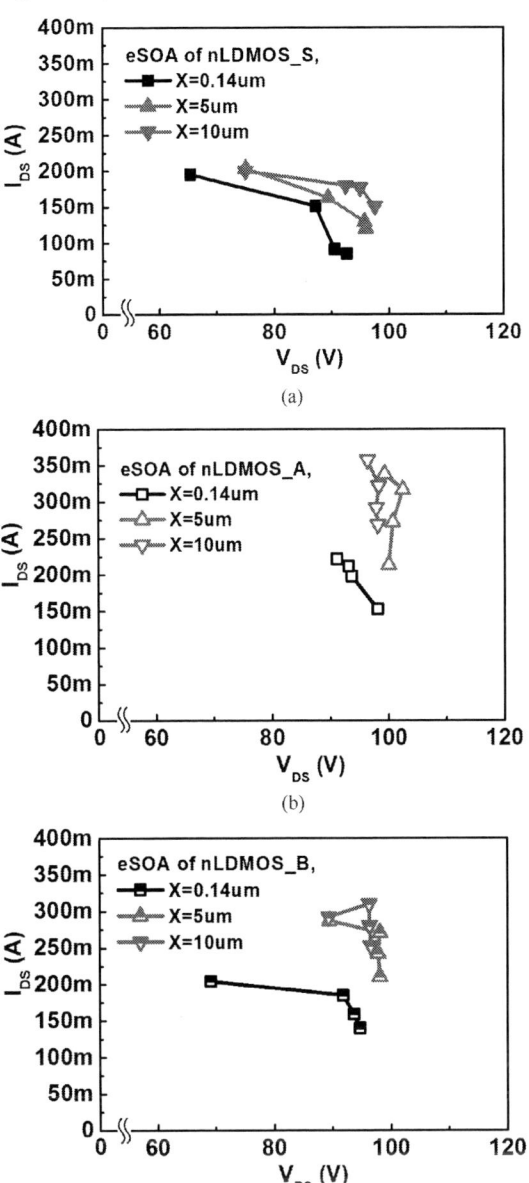

Fig. 8. The measured eSOA boundary of (a) nLDMOS_S, (b) nLDMOS_A, and (c) nLDMOS_B, with different distances X of 0.14, 5, and 10 μm, respectively.

B. TLP-Measured Results and ESD Robustness

Fig. 9 shows the TLP-measured characteristics under the same condition of V$_{GS}$ = 0 V. The data of secondary breakdown current (I$_{t2}$) are extracted from the TLP-measured *I-V* curves. The TLP-measured results and ESD robustness among the three HV test devices are summarized in Table II. From the measurement results, the test devices immediately failed as the snapback happened. It indicated that the parasitic n-p-n BJT in test devices were not triggered on during ESD stress. Thus, the ESD current was totally discharged through the reverse diode path. Although the BJT path was not triggered on during ESD stress, the ESD robustness of all test devices can still be improved with a large distance X due to reduction of power density. Moreover, the ESD performance of nLDMOS_A is the greatest, while that of nLDMOS_S is

978-1-4673-4845-4/13 $31.00 © 2013 IEEE

the worst. It demonstrates that the ESD current can be spread underneath the device, therefore reducing the power density and improving the ESD robustness. Similarly, the nLDMOS_B with slotted DPW structure can have better ESD robustness than that of nLDMOS_S.

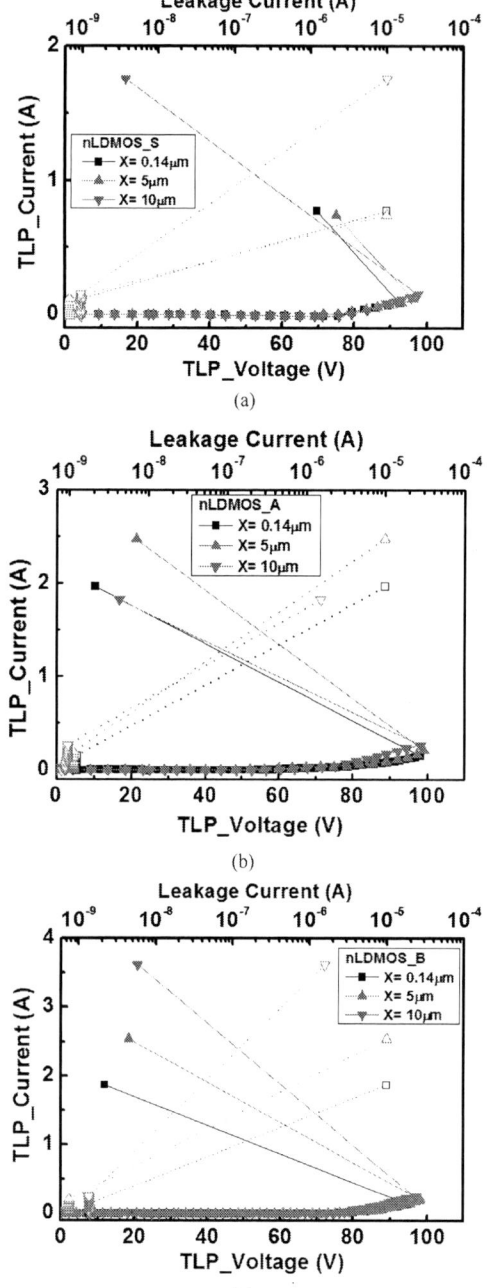

Fig. 9. The TLP-measured *I-V* characteristics of (a) nLDMOS_S, (b) nLDMOS_A, and (c) nLDMOS_B, with different distances X of 0.14, 5, and 10 μm, respectively.

IV. CONCLUSION

The test devices for self-protected HV MOSFET have been investigated with different DPW test structures in a 0.25-μm 60-V BCD process. According to the measurement results, the nLDMOS_B with slotted DPW structure can maintain high breakdown voltage for wide eSOA and get better ESD robustness at the same time. Therefore, nLDMOS with the appropriate drain-side engineering is a useful technique for self-protection ESD design in HV integrated circuits.

TABLE II
THE MEASURED RESULTS AMONG DIFFERENT TEST DEVICES

Device (V_{GS}=0V)	TLP measurement I_{t2} (A)	ESD tester	
		HBM (kV)	MM (V)
nLDMOS_S (X=0.14μm)	0.09	0.3	< 50
nLDMOS_S (X=5μm)	0.12	0.4	50
nLDMOS_S (X=10μm)	0.15	0.5	50
nLDMOS_A (X=0.14μm)	0.16	0.5	50
nLDMOS_A (X=5μm)	0.21	0.6	100
nLDMOS_A (X=10μm)	0.27	0.8	100
nLDMOS_B (X=0.14μm)	0.14	0.5	50
nLDMOS_B (X=5μm)	0.21	0.6	100
nLDMOS_B (X=10μm)	0.25	0.7	100

ACKNOWLEDGMENT

The authors would like to thank the National Chip Implementation Center (CIC), Hsinchu, Taiwan, for the support of chip fabrication, and the Hanwa Electronic Ind. Company, Ltd., Japan, for the support of transmission line pulse (TLP) equipment. This work was partially supported by National Science Council (NSC), Taiwan, under Contracts of NSC 101-2220-E-009-020, NSC 101-2221-E-009-141, and NSC 101-3113-P-110-004.

REFERENCES

[1] B. Murari, F. Bertoti, and G. A. Vignola, *Smart Power ICs: Technologies and Applications.* Berlin, Germany: Springer-Verlag, 2002.

[2] J. Webster, *Reliability Issues of Power Devices*, Wily Encyclopedia of Electrical and Electronics Engineering, 2007.

[3] M. P. J. Mergens, W. Wilkening, S. Mettler, H.wolf, A.Stricker, and W. Fichtner, "Analysis of lateral DMOS power devices under ESD stress condition," *IEEE Trans. on Electron Devices*, vol. 47, no.11, pp. 2128-2137, Nov. 2000.

[4] W.-Y. Chen and M.-D. Ker, "Improving safe operating area of an nLDMOS array with an embedded silicon controlled rectifier for ESD protection in a 24-V BCD process," *IEEE Trans. Electron Devices*, vol. 58, no. 9, pp. 2944-2951, Sep. 2011.

[5] W.-Y. Chen and M.-D. Ker, "Characterization of SOA in time domain and the improvement techniques for using in high-voltage integrated circuits," *IEEE Trans. Device and Mater. Reliab.*, vol. 12, no. 2, pp. 382-390, Jun. 2011.

[6] S. Voldman, "Smart power, LDMOS, and BCD technology," in *ESD: Failure Mechanisms and Models.* Hoboken, NJ: Wiley, 2009.

[7] C.-T. Dai, P.-Y. Chiu, M.-D. Ker, F.-Y. Tsai, Y.-H. Peng, and C.-K. Tsai, "Failure analysis on gate-driven ESD clamp circuit after TLP stresses of different voltage steps in a 16V CMOS process," in *Proc. of International Symp. on Physical and Failure Analysis of Integrated Circuits (IPFA)*, 2012.

[8] A. W. Ludikhuize, "A review of RESURF technology," in *Proc. IEEE Int. Symp. Power Semiconductor Devices and ICs*, 2000, pp. 11-18.

[9] V. Parthasarathy, V. Khemka, R. Zhu, J. Whitfield, A. Bose, and R. Ida, "A double RESURF LDMOS with drain profile engineering for improved ESD robustness," *IEEE Trans. Electron Device Lett.*, vol. 23, no.4, pp. 212-214, Apr. 2002.

978-1-4673-4845-4/13 $31.00 © 2013 IEEE

A Test Structure for Analysis of Temperature Distribution in CMOS LSI with Sensing Device Array

T. Matsuda[1], *Member, IEEE*, H. Hanai[1], H. Iwata[1], D. Kondo[1], T. Hatakeyama[1], M. Ishizuka[1]
and T. Ohzone[2]

[1]Department of Information Systems Engineering, Toyama Prefectural University, Imizu, Japan.
matsuda@pu-toyama.ac.jp
[2]Dawn Enterprise, Nagoya, Japan.

Abstract— **A test structure for analysis of temperature distribution in CMOS LSI is presented. Fundamental thermal properties of LSI chip were measured and discussed with simulation results. The test structure consists of 24 sensor blocks, each of which has a resistor as an on-chip heater, a p-n diode array for temperature sensing and selector switches. Dependence of heating time and distance from the resistor were analyzed as well as transient phenomena. The test structure can provide an effective methodology for analysis of fundamental thermal properties in LSIs packaged in various ways.**

Keywords—CMOS; temperature distribution; on chip sensor

I. INTRODUCTION

With progress of CMOS device and packaging technology, the increase of device density in LSI systems causes the larger power consumption and thus the higher temperature in a chip [1]. Since the temperature-related effects have an impact on performance and reliability of LSI circuits, analysis of temperature distribution in CMOS LSI becomes an important issue [2-3]. Though thermal simulators provide a powerful LSI design tool, verification of simulation results with experimental measurements is essential to accurate analysis. In this paper, a test structure for analysis of temperature distribution in CMOS LSI is presented and fundamental thermal properties of LSI chip are measured and discussed.

II. EXPERIMENTS

A. Test Structure

Fig. 1 shows a schematic of a sensing block for temperature distribution measurement. Each block, of which size is 260 x 60 μm^2, has a resistor of about 500 Ω as an on-chip heater, 32 p-n diodes for temperature sensing and sensor selector switches. P type diffusion resistor and p type polysilicon resistor are used for the heaters, and their cross sections are given in Fig. 2 (a) and (b), respectively. A photograph of a sensor block in a test chip is given in Fig. 3. According to the addressing signals from X and Y decoders, the selected switch connects force and sense lines to one of the 32 sensor diodes. Since a heater resistor has a certain area rather than a thermal point source, distance *L* is defined as an average of distances from four points placed evenly in the resistor as shown in Fig. 4. Fig. 5 shows a circuit diagram of the selector switch and the sensor diode, which is formed between p-diffusion Source/Drain and n-Well regions.

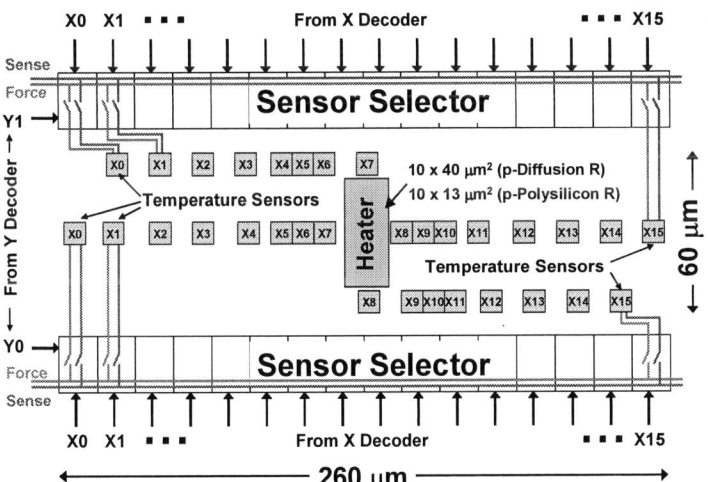

Fig. 1. Schematic of a sensing block for temperature distribution measurement. It consists of a resistor as an on-chip heater, 32 p-n diodes for temperature sensing and selector switches.

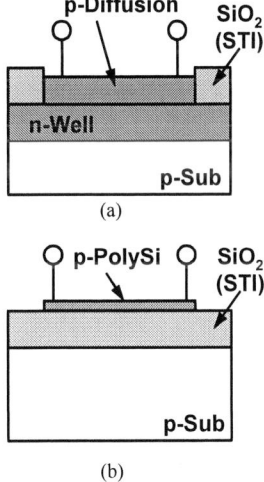

Fig. 2. Cross section of (a) p type diffusion resistor and (b) p type polysilicon resistor.

978-1-4673-4845-4/13 $31.00 © 2013 IEEE

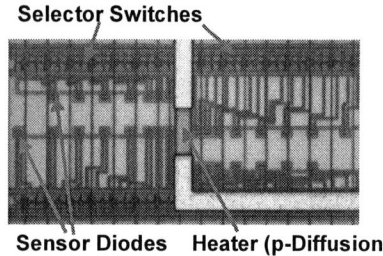

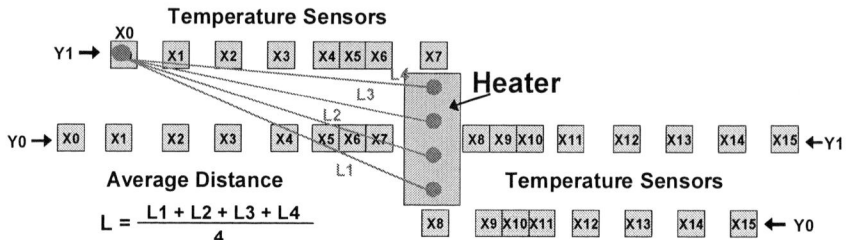

Fig. 3. Photograph of a sensor block in a test chip.

Fig. 4. Definition of distance L is an average of distances from four points placed evenly in the resistor.

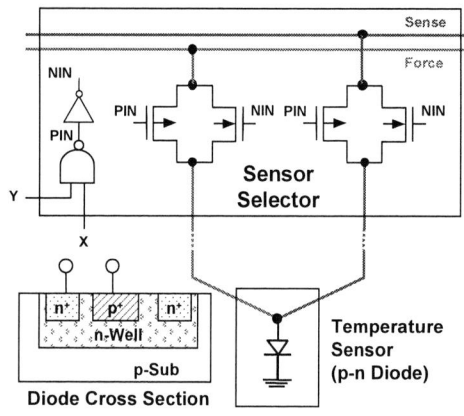

Fig. 5. Circuit diagram of the selector switches and the sensor diode with force and sense lines for accurate I - V measurements.

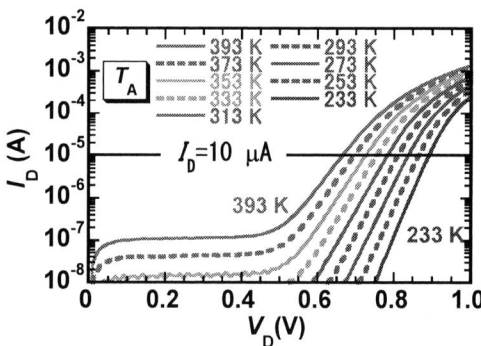

Fig. 7. Example of I_D - V_D characteristics of sensor diode measured without on-chip heating in a constant temperature chamber over a range from 233 to 393 K.

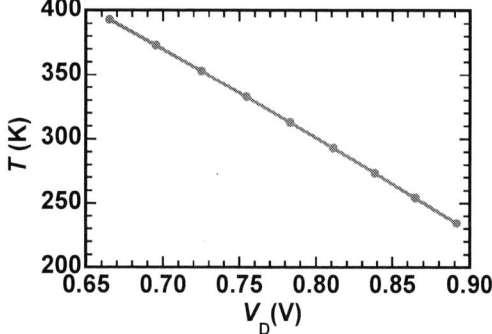

Fig. 8. Sensor diode temperature T versus diode voltage V_D at the constant current of 10 µA.

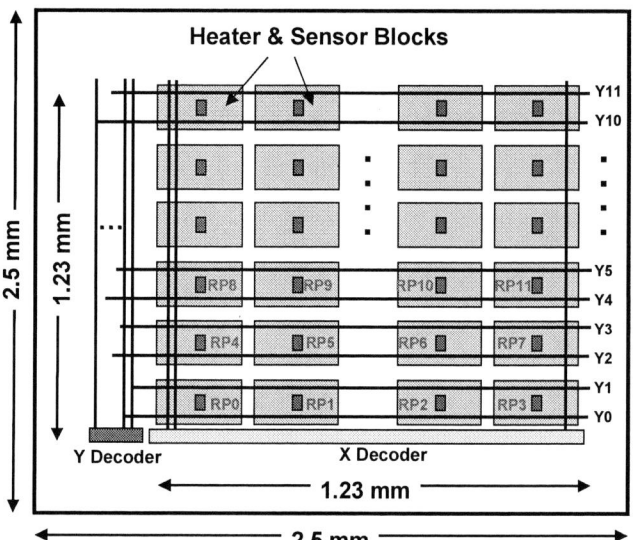

Fig. 6. Test structure chip floor plan, which has 24 blocks (4 rows by 6 columns) of 1.23 x 1.23 mm² in size

Transmission gates switch the force and sense lines, which are shorted at the anode electrode of the diode for accurate I - V measurements. Fig. 6 shows a chip floor plan, which has 24 blocks (4 rows by 6 columns) of 1.23 x 1.23 mm² in size. The test chips were fabricated by a standard 0.18 µm CMOS process and assembled into ceramic 80 pin QFPs with metal covers.

B. Temperature Measurement Procedure

Fig. 7 shows an example of current - voltage (I_D - V_D) characteristics of sensor diode measured without on-chip heating in a constant temperature chamber over a range from 233 to 393 K. The temperature step is 20 K, and the wait time for the measurements at each step is 20 min. to keep stable thermal environment. The sensor diode temperature T is proportional to the diode voltage V_D at the constant current I_D of 10 µA as shown in Fig. 8. The diode power consumption of about 7 - 9 µW is low enough to suppress self-heating influence. The specific sensor temperature under the on-chip heating condition can be determined by the T - V_D curves, which were obtained by the measurement of every diode in the sensor blocks in advance.

978-1-4673-4845-4/13 $31.00 © 2013 IEEE

III. Experimental Results

Fig. 9 shows time dependent temperature changes of the sensor diode adjacent to the heater at the different values of dissipated power. ΔT (difference from the ambient temperature T_A) in Fig. 10 gives similar time dependence regardless of T_A, which is equal to the chamber temperature of 233 to 373 K. Since the temperature change from the start of on-chip heating becomes saturated after 200 s, the temperature for the steady state was measured at 250 s.

Temperature distributions in a chip are analyzed under various conditions. Fig. 11 summarizes distance dependences of temperature for both diffusion and polysilicon resistors with 110 mW power dissipation at $T_A = 293$ K. The temperature decreases with the distance L rapidly, and saturates around 100 μm. Note that the polysilicon resistor formed on the SiO_2 layer has the same L dependence as the diffusion resistor in our test structure. The temperature T is proportional to the reciprocal of L regardless of the heater power as shown in Fig. 12, and is expressed by an empirical equation: $T = T_0 + \alpha \cdot (1/L)$, where T_0 and α are the temperature at $L = \infty$ and a constant, respectively. Since α is also proportional to the heater power P_R as given in Fig. 13, T can be rewritten in the form of $T = T_0 + \beta \cdot P_R \cdot (1/L)$, where β is a constant of 1.26 (K·μm)/mW.

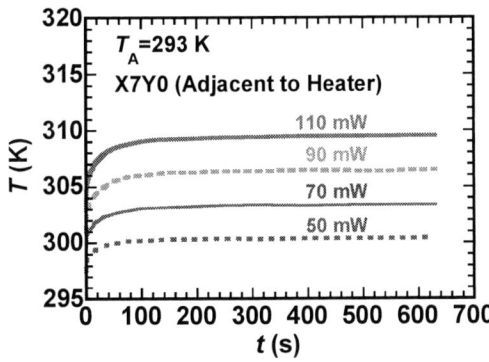

Fig. 9. Time dependent temperature changes of the sensor diode adjacent to the heater at the different values of dissipated power.

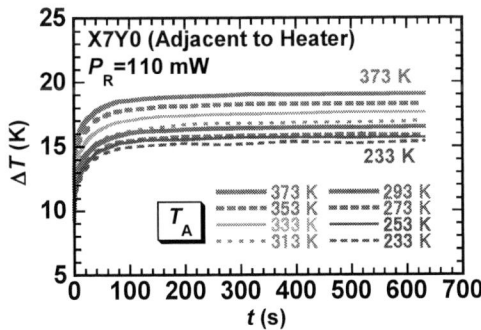

Fig. 10. Time dependence of ΔT, which is a temperature difference from the ambient temperature T_A

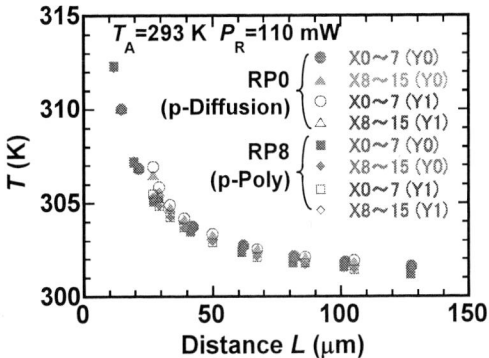

Fig. 11. Distance dependences of temperature T for both diffusion and polysilicon resistors at T_A=293 K.

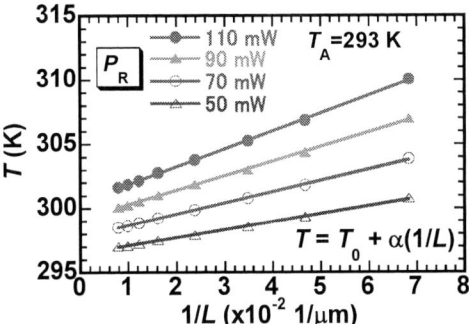

Fig. 12. T versus $(1/L)$ with an empirical equation : $T = T_0 + \alpha(1/L)$.

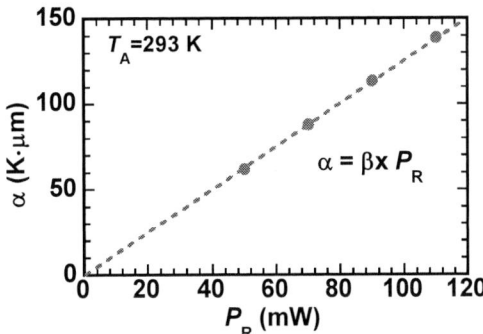

Fig. 13. Heater power P_R dependence of proportional constant α in Fig. 12.

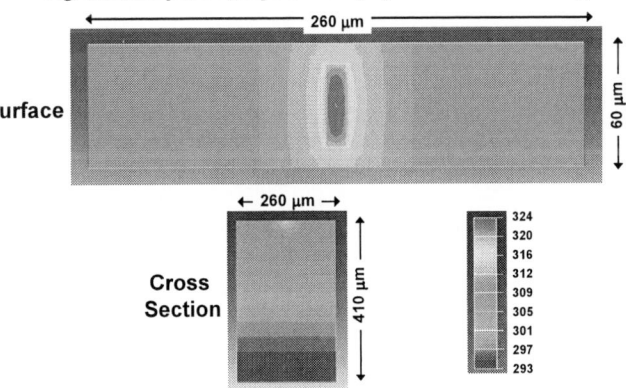

Fig. 14. Temperature distribution with 110 mW heating of p-diffusion resistor simulated by Ansys Icepak ver. 14.0.

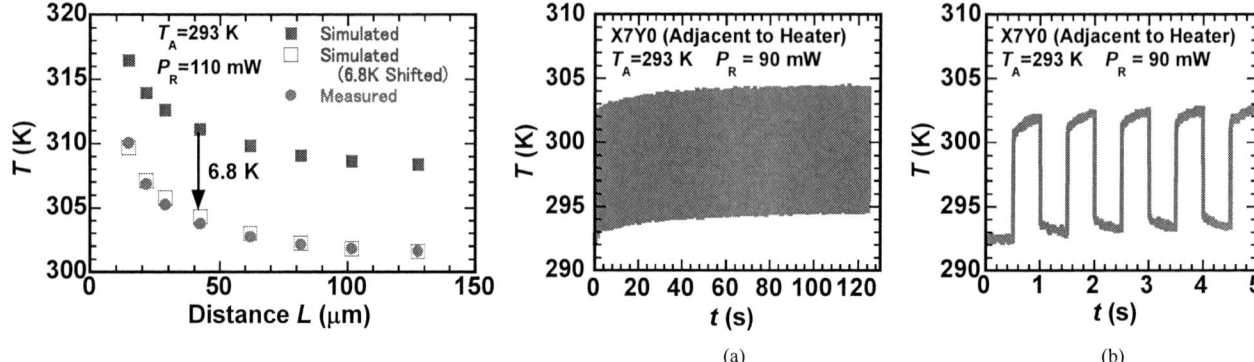

Fig. 15. Comparison of simulated and measured results of the temperature change dependent on L.

Fig. 16 Time dependent temperature change under the pulsed heating in the time range of (a) 0 to 125 s, and (b) the first 5 s.

Fig. 14 shows a temperature distribution with 110 mW heating of p-diffusion resistor simulated by Ansys Icepak ver. 14.0. The red portion indicates resistor region and the temperature propagates in the surrounding area. Fig. 15 gives a comparison of simulated and measured results of the temperature change dependent on L. In spite of its simple modeling, the simulation produces an L dependence of temperature similar to the measured result except for the higher temperature of about 6.8 K as shown by open/filled square markers in Fig. 15.

To analyze thermal transient phenomena in a LSI chip, voltage pulses were applied to the heater resistor and V_D changes were measured under the constant current condition of 10 μA. Fig. 16 shows a time dependent temperature change under the pulsed heating in the time range of (a) 0 to 125 s, and (b) the first 5 s. The pulse power of the heater during ON state was set to 90 mW, and the pulse period, width and rise/fall times were 1000, 500 and 1 ms, respectively. The sensor temperature changes abruptly at the heater switching, and then it increases (decreases) gradually while the heater power is ON (OFF). Fig. 17 shows a temperature change during the first 10 s at the different distance of about 15 to 1077 μm. The swing of the measured temperature pulse is proportional to (1/L) as shown in Fig. 18. The result of the pulse response suggests that the measured thermal transient phenomena comprise fast (much less than 1 ms) and relatively slow processes.

IV. CONCLUSIONS

The temperature distributions and transient phenomena in CMOS LSI were analyzed with the proposed test structure, which can provide an effective methodology for analysis of fundamental thermal properties in LSIs. The temperature is proportional to the reciprocal of the distance from the heat source, and an empirical equation is proposed. The pulse response suggests that the measured thermal transient phenomena comprise both fast and slow processes.

ACKNOWLEDGMENT

This work is supported by VLSI Design and Education Center (VDEC), the University of Tokyo in collaboration with Cadence Design Systems, Inc., Mentor Graphics, Inc., Rohm Corporation and Toppan Printing Corporation.

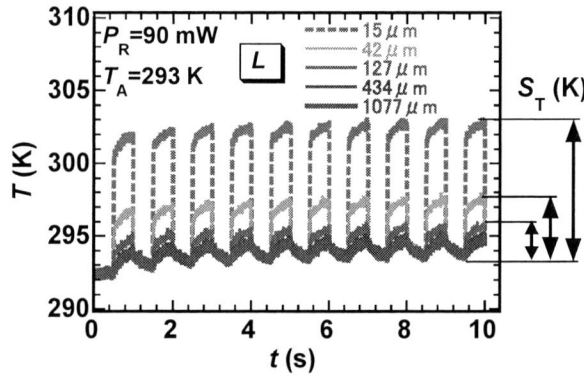

Fig. 17. Temperature change during the first 10 s at the different distance of about 15 to 1077 μm.

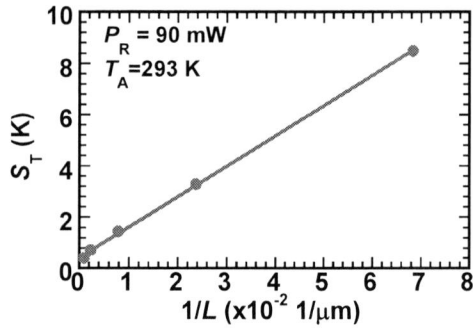

Fig. 18. (1/L) dependence of the swing S_T of the measured temperature pulse.

REFERENCES

[1] A. Shakouri, S.-M. Kang, A. Bar-Cohen, and B. Courtois, "Scanning the Special Issue on On-Chip Thermal Engineering," Proceedings of the IEEE, vol. 94, No. 8, pp. 1473-1475, 2006.

[2] N. Minas, G. Van der Plas, H. Oprins, Y. Yang, C. Okoro, A. Mercha, et al, "Test Structures for Characterization of Thermal-Mechanical Stress in 3D Stacked IC for Analog Design," Proc. IEEE Int. Conf. on Microelectronic Test Structures, pp. 140-144, 2010.

[3] R. Koh, and T. Iizuka, "Self-heating Parameter Extraction of Power MOSFETs Based on Transient Drain Current Measurements and on the 2-cell Self-heating Model," IEEE Int. Conf. on Microelectronic Test Structures, pp. 191-195, 2012.

SESSION 7: Parameter Extraction

978-1-4673-4845-4/13 $31.00 © 2013 IEEE

978-1-4673-4845-4/13 $31.00 © 2013 IEEE

Analysis of Narrow Gate to Gate Space Dependence of MOS Gate-Source/Drain Capacitance by Using Contact-less and Drawn-out Source/Drain Test Structure

Yasuhisa Naruta[1] and Shigetaka Kumashiro[2]

Renesas Electronics Corporation

[1]5-20-1 Josuihon-cho, Kodaira, Tokyo 187-8588, Japan

[2]1753 Shimonumabe, Nakahara-ku, Kawasaki, Kanagawa 211-8668, Japan

[1]yasuhisa.naruta.xk@renesas.com, [2]shigetaka.kumashiro.zj@renesas.com

Abstract—**A new test structure which can provide voltage to the very narrow source/drain region between adjacent gates by drawing out the source/drain silicide layer has been developed. By using the test structure, the dependence of the gate-drain capacitance (*Cgd*) on the gate-gate space (*Lsp*) has been successfully measured until the minimum gate pitch where no contact can be placed. Decrease of *Cgd* with respect to the decrease of *Lsp* has been observed and its main cause is identified as the decrease of the gate-drain overlap length.**

Keywords—test structure; gate-drain capacitance; gate to gate space; contact-less; drawn-out drain; gate-drain overlap length

I. INTRODUCTION

As process technology node proceeds, not only MOSFET gate length but also gate-gate space becomes very short due to device structure scaling. The decrease of gate-gate space inevitably leads to the decrease of the distance between the gate and the source/drain contacts, hence the parasitic capacitance between gate and source/drain increases. Parasitic capacitances such as gate-contact capacitance should be separated from elemental gate-source/drain capacitance in a compact MOSFET model and should be handled by LPE in order to maintain the generality of the compact MOSFET model. On the other hand, extraction of the elemental gate-source/drain capacitance for advanced process technology is difficult because the gate-contact capacitance becomes significant [1]. Moreover, in NAND and/or/ NOR circuits, measurement of the elemental gate-source/drain capacitance is also difficult because the voltage of the shared source/drain between the series transistors cannot be monitored through contact.

In this paper, a new contact-less and drawn-out source/drain test structure has been developed to measure elemental gate-source/drain capacitance without any contact. By using the test structure, gate pitch dependence of the elemental gate-source/drain capacitance has been successfully measured. Furthermore, origin of the gate pitch dependence of the gate-source/drain capacitance has been clarified by measuring the gate leak current at the gate-source/drain overlap.

II. TEST STRUCTURE DESIGN

As shown in Fig. 1, the test structure has source/drain silicide layer drawn out longer than the poly gate edge. The measurement terminals are the gate, the shorted source/drain and the substrate. Since there is no need to deploy any contact plug between the gates, elemental gate-source/drain capacitance can be measured even for minimum gate pitch layout.

III. MEASUREMENT & EXTRACTION

Gate-source/drain capacitance (*Cgs*+*Cgd*) is measured by using LCR-meter (Agilent 4284A) as shown in Fig. 2 To evaluate the gate-source/drain overlap length, gate leak current between the gate and the shorted source/drain (*Igs*+*Igd*) is also measured as shown in Fig. 3.

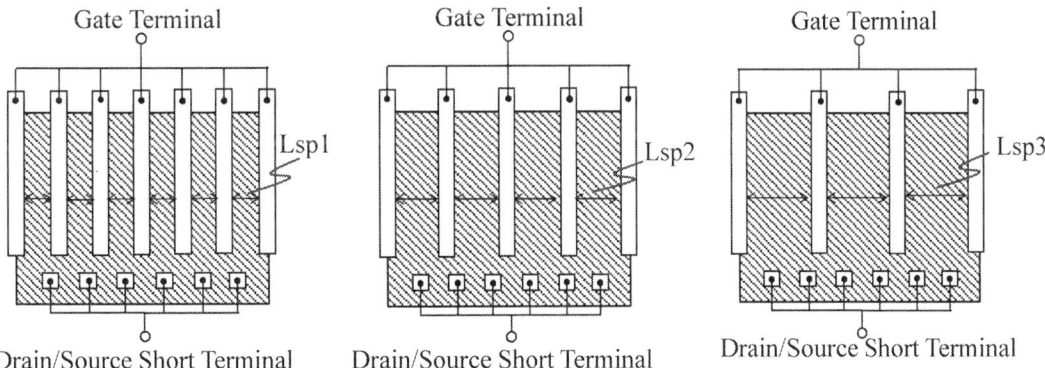

Fig. 1 Test structure for gate space dependence (Lsp1 < Lsp2 < Lsp3). This structure has source/drain silicide layer drawn out longer than the poly gate.

The gate-drain capacitance (*Cgd*) can be decomposed into three components as (1), i.e. the gate-to-drain overlap capacitance (*Cov*), the gate-to-diffusion outer-fringing capacitance (*CoutF*), and the gate-to-diffusion inner-fringing capacitance (*CinF*) as shown in Fig. 4.

$$Cgd = CoutF + Cov + CinF \big|_{Vg=Vd=Vb=0} \quad (1)$$

As shown in Fig. 5, since *CinF* appears together with the expansion of the depletion layer at the bulk surface, it can be decreased to a negligible value by applying flat-band voltage (-*Vfb*) to the gate as (2).

$$Cgd = CoutF + Cov \big|_{Vg=-Vfb,Vd=Vb=0} \quad (2)$$

As shown in Fig. 6, *CoutF* is expected to decrease with respect to the decrease of the gate-gate space. As for *Cov*, it was reported that overlap length (i.e. *Cov*) increases with respect to the decrease of the gate-gate space [2]. On the other hand, *CinF* is expected to be invariant with respect to the gate-gate space as long as the source/drain junction depth is kept constant. Fig. 7 shows *Cgd* dependence on gate-gate space. *Cgd* smoothly decreases with the decrease of the gate-gate space until minimum gate pitch.

IV. SEPARATION METHOD OF COV, COUTF AND CINF

Fig. 8 shows an extraction flow of *Cov*, *CoutF* and *CinF*. Here, the flow is explained step by step.

Step1: Extraction of *TOXE* and *CoutF@(Lsp=Lsp_ref)*

First, the electrical gate oxide thickness (*TOXE*) is extracted from large gate area pattern. Then, CoutF at a standard gate-gate space (*Lsp=Lsp_ref*) is calculated by using 3D capacitance simulator.

Step 2: Extraction of *Lov@(Lsp=Lsp_ref)*

Cov is expressed as (3), where *Lov* is the gate-drain overlap length, ε_r is the relative permittivity of SiO_2, ε_o is the permittivity of vacuum.

$$Cov = Lov \cdot \frac{\varepsilon_r \cdot \varepsilon_o}{TOXE} \quad (3)$$

From (1) and (3), *Lov* when *Vg=-Vfb* is expressed as (4).

$$Lov = (Cgd - CoutF) \cdot \frac{TOXE}{\varepsilon_r \cdot \varepsilon_o} \quad (4)$$

Lov@(Lsp=LSP_ref) can be calculated by using *CoutF* already

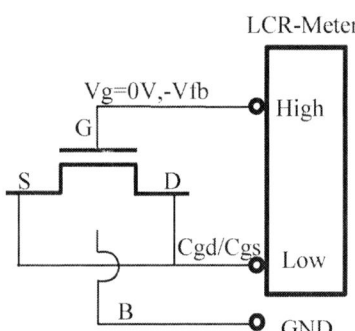

Fig. 2 Simplified Gate to Drain/Source capacitance (*Cgd/Cgs*) measurement schematics for MOSFET.

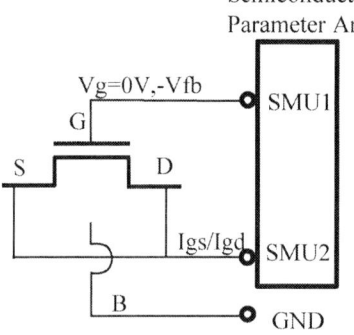

Fig. 3 Simplified Gate to Drain/Source leak current (*Igd/Igs*) measurement schematics for MOSFET.

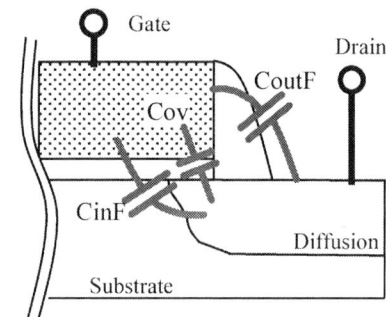

Fig. 4 Components of the Gate to Drain capacitance (*Cov, CoutF, CinF*)

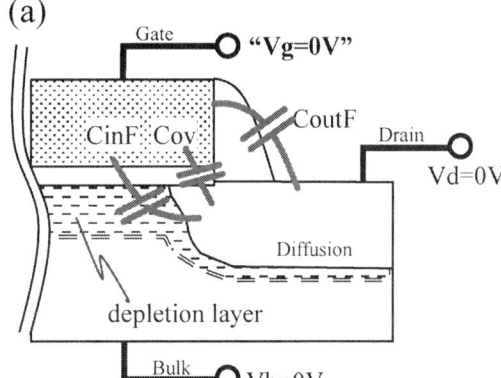

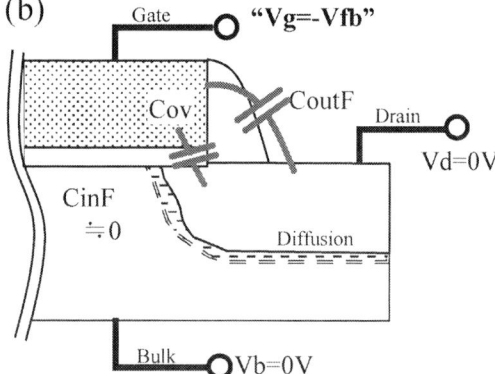

Fig. 5 Capacitance change in three-terminal MOS structure. (a) *CinF* appears together with the expression of the depletion layer at the bulk surface at "*Vg=0V*". (b) *CinF* decreases to a negligible value at "*Vg=-Vbf*".

978-1-4673-4845-4/13 $31.00 © 2013 IEEE

(a) Narrow Lsp

(b) Wide Lsp

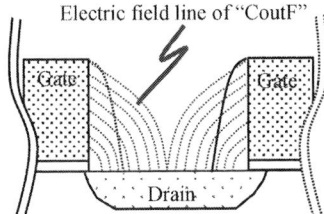

Fig. 6 Sectional views of outer-fringing (*CoutF*) electric field line of MOSFET. (a) Narrow *Lsp*; (b) Wide *Lsp*.

obtained.

Step 3: Extraction of *Lov@(Lsp=x)*

Since *Igd* is proportional to *Lov*, *Lov* at arbitrary *Lsp* (i.e. *Lov(x)*, where *Lsp=x*) can be extracted from *Lov(Lsp_ref)* and *Igd(Lsp_ref)* as (5) [3].

$$Lov(x) = Lov(Lsp\_ref) \cdot \frac{Igd(x)}{Igd(Lsp\_ref)} \qquad (5)$$

Gate-gate space dependence of *Lov* obtained by (5) is shown in Fig. 9. From *Lsp_ref* to minimum gate pitch, *Lov* shrinks by 2.8nm. The shrink is opposite to the expansion of *Lov* observed in [2].

Step 4: Extraction of *CoutF, CinF*

From *Lov* at each *Lsp*, *CoutF* is calculated from (1) and (3) as (6).

$$CoutF = Cgd - Lov \cdot \frac{\varepsilon_r \cdot \varepsilon_o}{TOXE} \qquad (6)$$

CinF@(Vg=0) can be calculated as a difference between (1) and (2).

Gate-gate space dependence of *Cov*, *CoutF* and *CinF* obtained through these steps are shown in Fig. 10. The reason why *Cgd* decreases with respect to the decrease of *Lsp* is not due to the decrease of *CoutF* but due to the decrease of *Cov* (i.e.

Lov). The origin of this phenomenon is probably due to the decrease of total source/drain-extension implantation dose as *Lsp* decreases. The smaller the dose becomes, the shorter the *Lov* becomes.

Step 1: (i) Extract the electrical gate oxide thickness (TOXE) by using large gate area pattern.

:(ii) Calculate CoutF at a standard gate-gate space (Lsp=Lsp_ref) by using 3D capacitance simulator.

Step 2: Calculate "Lov@(Lsp=Lsp_ref)" by using (4).

Step 3: Extract "Lov(x)" at Lsp=x by using (5). And calculate" Cov(x)" by using (3) from Lox(x).

Step 4: Calculate "CoutF(x)" at Lov(x) by using (6).

Step 5: Calculate CinF as a difference between (1) and (2).

Fig. 8 Flow-chart of extracting *Lov, Cov, CoutF* and *CinF*.

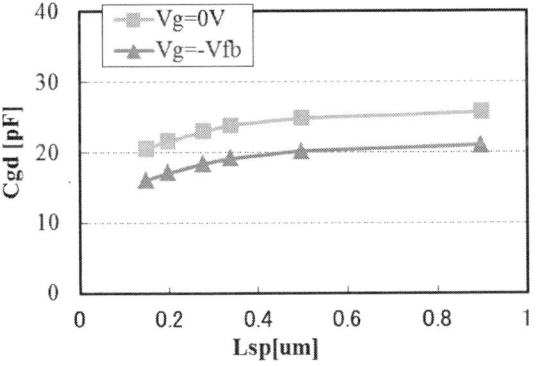

Fig. 7 The Gate space dependence of *Cgd@(Vg=0 and Vg=-Vfb)* of NMOSFET.

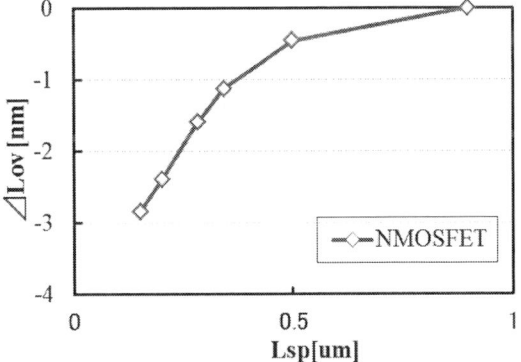

Fig. 9 Gate space dependence of *ΔLov* of NMOSFET.

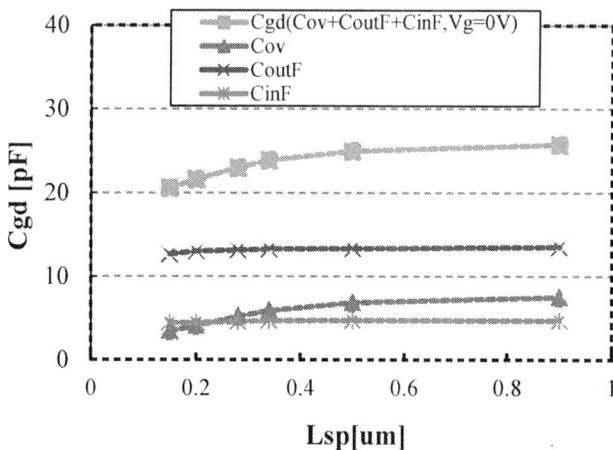

FIG. 10 Gate space dependence of *Cgs@(Vg=0), Cov, CoutF, CinF* of NMOSFET.

V. CONCULUSIONS

A new test structure which can provide voltage to the very narrow source/drain region between adjacent gates by drawing out the source/drain silicide layer has been developed. By using the test structure, *Cgd* dependence on *Lsp* has been successfully measured until the minimum gate pitch where no contact can be placed. Decrease of *Cgd* with respect to the decrease of *Lsp* has been observed and its main cause is identified as the decrease of *Lov*.

ACKNOWLEDGMENT

The authors would like to thank Mr. Takahiro Iizuka of Hiroshima University for his help and valuable discussions held while he was in Renesas Electronics Corporation.

REFERENCES

[1] A. Khakifrooz and D. Antoniadis, "MOSFET Performance Scaling—Part II: Future Directions," IEEE Trans. Electron Devices, vol. 55, pp. 1401–1408, 2008.

[2] H. Tsuno, K. Anzai, M. Matsumura, S. Minami, A. Honjo, H. Koike, Y. Hiura, A. Takeo, W. Fu, Y. Fukuzaki, M. Kanno, H. ansai and N. Nagashima, "Advanced Analysis and Modeling of MOSFET Characteristics Fluctuation Caused by Layout Variation," Symp. on VLSI Technology, pp. 204-205, Kyoto, 2007.

[3] S. Kim, S. Narasimha, and K. Rim, "A New Method to Determine Effective Lateral Doping Abruptness and Spreading-Resistance Components in Nanoscale MOSFETs," IEEE Trans. Electron Devices, vol. 55, pp. 1035–1041, 2008.

Three- and four-point Hamer-type MOSFET parameter extraction methods revisited

Kjell O. Jeppson, *Senior Member, IEEE*
Chalmers University of Technology
Department of Microtechnology and Nanoscience
Gothenburg, Sweden
jeppson@chalmers.se

Abstract— In this paper the three-point Hamer type and four-point Karlsson & Jeppson type MOSFET parameter extraction methods are revisited concerning robustness and selection of data points. The method for fitting models described by rational functions to measured data proposed by Hamming is also discussed and it is shown how this method calculates its weighted data points. An alternative method where MOSFET resistance values are used instead of current values for the extraction procedure is also investigated in an attempt to increase extraction method robustness. Finally, it is shown how the three point extraction method can be applied not only to the triode region but also to the MOSFET saturation region for separating parameters for the body effect and the velocity saturation.

Keywords—Direct parameter extraction, MOSFET models;

I. INTRODUCTION

The three-point model parameter extraction method proposed by Hamer [1] is a popular method for rapid MOSFET triode region extraction of model parameter. By this method, values for the three main triode region model parameters can be extracted from as few as three experimental data points. Guidelines for how to choose the data points for most reliable extraction has been empirically established by many workers. The method has been extended by Karlsson and Jeppson to include deep submicron models by adding a second-order mobility roll-off parameter [2, 3]. On previous ICMTS conferences, Karlsson and Jeppson have presented studies on how data point selection and measurement noise could possibly influence the reliability of the extracted model parameter values [4]. They also showed how the method proposed by Hamming [5] for fitting models described by rational functions to measured data could be used. This method yields a result that is very close to a least-square fitting of model to data.

A least-square fitting procedure includes a process of reducing the N equations described by N data points obtained from measurements to the number of model parameters. This paper contains a study of the result of this process, i.e. studies of where these weighted "data points" are located on the model curve. The result for the three-point method is very close to the expected with two closely related data points just above the threshold voltage and one data point close to V_{DD}. However, somewhat unexpected there is also a second solution with only one data point close to V_T and the two closely located data points near V_{DD}. For the four-point method the weighted data points first appear to be evenly distributed between V_T and V_{DD}. However, a closer study reveals two different scenarios: in the first scenario two data points appear close to V_T and two data points appear close to V_{DD}, while in the second scenario three data points appear very close either just above V_T or just below V_{DD}. This finding suggests that the method could be sensitive to measurement noise, a finding that is confirmed in the study presented in section F.

Finally, a new procedure is presented allowing the three-point method to be used not only in the triode region, but also in the saturation region. The suggested method provides an attractive means for obtaining drain-induced barrier-lowering parameters for the threshold voltage, and for separating the body effect described by the body parameter α from the velocity saturation parameter $V_C=2LE_{sat}$. The method has been used in microelectronics and integrated circuit courses providing students with a simple yet surprisingly accurate tool for obtaining MOSFET models accurate enough for propagation delay modeling and first estimate analog designs.

II. PARAMETER EXTRACTION: THEORY AND PROCEDURE

A. Background of three-point extraction method

The three-point direct extraction method was first proposed in 1986 by Hamer [1]. It relies on a simple three-parameter MOSFET model for the linear region

$$I_{DS} = k\frac{V_{GS} - V_T - V_{DS}/2}{1+\theta(V_{GS}-V_T)}V_{DS}, \tag{1}$$

where k, V_T and θ are the three parameters to be extracted. The general form of this equation is a rational function,

$$I_{DS} = \frac{a+bV_{GS}}{1+cV_{GS}}, \tag{2}$$

where

$$\begin{cases} a = -b(V_T + V_{DS}/2) \\ b = kV_{DS}/(1-\theta V_T) \;, \\ c = \theta/(1-\theta V_T) \end{cases} \tag{3}$$

The parameters a, b, and c can be determined from three measurements of (I_{DS}, V_{GS}) by solving the following linear system of equations

$$\begin{pmatrix} 1 & V_{GS1} & -I_{DS1}V_{GS1} \\ 1 & V_{GS2} & -I_{DS2}V_{GS2} \\ 1 & V_{GS2} & -I_{DS3}V_{GS3} \end{pmatrix}\begin{pmatrix} a \\ b \\ c \end{pmatrix} = \begin{pmatrix} I_{DS1} \\ I_{DS2} \\ I_{DS3} \end{pmatrix}. \tag{4}$$

978-1-4673-4845-4/13 $31.00 © 2013 IEEE

By using built-in matrix inversion and multiplication formulas in for instance Microsoft Excel this is a simple task. We obtain

$$\begin{pmatrix} a \\ b \\ c \end{pmatrix} = \begin{pmatrix} 1 & V_{GS1} & -I_{DS1}V_{GS1} \\ 1 & V_{GS2} & -I_{DS2}V_{GS2} \\ 1 & V_{GS2} & -I_{DS3}V_{GS3} \end{pmatrix}^{-1} \begin{pmatrix} I_{DS1} \\ I_{DS2} \\ I_{DS3} \end{pmatrix}. \tag{5}$$

From a, b, and c the MOSFET model parameters can be extracted according to the following

$$\begin{cases} k = \dfrac{b/V_{DS}}{1+cV_T} \\ V_T = -a/b - V_{DS}/2, \\ \theta = \dfrac{c}{1+cV_T} \end{cases} \tag{6}$$

Many authors have discussed the importance of selecting the three data points for minimum error and best fit between model and data. For mismatch and variability studies, for instance, it is important that V_T variations are not due to the selection of data points nor to model imperfections [6].

B. Background of four-point extraction method

For deep sub-micron (DSM) devices a second order mobility roll-off parameter was introduced to improve modeling in the BSIM2 MOSFET model [7]. For rapid extraction of the four model parameters, Karlsson and Jeppson introduced the four-point extraction method [2, 3]. The new model was written on the following form:

$$I_{DS} = k \frac{V_{GS} - V_T - V_{DS}/2}{1 + \theta_1(V_{GS} - V_T) + \theta_2(V_{GS} - V_T)^2}. \tag{7}$$

The generic form of this model is given by

$$I_{DS} = \frac{a + bV_{GS}}{1 + cV_{GS} + dV_{GS}^2}, \tag{8}$$

where parameters a, b, c, and d can be solved from four experimental data points (I_{DS}, V_{GS}) by solving the following linear system of equations:

$$\begin{pmatrix} 1 & V_{GS1} & -I_{DS1}V_{GS1} & -I_{DS1}V_{GS1}^2 \\ 1 & V_{GS2} & -I_{DS2}V_{GS2} & -I_{DS2}V_{GS2}^2 \\ 1 & V_{GS3} & -I_{DS3}V_{GS3} & -I_{DS3}V_{GS3}^2 \\ 1 & V_{GS4} & -I_{DS4}V_{GS4} & -I_{DS4}V_{GS4}^2 \end{pmatrix} \begin{pmatrix} a \\ b \\ c \\ d \end{pmatrix} = \begin{pmatrix} I_{DS1} \\ I_{DS2} \\ I_{DS3} \\ I_{DS4} \end{pmatrix}. \tag{9}$$

From the extracted values of parameters a, b, c, and d the MOSFET model parameters k, V_T, θ_1, and θ_2 can easily be obtained.

$$\begin{cases} a = -b(V_T + V_{DS}/2) & \Rightarrow k = \dfrac{b/V_{DS}}{1 + cV_T + dV_T^2} \\ b = \dfrac{kV_{DS}}{1 - \theta V_T + \theta_2 V_T^2} & \Rightarrow V_T = -a/b - V_{DS}/2 \end{cases}, \tag{10a}$$

$$\begin{cases} c = \dfrac{\theta - 2\theta_2 V_T}{1 - \theta V_T + \theta_2 V_T^2} & \Rightarrow \theta = \dfrac{c + 2dV_T}{1 + cV_T + dV_T^2} \\ d = \dfrac{\theta_2}{1 - \theta V_T + \theta_2 V_T^2} & \Rightarrow \theta_2 = \dfrac{d}{1 + cV_T + dV_T^2} \end{cases}. \tag{10b}$$

C. Selection of data points

An extensive study of how to select the three gate voltages for drain current measurements for the most reliable parameter extraction has been presented by Karlsson and Jeppson [4]. At several occasions, Tuinhout has pointed out the importance of appropriate data point selection for extracting reliable parameter values useful in mismatch studies [8]. In his work the fixed linear region three-point extraction method has become the standard technique [9]. In this method, an initial value of the threshold voltage must be used for ensuring appropriate gate voltages for drain current measurements. Some implementations of the three- and four-point extraction schemes use an iterative procedure where the last extracted V_T is used to suggest the next set of gate voltage bias points.

D. Least-square fitting techniques

Another approach is to use least-square fitting of the model to a large number of data points. A strategy for fitting rational function models to measured data was proposed by Hamming [5]. The sum of the square errors in each of N data points can be written

$$\varepsilon = \sum_{i=1}^{N} \left[I_{DS} - \frac{a - bV_{GS}}{1 + cV_{GS}} \right]^2 = \sum_{i=1}^{N} \frac{\left[I_{DS}(1 + cV_{GS}) - a + bV_{GS} \right]^2}{\left[1 + cV_{GS} \right]^2} \tag{11}$$

To obtain the parameter values, a, b, and c that minimizes the error, the three derivatives with respect to these parameters are taken and set to zero. This results in a system of three linear equations. Defining the denominator N_i as

$$DN_i = \left[1 + c_{start}V_{GS} + d_{start}V_{GS}^2 \right]^2, \tag{12}$$

we obtain the following system of equations

$$\begin{pmatrix} \sum\limits_{i=1}^{N}\dfrac{1}{DN_i} & \sum\limits_{i=1}^{N}\dfrac{V_{GSi}}{DN_i} & -\sum\limits_{i=1}^{N}\dfrac{I_{DSi}V_{GSi}}{DN_i} \\ \sum\limits_{i=1}^{N}\dfrac{V_{GSi}}{DN_i} & \sum\limits_{i=1}^{N}\dfrac{V_{GSi}^2}{DN_i} & -\sum\limits_{i=1}^{N}\dfrac{I_{DSi}V_{GSi}^2}{DN_i} \\ \sum\limits_{i=1}^{N}\dfrac{I_{DSi}}{DN_i} & \sum\limits_{i=1}^{N}\dfrac{I_{DSi}V_{GSi}}{DN_i} & -\sum\limits_{i=1}^{N}\dfrac{I_{DSi}^2V_{GSi}}{DN_i} \end{pmatrix} \begin{pmatrix} a \\ b \\ c \end{pmatrix} = \begin{pmatrix} \sum\limits_{i=1}^{N}\dfrac{I_{DSi}}{DN_i} \\ \sum\limits_{i=1}^{N}\dfrac{I_{DSi}V_{GSi}}{DN_i} \\ \sum\limits_{i=1}^{N}\dfrac{I_{DSi}^2}{DN_i} \end{pmatrix}. \tag{13}$$

In a first round $c_{start}=d_{start}=0$ could be used, only to be replaced by the extracted values for c and d in a second, and most often, final round. From this system of equations, the solutions for parameters a, b, and c are easily found. If (13) is rewritten as

$$\begin{pmatrix} 1 & \sum\limits_{i=1}^{N}\dfrac{V_{GSi}}{DN_i}/\sum\limits_{i=1}^{N}\dfrac{1}{DN_i} & -\sum\limits_{i=1}^{N}\dfrac{I_{DSi}V_{GSi}}{DN_i}/\sum\limits_{i=1}^{N}\dfrac{1}{DN_i} \\ 1 & \sum\limits_{i=1}^{N}\dfrac{V_{GSi}^2}{DN_i}/\sum\limits_{i=1}^{N}\dfrac{V_{GSi}}{DN_i} & -\sum\limits_{i=1}^{N}\dfrac{I_{DSi}V_{GSi}^2}{DN_i}/\sum\limits_{i=1}^{N}\dfrac{V_{GSi}}{DN_i} \\ 1 & \sum\limits_{i=1}^{N}\dfrac{I_{DSi}V_{GSi}}{DN_i}/\sum\limits_{i=1}^{N}\dfrac{I_{DSi}}{DN_i} & -\sum\limits_{i=1}^{N}\dfrac{I_{DSi}^2V_{GSi}}{DN_i}/\sum\limits_{i=1}^{N}\dfrac{I_{DSi}}{DN_i} \end{pmatrix} \begin{pmatrix} a \\ b \\ c \end{pmatrix} = \begin{pmatrix} \sum\limits_{i=1}^{N}\dfrac{I_{DSi}}{DN_i}/\sum\limits_{i=1}^{N}\dfrac{1}{DN_i} \\ \sum\limits_{i=1}^{N}\dfrac{I_{DSi}V_{GSi}}{DN_i}/\sum\limits_{i=1}^{N}\dfrac{V_{GSi}}{DN_i} \\ \sum\limits_{i=1}^{N}\dfrac{I_{DSi}^2}{DN_i}/\sum\limits_{i=1}^{N}\dfrac{I_{DSi}}{DN_i} \end{pmatrix}.$$

the similarities with (4) are obvious. The second column corresponds to three weighted voltages, and the column on the right-hand side corresponds to three weighted current values based on the measurements. These three data points are not placed on the extracted model line, but they can easily be moved there after some manipulations of the three equations described by the matrix. It is interesting to note that there are two different solutions to this problem as illustrated in Fig. 1: we either find two data points just above V_T and a third close to the supply voltage V_{DD} (filled symbols), or just one data point close to V_T and two data points close by V_{DD} (open symbols).

The same method can be used for finding the four parameters in the DSM model described by (7). Depending on the MOSFET data, we have found some different examples of how this "least-square fitting" method yields its four data points on which the parameter extraction is based.

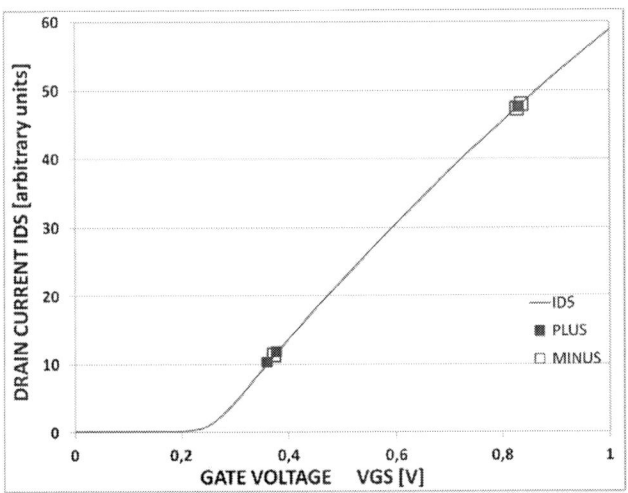

Fig. 1. Illustration of the three-point extraction method.

A good example is shown in Fig. 2 where the four data points are evenly distributed across the measurement range. A not so good example is shown in Fig. 3, where three data points are clustered more or less on the same values.

Here, we can see that the four "original" data points found in the V_{GS} and I_{DS} columns are evenly distributed. However, when the equations are manipulated to place the four data points on the model line, then three data points are clustered either close to V_T or close to V_{DD}. This makes this extraction method very sensitive to measurement noise, as will be illustrated in section F.

E. Extending the extraction method to the saturation region

So far, the three-point Hamer type extraction method has been mostly limited to extraction of model parameters in the linear region. However, similar methods have been shown to be useful for extracting the effective geometries and series resistances of MOS transistors [10, 11].

Nevertheless, up till now the method has not, to our knowledge, been used for extraction of model parameters in the saturation region. But such a method will now be presented in this section.

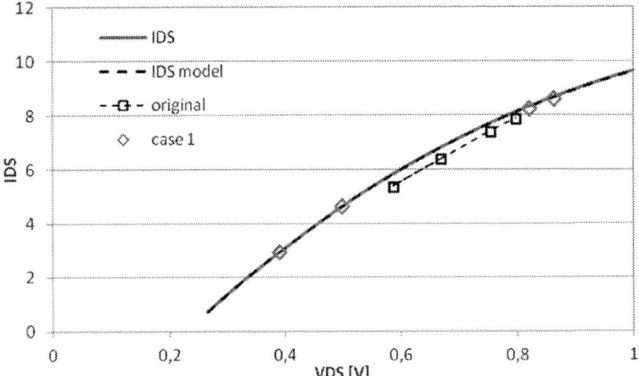

Fig. 2. Illustration of four-point parameter extraction.

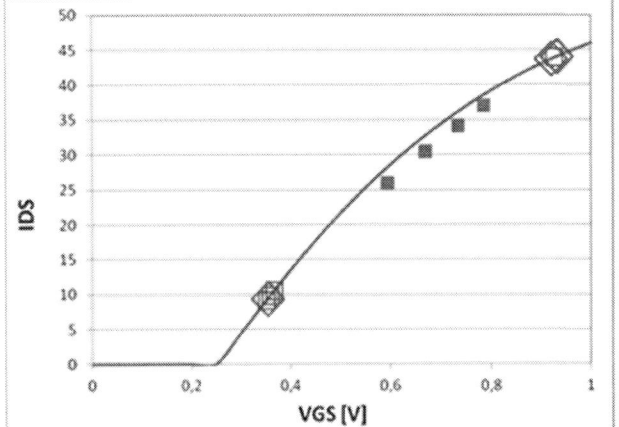

Fig. 3. Illustration of four-point parameter extraction.

In the saturation region, the drain current in a velocity-saturation limited MOSFET can be written according to the Berkeley model,

$$I_{DS} = \frac{k}{2}(V_{GS} - V_T)V_{DSAT}, \tag{15}$$

where the drain saturation voltage V_{DSAT} is given by

$$V_{DSAT} = \frac{(V_{GS} - V_T)V_C}{V_{GS} - V_T + \alpha V_C}, \tag{16}$$

and where $V_C = 2LE_{sat}$ as in the BSIM model; L being the channel length and E_{sat} the critical field for velocity saturation. Now, the parameter values for k and the linear region V_T are assumed to be known from a preceding extraction procedure performed in the linear region, while V_T, α, and V_C are the parameters to be determined in the saturation region. The threshold voltage extracted in the linear region could here be used as a start value for extracting a different value for V_T in saturation due to drain-induced barrier-lowering (DIBL).

978-1-4673-4845-4/13 $31.00 © 2013 IEEE   143

The saturation region model can now be written

$$I_{DS} = \frac{a + bV_{GS}}{1 + c/(V_{GS} - V_T)}, \tag{17}$$

where

$$\begin{cases} a = -bV_T & \Rightarrow & V_T = -a/b \\ b = kV_C/2 & \Rightarrow & V_C = 2b/k \ . \\ c = \alpha V_C & \Rightarrow & \alpha = c/V_C \end{cases} \tag{18}$$

Here, extraction can easily be handled by manually selecting three data points while assuming initially that for the denominator the V_T value from the linear region could be used for starters. If the least-square fitting technique previously described is used based on an arbitrary number of measurement data, $c=0$ could be used as the start value for initial parameter extraction. In a second round, the first round value for c is used. For most cases where we have tested this extraction scheme the convergence was relatively quick and required only two or three rounds, or even less. The results from one example extraction procedure are shown in Fig. 4. The results of fitting a very simple MOSFET model based on these model parameters, k, V_{T0}, roll-off theta, DIBL sigma, and velocity saturation V_C, are shown in Fig. 5 for a 65 nm p-channel device.

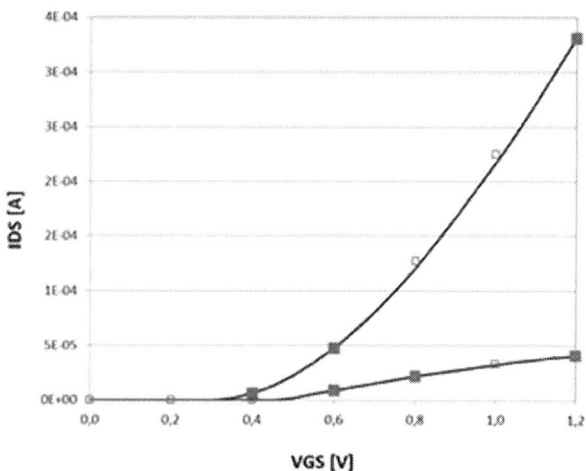

Fig. 4. Model to data fit based on model parameter extraction from the three filled data point in the two regions.

F. Robustness and Sensitivity to measurement noise

Since the robustness of the "least-square" fitting technique described for the four point model in section D could be questioned after the position of the "weighted" data points were revealed, the method was tested for robustness against measurement noise. For small noise levels at a few per cent the problem is negligible, but at 10-15% noise levels any negative slopes between two data points can have dramatic impact on the values of the extracted parameters. As expected, experiments with synthetic noise revealed the weakness of the proposed method as shown in Fig. 6. The main problem is that the model has two parameters in the denominator, two

parameters that can be extracted with different signs when they are supposed to be both positive.

To increase the robustness against parameter variations due to measurement noise or model imperfections, an alternative method was developed. In short, the method is based on extracting model parameters for the MOSFET ON resistance instead of extracting parameters for the ON conductance,

$$R_{ON} = \frac{a + bV_{GS} + cV_{GS}^2}{1 + dV_{GS}}. \tag{19}$$

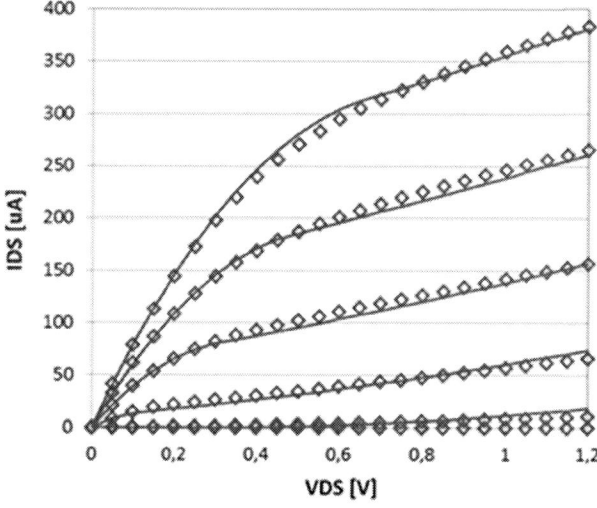

Fig. 5. Model and measurement data for a 65 nm p-channel MOSFET.

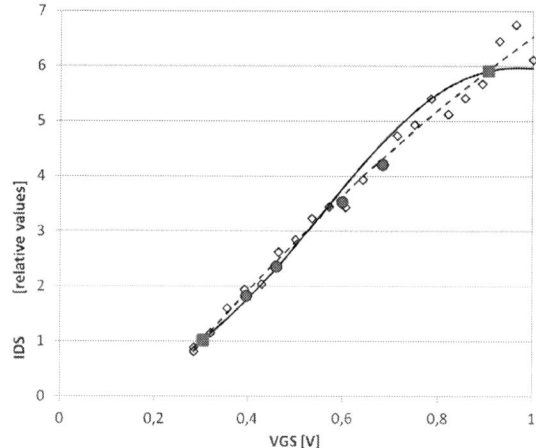

Fig. 6. Poor method robustness due to measurement noise.

As will be discussed more in detail at the conference, this method showed improved robustness, partly because it has only one unknown model parameter in the denominator. Results from this extraction method are shown in Fig. 7.

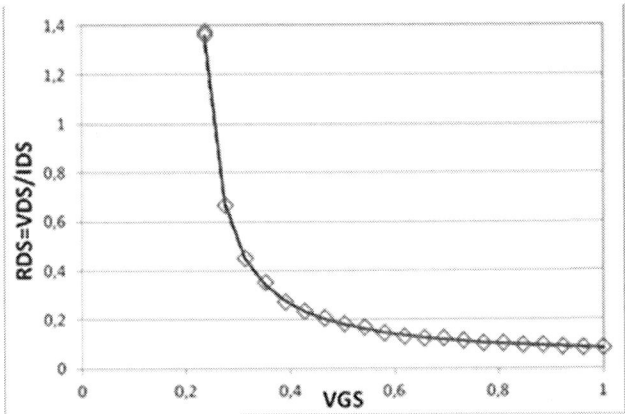

Fig. 7. Results from the proposed R_{ON} extraction method.

III. CONCLUSION

The three- and four-point Hamer-type model parameter extraction procedures have been revisited and their robustness and the selection of data points discussed. These methods based on a limited number of data points have also been compared to the "least-square" fitting technique proposed by Hamming; a technique that is not really a least-square fitting technique but in most cases good enough, at last for the three parameter model used originally by Hamer. The "least-square" fitting method is more questionable for the MOSFET model where the mobility roll-off model has been extended by a second-order term. In this case, extraction can result in two mobility roll-off parameters with different signs where one of the parameters is negative, a result that is not physically correct. To circumvent this problem, a more robust extraction method using R_{ON} values rather than G_{ON} values has been proposed and found to be more robust and insensitive to noise. Finally, the three-point extraction method was found to be useful also for extraction of saturation region parameters and for separating the substrate effect from the velocity saturation effect.

IV. REFERENCES

[1] M. F. Hamer, "First-order parameter extraction on enhancement silicon MOS transistors", IEE Proceeding on Solid-State Electron Devices, vol. 133, 1986, pp. 49-54

[2] P. R. Karlsson and K. O. Jeppson,, "A direct extraction algorithm for a submicron MOS transistor model", Proc. IEEE International Conference on Microelectronic Test Structures (ICMTS), vol 10, 1993, pp. 157-162

[3] P. R. Karlsson and K. O. Jeppson, "An Analytical Extraction Method for Submicron MOS Transistor Model Parameters in the Linear region", IEE Proceedings on Circuits, Devices and Systems, vol. 141, No. 6, 1994, pp. 457-461

[4] P. R. Karlsson and K. O. Jeppson, "Test Chip and Data Considerations for MOS Parameter Extraction", Proc. IEEE International Conference on Microelectronic Test Structures (ICMTS), Vol 10, 1997, pp. 159 – 164

[5] R. W. Hamming, Numerical Methods for Scientists and Engineers, McGraw-Hill, 1973, pp. 497-498

[6] H. P. Tuinhout, *Electrical characterization of matched pairs for evaluation of integrated circuit technologies*, Ph. D thesis, TU Delft, 2005

[7] M-C. Jeng, P. K. Ko, and C. Hu, A Deep Submicron MOSFET Model for Analog/Digital Circuit Simulations", Technical Digest of Electron Devices Meeting, 1988, pp. 114-117

[8] J. A. Croon, H. P. Tuinhout, R. Difrenza, J. Knol, A.J. Moonen, S. Decoutere, H. E. Maes, and W. Sansen, "A comparison of extraction techniques for threshold voltage mismatch", Proc. IEEE International Conference on Microelectronic Test Structures (ICMTS), vol 15, 2002, pp. 235 - 240

[9] J. P. de Gyvez, and H. P. Tuinhout, "Threshold Voltage Mismatch and Intra-Die Leakage Current in Digital CMOS Circuits", IEEE JOURNAL OF SOLID-STATE CIRCUITS, VOL. 39, NO. 1, 2004, pp. 157-168.

[10] P. R. Karlsson and K. O. Jeppson, A Direct Method to Extract Effective Geometries an Series Resistances of MOS Transistors", Proc. IEEE International Conference on Microelectronic Test Structures (ICMTS), vol 11, 1994, pp. 184-189

[11] P. R. Karlsson and K. O. Jeppson , "A new method of determining the effective channel width and its dependence on the gate voltage", Proc. IEEE International Conference on Microelectronic Test Structures (ICMTS), vol 13, 1996, pp. 151-156

Die-to-Die and Within-Die Variation Extraction for Circuit Simulation with Surface-Potential Compact Model

Y. Ohnari, A.A. Khan, A. Dutta, M. Miura-Mattausch, and H. J. Mattausch

Research Institute for Nanodevice and Bio Systems, Hiroshima University

Kagamiyama 1-5-2, Higashi-Hiroshima, Japan

Abstract— **A 65nm CMOS TEG for die-to-die and within-die variation analysis is reported. From measured Vth and Ion variation data of transistor pairs, die-to-die and within-die microscopic-parameter variations of a surface-potential model are extracted. Consideration of only five microscopic parameters is found sufficient to capture the channel-length dependence of these variations.**

Keywords— *Variation; Die-To-Die; Within-Die; CMOS; Compact Model; Surface Potential; Microscopic Parameter; Macroscopic Parameter*

I. INTRODUCTION

Down scaling of transistors has been the main method for miniaturization and speeding up of CMOS LSIs. However, in recent years, increasing variation in device characteristics has become a major problem when transistor sizes are scaled down further. With the larger variation of transistor characteristics, the circuit-performance variation is also increased, so that IC yield can be degraded substantially. Therefore, it is important to construct a design method which allows the design of circuits having robustness in their functionality under process variations [1].

Here we report a TEG, designed in 65 nm CMOS technology, for the purpose of developing and testing a surface-potential-based compact-model method to take into account the manufacturing variation at the time of circuit design.

II. TEST-STRUCTURE (TEG) DESCRIPTION

Fig.1 shows the photomicrograph of the prototype TEG for die-to-die (DTD) and within-die (WID) variation analysis. This TEG is equipped with transistor pairs for extracting the size dependence (channel length and width) of DTD and WID variation. Low complexity circuits (inverters, NANDs, NORs, ring oscillators) are also provided for analysis and verification of the prediction capability of circuit variation.

The surface-potential transistor model HiSIM [2] is used to establish the correlation between microscopic transistor-parameter variation and macroscopic variation of measured electrical transistor characteristics [3].

III. CLASSIFICATION OF VARIATION

There are various types of manufacturing variations, which are different in origin, and affect the transistor and circuit

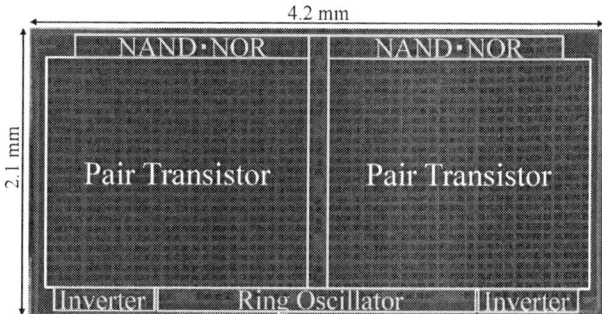

Figure 1. Photomicrograph of TEG in 65nm CMOS for die-to-die (DTD) and within-die (WID) variation extraction, as well as verification circuits for circuit-variation prediction.

characteristics in different ways. DTD and WIN variations are an important concept when considering manufacturing variations, in particular to capture the distance dependence of variations. The term DTD variation refers to process variations between different dies normally within a wafer [4]. On the other hand the term WID variation refers to process variations between identically designed devices or circuits within a die [5].

As a characteristic feature, DTD variations additionally contain a systematic factor in comparison to WID variation. Systematic variations are caused by changing manufacturing conditions in the used equipment, and lead to a characteristic that varies gradually over the wafer. On the other hand, WID variation mainly gives the independent variation for each element in the same die, causing a random type of change in the transistor characteristics.

Conventional statistical circuit analysis, that takes into account the variation of transistor characteristics due to process variations, often considers mainly DTD variations and assumes that all transistor characteristics change uniformly within the die or circuit. In reality, each transistor characteristic additionally varies independently within the die under the influence of the random WID-variation boundaries. So, WID variations greatly affect the yield of circuits in addition to DTD variations. It is expected that this will become more pronounced with further scaling down of transistor sizes. Thus, in the circuit simulation considering manufacturing variations, it is necessary to include the effects of both DTD and WID variation.

978-1-4673-4845-4/13 $31.00 © 2013 IEEE

IV. EXTRACATION METHOD OF DIE-TO-DIE AND WITHIN-DIE VARIATION FROM TRANSISTOR PAIRS

The term "transistor pair" refers to two adjacent identically designed single MOSFETs. Fig.2 shows a layout example. Dummy transistors are placed around each transistor pair to prevent additional variation due to the layout style. The change of MOSFET characteristics among different dies gives the DTD variation.

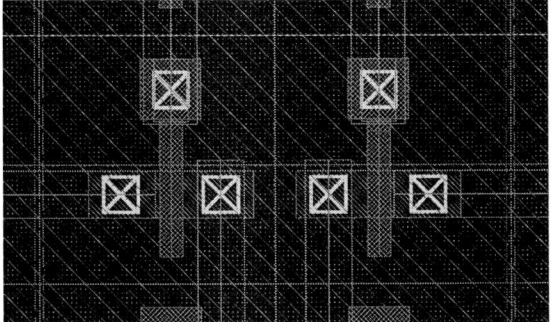

Figure 2. Layout of a transistor pair.

The extraction of the WID variation is more difficult, because there is only one transistor pair of the same size on each fabricated TEG chip. Therefore, the transistor pair data from all fabricated chips must be included in the WID-variation extraction method. For this purpose the random characteristic of the WID variation can be exploited, which has to a normal variation distribution. It can also be assumed that there is no inter-correlation between DTD and WID variations. In the following, the threshold voltages of left and right transistor in each transistor pair are referred to as Vth,right and Vth,left, respectively. Since left and right transistors are designed exactly equal, their WID variations must be equal and correspond to the WID variation of the studied technology as shown in Eq. (1).

$$\sigma_{V_{th, WID}} = \sigma_{V_{th, right,WID}} = \sigma_{V_{th, left,WID}} \tag{1}$$

Furthermore, the threshold-voltage difference ($\Delta V_{th,measured}$) between left and right transistor is equal to their WID-variation differences as shown in Eq. (2).

$$\Delta V_{th, measured} = V_{th, right} - V_{th, left}$$

$$= (V_{th, right, nominal\&DTD} + \Delta V_{th, right, WID})$$

$$- (V_{th, left, nominal\&DTD} + \Delta V_{th, left, WID})$$

$$= \Delta V_{th, right,WID} - \Delta V_{th, left,WID} \tag{2}$$

The nominal Vth and its DTD-variation component cancel, because both transistors are located in close proximity on the same die. From Eq. (2) and the statistical independence of the variation of left and right transistor, we obtain the variance of ΔV_{th},measured and consequently the standard deviation σV_{th},WID of the WID threshold-voltage variation as:

$$(\sigma_{V_{th, measured}})^2 = (\sigma_{V_{th, right, WID}})^2 + (\sigma_{V_{th, left, WID}})^2$$

$$= 2(\sigma_{V_{th, WID}})^2$$

$$\therefore \sigma_{V_{th, WID}} = \frac{\sigma_{V_{th, measured}}}{\sqrt{2}} \tag{3}$$

Eq. (2), (3) are used to extract the WID variation of Vth including its transistor-size dependence from the measured data of the transistor pairs. The same method also holds for the WID variation extraction of the transistor's on-current Ion.

V. VARIATION EXTRACTION WITH HISIM2

In this study, the HiSIM model is used for modeling the manufacturing variations. HiSIM is a transistor model based on the surface potential, and can reproduce electrical properties with high accuracy by adjusting a plurality of model parameters based on physical principles to match the electrical characteristics of the measured transistors as e.g. the threshold voltage (Vth) and the on-current (Ion) [6]. In addition, the result of reproducing DTD manufacturing variations with only four physical parameters, which are substrate doping concentration (NSUBC), interface-roughness mobility (MUESR1), pocket doping concentration (NSUBP) and difference between manufactured and designed channel length (XDL), has been obtained [3].

First, in order to reproduce the manufacturing variations, it is necessary to extract the model parameters of a nominal chip serving as the center of variation. Then, the values of these nominal parameters are adjusted for extracting the microscopic parameter variation which reproduces the macroscopic Vth and Ion variation of the manufactured transistors.

A. Reproduction of the measured PMOSFET variation

Table I and Figs. 3 - 6 show the results of the DTD and WID variation reproduction for Vth and Ion by changing only the above mentioned four microscopic HiSIM-model parameters from their values extracted for the PMOSFETs on the nominal die. The applied extraction method is the same as the extraction method described in reference [3]. Table I lists the maximum deviation in percent of each microscopic HiSIM parameter from its nominal value, which is necessary to reproduce the measured DTD and WID variation of Vth and Ion as a function of channel length (L). In the Figs.3, 4, 5 and 6, where the data is normalized to the values of the nominal chip, red circles indicate nominal-chip values, blue diamonds are the measured variation data and green circles are the simulated data for the 16 combinations with the variation boundaries of Table I. These figures verify that DTD and WID variations of PMOSFETs can be reproduced well by changing the values of only these four microscopic HiSIM-model parameters. Furthermore, the extracted microscopic parameter variations of Table I, which is necessary to reproduce the measured macroscopic DTD and WID variations of Vth and Ion for the PMOSFETs, indicates that the pocket implant doping (NSUBP) variation has a dominating influence for both DTD and WID variation of Vth and Ion.

978-1-4673-4845-4/13 $31.00 © 2013 IEEE

TABLE I. EXTRACTION RESULTS OF PMOSFET VARIATION BOUNDARIES FOR THE MICROSCOPIC HiSIM2 PARAMETERS

Variation type	NSUBC [%]	MUESR1 [%]	NSUBP [%]	XL [nm]
Die-to-Die	-0.00/+0.19	-0.51/+0.61	-14.7/+23.7	-0.54/+0.58
Within-Die	-0.00/+0.10	-0.38/+0.39	-6.67/+11.2	-0.13/+0.31

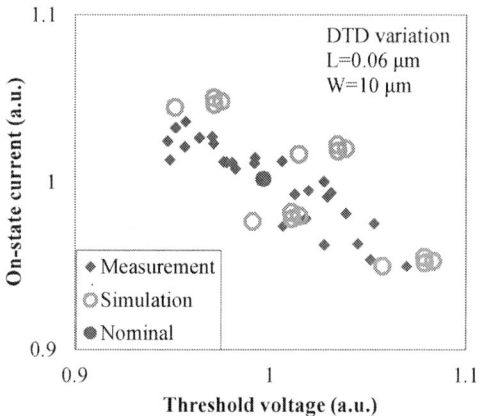

Figure 3. Extraction results of die-to-die (DTD) variation from fabricated pair transistors (PMOSFETs) on the TEG (L=0.06μm,W=10μm).

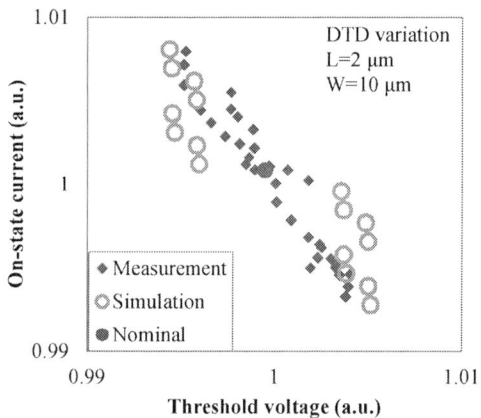

Figure 4. Extraction results of die-to-die (DTD) variation from fabricated pair transistors (PMOSFETs) on the TEG (L=2μm,W=10μm).

B. Reproduction of the measured NMOSFET variation

Fig.7 shows an example of measured DTD variations of long channel NMOSFETs (L=2μm, channel width W=10μm) for Vth and Ion. As can be seen from comparison to Fig.4 the shape of the plotted data for the NMOSFET variation has changed to a more circular form, meaning that there is much less correlation between Vth and Ion for the NMOSFETs than for the PMOSFETs. As a consequence the spread of measured data in the vertical direction becomes much larger for the NMOSFET case. This indicates that only usage of the same four microscopic parameters as in the PMOSFET case may not be sufficient to reproduce the NMOSFET variation data.

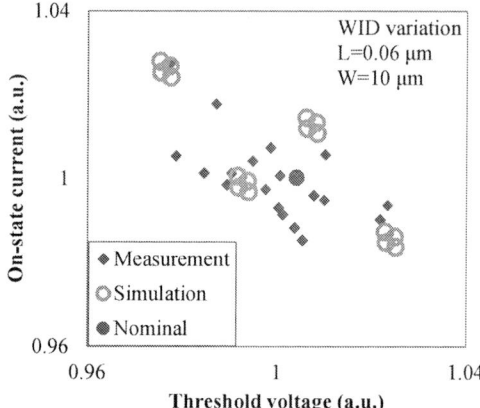

Figure 5. Extraction results of within-die (WID) variation from fabricated pair transistors (PMOSFETs) on the TEG (L=0.06μm,W=10μm).

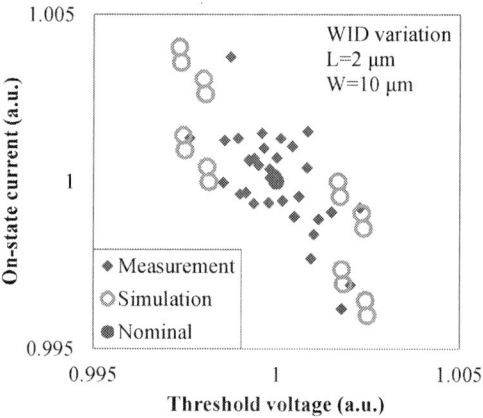

Figure 6. . Extraction results of within-die (WID) variation from fabricated pair transistors (PMOSFETs) on the TEG (L=2μm,W=10μm).

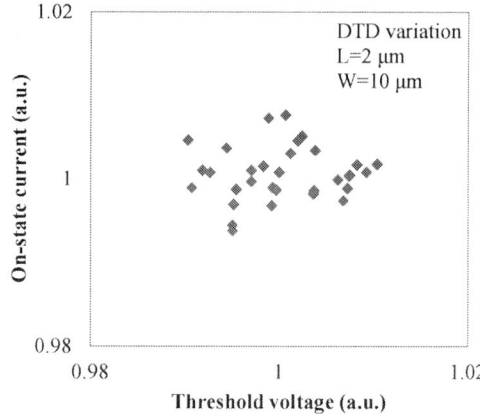

Figure 7. Die-to-die (DTD) variation from fabricated pair transistors (NMOSFETs) on the TEG (L=2μm,W=10μm).

We found that consideration of the microscopic variation of the electron mobility due to Coulomb scattering, represented by the microscopic HiSIM-parameter MUECB0, is necessary and sufficient to reproduce the experimental macroscopic variation data.

Fig. 8 and Table II explain the method for extracting the microscopic parameter variation from the measured DTD and WID variation of Vth and Ion in the case of NMOSFETs. The microscopic variation of MUECB0 expands nearly exclusively the macroscopic Ion variation and has a much larger effect for long-channel transistors than for short-channel transistors. The corresponding extraction method becomes therefore somewhat different as illustrated with Fig. 8 and Table II. Firstly, the Vth and Ion variation of long channel transistors is reproduced with the three parameters MUESR1, NSUBC and MUCEB0. Then, the variation of short channel transistors is reproduced by adding the variation of the two parameters NSUBP and XDL. The reproduction of the channel-length-dependent macroscopic variation of the NMOSFETs becomes possible in this way.

variation source in the case of NMOSFETs. In particular, the Coulomb scattering related microscopic variation reduces the correlation between macroscopic Ion and Vth variations. As Table III verifies, macroscopic DTD and WID variations of the NMOSFETs show, as in the PMOSFET case, a dominant influence of the microscopic pocket implant doping (NSUBP) variation and that the Coulomb-scattering variation is the 2nd most important microscopic variation source.

TABLE III.   EXTRACTION RESULTS OF NMOSFET VARIATION BOUNDARIES FOR THE MICROSCOPIC HiSIM2 PARAMETERS

Variation type	NSUBC [%]	MUECB0 [%]	MUESR1 [%]	NSUBP [%]	XL [nm]
Die-to-Die	-0.43/ +0.48	-9.58/ +6.87	-1.75/ +1.10	-14.4/ +1.10	-0.75/ +1.30
Within-Die	-0.14/ +0.19	-1.54/ +3.21	-1.55/ +0.10	-8.12/ +2.34	-1.28/ +1.21

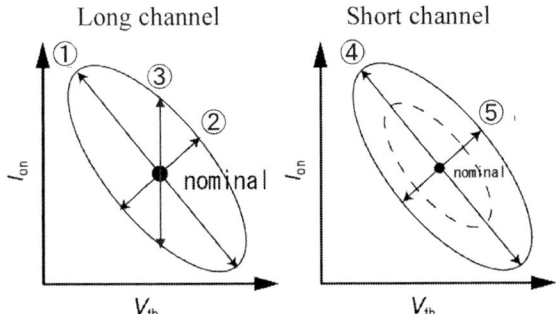

Figure 8.   Method to reproduce the die-to-die (DTD) and within-die (WID) channel-length-dependent variations of the NMOSFET by additionally considering the mobility variation due to Coulomb scattering which is represented by the HiSIM parameter MUCEB0.

TABLE II.   5 PARAMETER BASED METHOD TO REPRODUCE THE DIE-TO-DIE (DID) AND WITHIN-DIE (WID) VARIATIONS OF THE NMOSFETS

Step	Device	Parameter	Direction
1	Long	NSUBC	1
2		MUESR1	2
3		MUECB0	3
4	Short	NSUBP	4
5		XDL	5

Table III and Figs. 9 - 12 show the results of the DTD and WID variation reproduction for Vth and Ion by changing only the above five microscopic HiSIM-model parameters from their values extracted for the NMOSFETs on the nominal die. As can be seen from the figures, the manufacturing variations for Vth and Ion of NMOSFETs can be reproduced well by adding the microscopic HiSIM-parameter MUECB0 for electron-mobility reduction due to the Coulomb-scattering related mobility to the four parameters used for extracting the microscopic PMOSFET variation. Consequently, the measured manufacturing variations of NMOSFETs verify that the microscopic variation of Coulomb scattering is an important

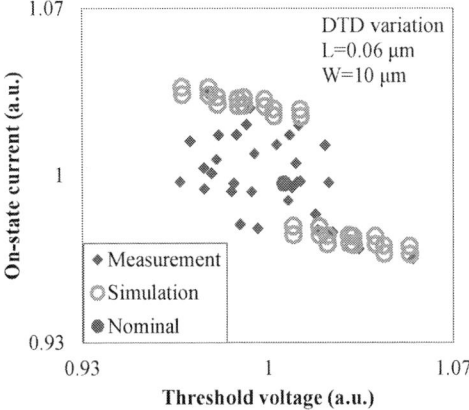

Figure 9.   Extraction results of die-to-die (DTD) variation from fabricated pair transistors (NMOSFETs) on the TEG (L=0.06μm, W=10μm).

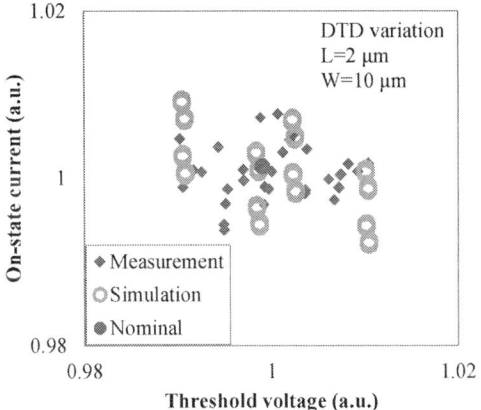

Figure 10. Extraction results of die-to-die (DTD) variation from fabricated pair transistors (NMOSFETs) on the TEG (L=2μm, W=10μm).

978-1-4673-4845-4/13 $31.00 © 2013 IEEE

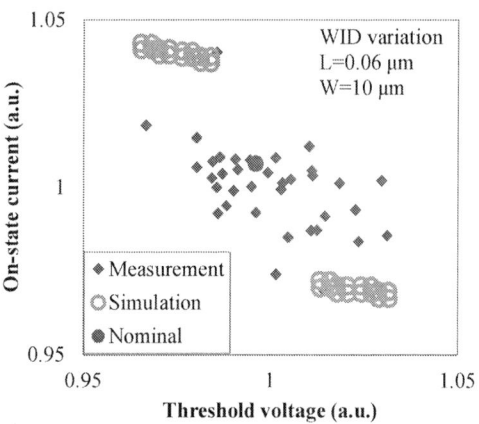

Figure 11. Extraction results of within-die (WID) variation from fabricated pair transistors (NMOSFETs) on the TEG (L=0.06μm,W=10μm).

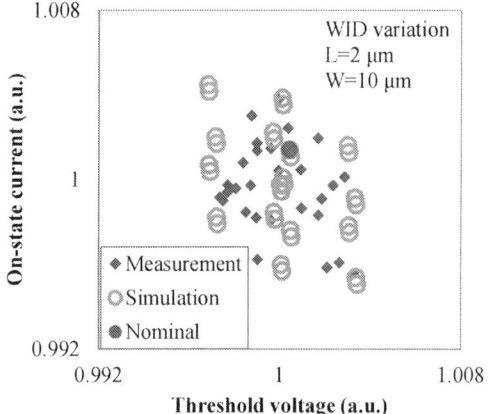

Figure 12. Extraction results of within-die (WID) variation from fabricated pair transistors (NMOSFETs) on the TEG (L=2μm,W=10μm).

VI. SUMMARY

A TEG for die-to-die (DTD) and within-die (WID) variation analysis is designed and microscopic parameter variations for reproducing macroscopic DTD and WID variations are extracted from measurements for Vth and Ion of transistor pairs on a TEG for 65 nm CMOS technology. Macroscopic DTD and WID variations of Vth and Ion for PMOSFETs can be reproduced by changing the values only four microscopic parameters in the surface-potential MOSFET model HiSIM. The measured DTD and WID variations for Vth and Ion of NMOSFETs reveal that a fifth microscopic parameter MUECB0, representing the influence of the Coulomb scattering variation on mobility variation, becomes necessary to reproduce the experimental data. Detailed DTD and WID variation effects for NMOSFETs and PMOSFETs are different in many ways. However, the largest microscopic impact on the macroscopic Vth and Ion variation comes for this 65 nm CMOS technology in all cases from the impurity-concentration variation of the pocket implantation.

Our future research will concentrate on the comparison of simulated and measured circuit variation. In particular, we hope to be able to predict measured circuits variations based on the microscopic parameter variations of the used PMOSFETs and NMOSFETs as extracted in the reported work.

References

[1] K. Kuhn, "Reducing variation in advanced logic technologies: approaches to process and design for manufacturability of nanoscale CMOS," Technical Digest of the 2007 International Electron Devices Meeting, pp. 471–474, 2007.

[2] HiSIM compact model information. [Online]. Available. http://www.hisim.hiroshima-u.ac.jp/index.php?id=87

[3] H.J.Mattausch, et. al., "Correlating microscopic and macroscopic variation with surface-potential compact model," IEEE Electron Device Letters, Vol.30, No.8, pp. 873-875, 2009.

[4] M. Pelgrom, et. al., "Matching properties of MOS transistors," IEEE Journal of Solid-State Circuits, pp. 1433-1439, 1989.

[5] J.Bastos, et. al., "Influence of die attachment on MOS transistor matching," Proceedings of IEEE International Conference.

[6] M. Miura-Mattausch, et. al., "The physics and modeling of MOSFETs: surface-potential model HiSIM," World Scientific, 2008.

978-1-4673-4845-4/13 $31.00 © 2013 IEEE

BSIM4 Parameter Extraction for Tri-gate Si Nanowire Transistors

Chika Tanaka, Masumi Saitoh, Kensuke Ota and Toshinori Numata

Advanced LSI Technology Laboratory, Corporate R&D Center,

Toshiba Corporation

8, Shinsugita-cho, Isogo-ku, Yokohama, Kanagawa 235-8522, Japan

E-mail:chika.tanaka@toshiba.co.jp

Abstract— **We investigated the BSIM4 parameter extraction procedure for tri-gate Si nanowire transistors with different geometries and fabrication processes using measurement data. Dependence of source/drain parasitic resistances on transistor geometry and fabrication process can be observed on the extracted parameters. Single sets of parameters can reproduce *I-V* characteristics with L_g down to 35nm.**

Keywords—parameter extraction; nanowire transistor; BSIM4

I. INTRODUCTION

Si nanowire transistors (NW Tr.) are promising device structures for further scaling [1]. The excellent gate controllability over the channel in the multi-gate structure leads to the ideal subthreshold slope and the strong immunity to short-channel effects [2]. Therefore, further voltage scaling with minimizing off-state leakage current (I_{off}) is expected for CMOS circuit with ultra-low power operation [3][4].

For circuit simulation, the Berkeley Short-Channel IGFET Model 4 (BSIM4) has been widely recognized as a scalable *IV* model for conventional planer MOSFETs [5]. Since BSIM4 addresses the physical effects in MOSFETs scaled down to sub-100nm regime, it should also be adequate model for NW Tr. In this study, we investigate the BSIM4 parameter extraction procedure for NW Tr. with different geometries and fabrication processes on the basis of measurement data.

II. DEVICE STRUCTURE AND CHARACTERISTICS OF NW TR.

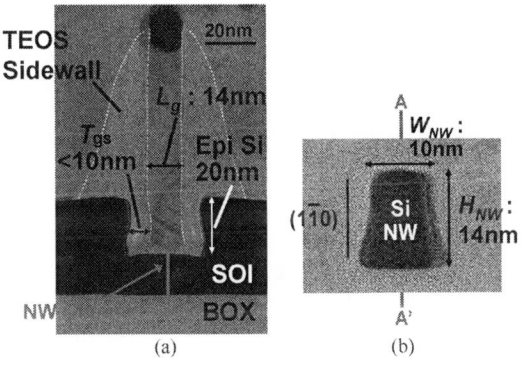

Fig.1 Cross-sectional TEM of NW Tr. (a) gate-length direction and (b) NW width direction.

We fabricated high performance tri-gate NW Tr. [2]. Fig. 1(a) and 1(b) show the cross-sectional TEM image of the NW Tr. Tri-gate NW Tr. with various channel length (L_g), nanowire width (W_{NW}) and gate sidewall thickness (T_{gs}) were fabricated on 300mm SOI wafers. The thickness of the buried oxide (BOX) of SOI wafer is 145nm and the NW Tr. were isolated from Si substrate. Poly-Si gate/SiO$_2$ (T_{ox}=3nm) gate stack and undoped nanowire channel were adopted in the fabrication.

No.	(T_{gs}, W_{NW}) [nm]
1	(10, 20)
2	(10, 30)
3	(10, 50)
4	(10, 100)
5	(30, 20)
6	(30, 30)
7	(w/o raised S/D, 30)

Table 1 Lists of the targeted device geometries for BSIM4 parameter extraction.

Table 1 lists the targeted device geometries, such as T_{gs} and W_{NW}, for parameter extraction in this study. T_{gs} of the device without raised source/drain (S/D) corresponds to zero.

Si-epitaxial growth was formed with thin gate sidewall to shorten the bottleneck region of S/D extension as shown in Fig.1. In the NW Tr. of this study, the S/D parasitic resistance (R_{sd}) was reduced by raised S/D extension structure [2]. Fig. 2 shows the dependence of measured R_{sd} on T_{gs} and W_{NW} for both *n*- and *p*-channel NW Tr. Larger R_{sd} is obtained in thin-T_{gs} devices than thick-T_{gs} devices. R_{sd} increases with a decrease in W_{NW}, namely larger R_{sd} is obtained in narrow-W_{NW} devices than wide-W_{NW} devices.

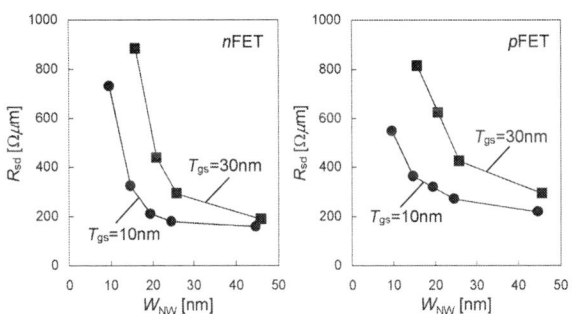

Fig. 2 Dependence of measured R_{sd} on W_{NW} and T_{gs} for both *n*- and *p*-channel NW Tr.

Fig. 3 shows the measured I_d-V_g characteristics of *n*-channel NW Tr. with T_{gs}=10nm, W_{NW}=20nm and H_{NW}=20nm. Good *IV* characteristics were obtained in L_g down to 35nm.

978-1-4673-4845-4/13 $31.00 © 2013 IEEE

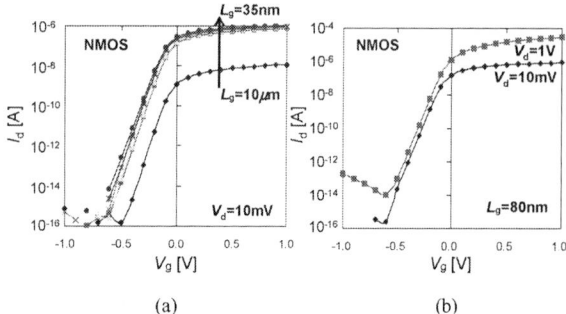

(a)   (b)

Fig. 3 The measured I_d-V_g characteristics of *n*-channel NW Tr. with T_{gs}=10nm, W_{NW}=20nm, H_{NW}=20nm: (a) L_g dependence at V_d=10mV and (b) V_d dependence with L_g=80nm.

Fig.4 shows the measured *S*-factor and drain induced barrier lowering (DIBL) for *n*-channel NW Tr. *S*-factor and DIBL are sufficiently suppressed in 20nm-W_{NW} NW Tr. due to the strong immunity against short-channel effects by thin nanowire channel. The ideal subthreshold slope of 60mV/dec. was obtained in narrow NW Tr. around 20nm. This is because of the high gate electrostatic controllability against the channel by thin NW and tri-gate structure.

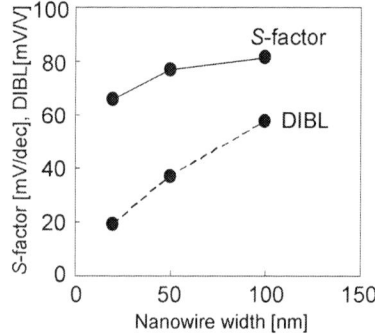

Fig. 4 Measured *S*-factor and DIBL for NW Tr. with T_{gs}=10nm, H_{NW}=20nm, L_g=80nm. V_d=1V.

III. SPICE PARAMETER EXTRACTION FOR NW TR.

A. DC model parameter extraction

SPICE modeling software, UTMOST III [6] was used to extract the SPICE model parameters of BSIM4 from the measurement data of NW Tr. The input measurement data includes I_d-V_g (linear region), I_d-V_g (saturation region), and I_d-V_d (at varying V_g). All data have no body bias (V_b) dependence because of the isolated substrate in a thick BOX of 145nm. Therefore, simulated *IV* and *CV* characteristics by using the extracted parameters have no V_b dependence. Three local optimization strategies in UTMOST III were used in this work: 1) long channel I_d-V_g, 2) short channel I_d-V_g, 3) I_d-V_d. Strategy 1) is used to optimize the I_d-V_g characteristic of long channel MOSFETs, mainly concentrating on the linear region. This strategy is activated for device with L_g=10μm. Strategy 2) is used to optimize the I_d-V_g characteristic of short channel MOSFETs, also concentrating on the linear region. This strategy is activated for devices with L_g=35~100nm. Strategy

3) is used to optimize the I_d-V_d of both long and short channel MOSFETs. The entire range of V_d, including both the linear and saturation regions, are considered. This strategy is activated for the I_d-V_d curves of all devices.

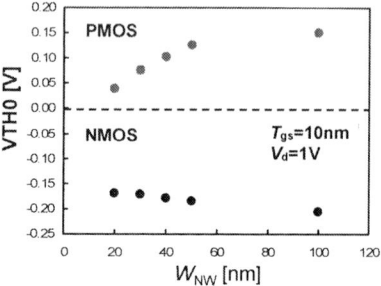

Fig. 5 Measured *S*-factor and DIBL for NW Tr. with T_{gs}=10nm, H_{NW}=20nm, L_g=80nm. V_d=1V.

Initial values of VTH0 and U0 were obtained from the measurement data. Fig.5 plots the extracted VTH0 versus W_{NW} for devices with T_{gs}=10nm. A decrease in the absolute VTH0 with smaller W_{NW} is observed for both *n*- and *p*-channel NW Tr. Since all the transistors under studied have very low threshold voltage, this decrease corresponds to easier turning on. The transistor width in SPICE model was set to be the effective nanowire width (W_{eff}=W_{NW}+2H_{NW}). In I_d-V_d optimization procedure, R_{sd} was carefully derived from the W_{NW} and T_{gs} dependence. Moreover, because of the immunity of DIBL in the NW Tr., the parameters related to DIBL were manually tuned. Rubberband method was used to extract DIBL and Gate Induced Drain Leakage (GIDL) related parameters. Extracted parameters were verified by the convergent of SPICE simulation and the results.

B. Capacitance model parameter extraction

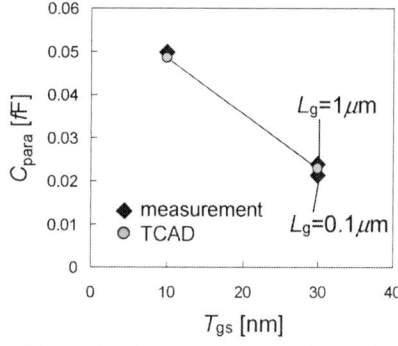

Fig.6 The simulated and measured parasitic capacitance.

The parasitic capacitance (C_{para}), especially fringe capacitance component from the gate to the raised S/D regions, is one of the key issues determining the CMOS circuit performance of multi-gate MOSFETs [7]. In order to derive the fringe capacitance in the SPICE model parameters, C_{para} was extracted from three-dimensional in-house TCAD simulation, as shown in Fig. 6. In the TCAD simulation, the device structure was formed in the same way as the fabricated two-finger NW Tr. with 200nm-NW-pitch. The overlap length between gate and S/D was set to be zero. Fig. 6 also shows

978-1-4673-4845-4/13 $31.00 © 2013 IEEE

C_{para} extracted from the gate-to-channel capacitance (C_{gc}) obtained from the Split-CV measurement [8]. Measured C_{para} is almost the same irrespective of the gate length (L_g=1μm and 0.1μm), as shown in Fig. 7, meaning that C_{para} was successfully extracted in this measurement.

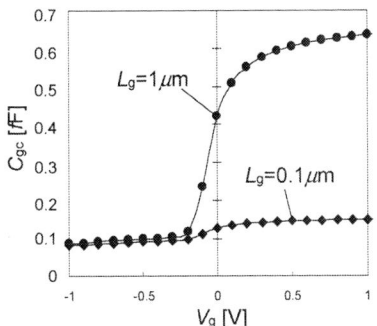

Fig. 7 The measured C_{gc}-V_g for NW Tr. with T_{gs}=10nm and H_{NW}=20nm.

The simulated values of C_{para} were manually put as fringe capacitances in SPICE netlists. The intrinsic capacitances at high V_g were equal to C_{ox}, where the effective oxide thickness was extracted from the measured CV data.

IV. RESULTS AND DISCUSSIONS

A. Parameter extraction results

Fig. 8 and 9 show the typical results of the simulated I_d-V_g and I_d-V_d curves, respectively. The behavior in subthreshold region and smaller DIBL effect were optimized to match the characteristics of NW Tr. Within the operation voltage range (V_g less than 1V), Good fitting with maximum error < 5% was achieved for both n- and p-channel NW Tr. with sub-100nm gate length.

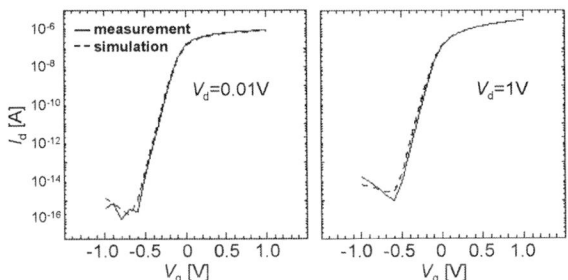

Fig. 8 The simulated versus measured I_d-V_g curves for n-channel NW Tr. with T_{gs}=10nm, W_{NW}=20nm, H_{NW}=20nm and L_g=80nm.

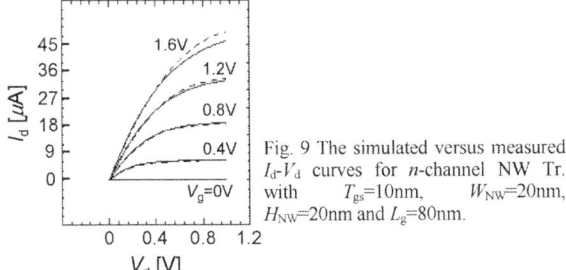

Fig. 9 The simulated versus measured I_d-V_d curves for n-channel NW Tr. with T_{gs}=10nm, W_{NW}=20nm, H_{NW}=20nm and L_g=80nm.

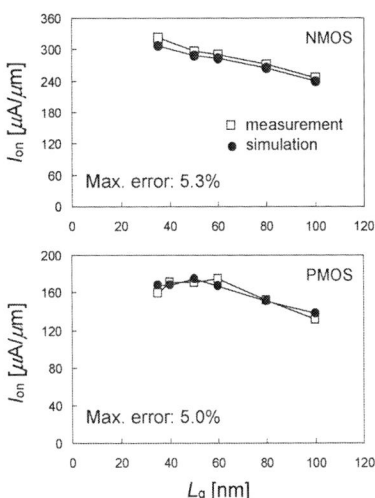

Fig. 10 The simulated and measured I_{on} for both n- and p-channel NW Tr. with T_{gs}=10nm and W_{NW}=20nm. I_{off}=50pA/μm and V_d=1V.

Fig. 10 shows the channel length dependence on the simulated and the measured on-current (I_{on}) for T_{gs}=10nm and W_{NW}=20nm. A maximum error of about 5% was observed. For all other device groups listed in Table 1, error less than 10% was observed. Single parameter sets for various L_g down to 35nm were obtained for n-channel NW Tr. that have short-channel effects immunity, while multiple parameter sets were extracted for p-channel NW Tr. of each L_g group.

B. SPICE simulation

The extracted parameters are imported into SPICE to simulate the delay time of inverter, in order to verify whether the dependence of R_{sd} and parasitic capacitance on transistor geometry and fabrication process is observed on the extracted parameters. Fig. 11 illustrates the circuit simulation structure for inverter. S/D fringe capacitances (CF) are included for each MOSFET. The output is assumed to have fan out equals to 1. Hence, the output loading capacitance (C_{load}) equals to the oxide capacitance plus CF of one nMOSFET and one pMOSFET.

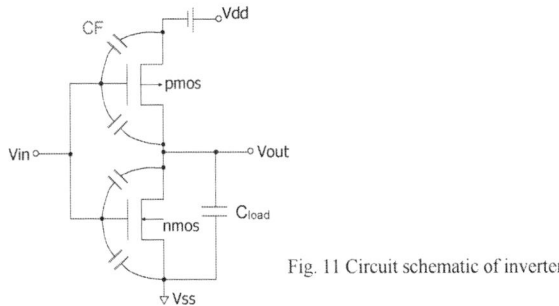

Fig. 11 Circuit schematic of inverter.

Fig. 12 shows the inverter delay versus L_g with and without CF. The inverter delay decreases with decreasing L_g, because shorter L_g device has higher I_{on}. With the fringe capacitance, both the rise time and fall time uniformly increase about 55%.

As a result, the fringe capacitances indeed affect the circuit performance seriously.

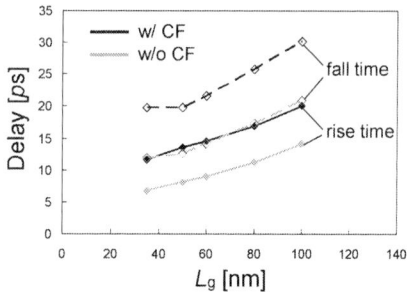

Fig. 12 Impact of fringe capacitance on the inverter delay. T_{gs}=10nm and W_{NW}=20nm.

Fig. 13 shows the process dependence of inverter delay. For LSTP application, devices with T_{gs}=10nm have larger delay than T_{gs}=30nm or w/o raised S/D (T_{gs}=0nm) devices, even with their smaller R_{sd}. This indicates that the parasitic capacitance is more important when considering the LSTP circuit performance. For HP application, inverter with T_{gs}=10nm and W_{NM}=20nm still has the largest delay, due to the fact that CF of thin T_{gs} is much larger than CF of thick T_{gs}. However, the delay of inverter with T_{gs}=30nm and W_{NM}=20nm is larger than that of inverter with T_{gs}=10nm and W_{NM}=30nm. This indicates that R_{sd} becomes important for HP applications.

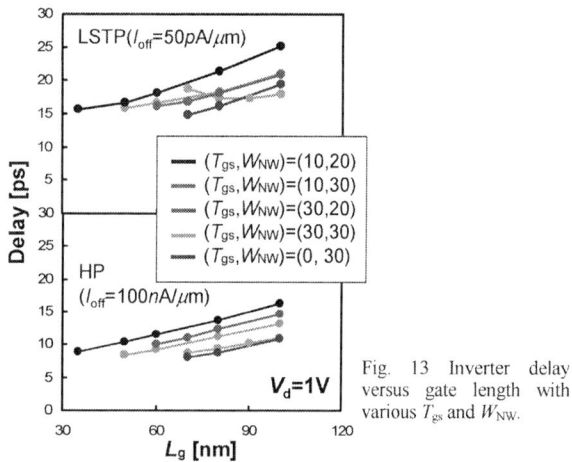

Fig. 13 Inverter delay versus gate length with various T_{gs} and W_{NW}.

The LSTP and HP inverter delays versus V_{dd} for various device groups with different W_{NW} are simulated and plotted in Fig. 14. The increase of the delay time is suppressed in narrower NW Tr. at low V_{dd}. Dramatic increase in the delay time with decreasing V_{dd} is observed for the LSTP inverter, while the delay time of HP inverter is less sensitive to V_{dd}. Delay is proportional to V_{dd}/I. The current flowing through the inverter decreases much faster than V_{dd} for LSTP conditions. Hence, even with decreasing V_{dd}, the delay increases. For HP conditions, the inverter current decreases more or less proportional to V_{dd}, leading to the very small change in delay. It is found that the extracted BSIM4 parameters reproduce the device characteristics of tri-gate NW Tr. well.

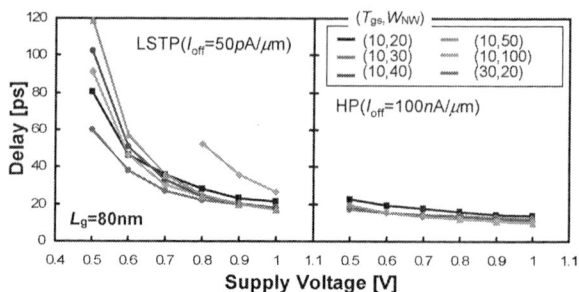

Fig. 14 Inverter delays versus V_{dd} for different W_{NW} devices.

V. CONCLUSIONS

SPICE model parameters of BSIM4 were successfully extracted for tri-gate NW Tr. using UTMOST III. Dependence of R_{sd} on transistor geometry and fabrication process can be observed on the extracted parameters. Single sets of parameters for L_g down to 35nm were obtained, as long as short-channel effects are well controlled. The extracted parameters will be a useful tool for characterizing the circuit performance of NW Tr. Our extraction procedure is applicable to extract the BSIM4 model parameters for NW Tr. as well as other multi-gate MOSFETs.

ACKNOWLEDGMENT

This work was partly supported by NEDO's Development of Nanoelectronic Device Technology. The authors would like to thank Dr. Min Chu (University of Florida) and Dr. Takayuki Ishikawa (Toshiba Corp.) for their great help with parameter extraction.

REFERENCES

[1] K. H. Yeo, S. D. Suk, M. Li, Y. -Y. Yeoh, K. H. Cho, K. -H. Hong, S. Yun, M. S. Lee, N. Cho, K. Lee, D. Hwang, B. Park, D. -W. Kim, D. Park, and B. -I. Ryu, "Gate-all-around (GAA) twin silicon nanowire MOSFET (TSNWFET) with 15 nm length gate and 4 nm radius nanowire," in *IEDM Tech. Dig.*, San Francisco, CA, Dec. 2006, p. 539.

[2] M. Saitoh, Y. Nakabayashi, K. Uchida, and T. Numata, "Short-channel performance improvement by raised source/drain extensions with thin spacers in trigate silicon nanowire MOSFETs" *IEEE Electron Device Lett.*, vol. 32, p. 273, 2011.

[3] C. Tanaka, M. Saitoh, K. Ota, K. Uchida, and T. Numata, "SPICE-Based Performance Analysis of Ultra-Low Voltage Si Nanowire CMOS Circuits" in Proc. of the European Solid-State Device Research Conference, 2011, p.159.

[4] C. Tanaka, M. Saitoh, K. Ota and T. Numata, " Analysis of Read Margin Improvement for Low Voltage SRAM Composed of Nano-Scale MOSFETs with Ideal Subthreshold Factor and Small Variability" in Ext. Abstr. Solid State Devices and Materials, 2012, p.819.

[5] BSIM Homepage, http://www-device.eecs

[6] Silvaco, http://www.silvaco.com/.

[7] M. Guillorn, J. Chang, A. Bryant, N. Fuller, O. Dokumaci, X. Wang, J. Newbury, K. Babich, J. Ott, B. Haran, R. Yu, C. Lavoie, D. Klaus, Y. Zhang, E. Sikorski, W. Graham, B. To, M. Lofaro, J. Tornello, D. Koli, B. Yang, A. Pyzyna, D. Neumeyer, M. Khater, A. Yagishita, H. Kawasaki and W. Haensch, "FinFET Performance Advantage at 22nm: An AC perspective", *VLSI Tech. Dig.*, 2008, p.12.

[8] C.G. Sodini, T.W. Ekstedt and J.L. Moll, "Charge accumulation and mobility in thin dielectric MOS transistors" , *Solid-State Electronics*, vol. 25, p.833, 1982.

SESSION 8: Emerging Technologies

978-1-4673-4845-4/13 $31.00 © 2013 IEEE

Benchmarking of a Surface Potential Based Organic Thin-Film Transistor Model against C_{10}-DNTT High Performance Test Devices

T. K. Maiti[*1], T. Hayashi[2], H. Mori[3], M. J. Kang[3], K. Takimiya[3],
M. Miura-Mattausch[1,2], and H. J. Mattausch[1,2]

[1]HiSIM Research Center, Hiroshima University, Higashi-Hiroshima, Japan
[2]Graduate School of Advanced Sciences of Matter, Hiroshima University, Higashi-Hiroshima, Japan
[3]Dept. of Applied Chemistry, Graduate School of Engineering, Hiroshima University, Higashi-Hiroshima, Japan
[*]E-mail: tkm@hiroshima-u.ac.jp

Abstract—**In this paper, a surface potential based compact model for organic thin-film transistors (OTFTs) including both tail and deep trap states across the band gap is presented and benchmarked against measured data from high-performance dinaphtho thieno thiophene (C_{10}-DNTT) based test devices. This model can accurately describe the OTFT test-structure current from week to strong inversion regime.**

Keywords— Organic Thin-Film Transistors; surface potential; traps; compact model

I. INTRODUCTION

Organic semiconductor devices have been extensively studied over the last few decades for the development of future organic electronics and optoelectronic applications such as light-emitting devices, solar cells, lasers, photodiodes, field-effect transistors (FETs), and integrated circuits [1]. These devices are now introduced for new applications in electronics, including bendable and flexible displays, wearable electronics, smart textiles, flexible bio-electronic, bio-optical sensors, smart skins and artificial muscles. To develop such new types of electronic products, organic thin-film transistor (OTFT) compact models are required as a link between device and circuit behaviour. Compact transistor models can be broadly classified into three categories, i.e. 1) threshold voltage (V_{th}) based, 2) inversion charge (Q_i) based and 3) surface-potential based. Several OTFT models for circuit simulation have been reported preciously [2-6] and are briefly described in the following. A V_{th}-based SPICE model for OTFTs, adapting the standard Berkeley Short-channel IGFET Model (BSIM) equations, was developed by Meixner *et al.* [2]. A DC model, using modified V_{th}-based model equations for amorphous silicon TFTs (AIM-SPICE Level 15), was developed by Horowitz's group [3]. A universal Q_i-based OTFT model (UOTFT) was developed by Mijalkovi *et al.* [4], applying the basic principles and ideas from M. Schur's model [5]. Li *et al.* developed a V_{th}-based compact model for DC current and capacitance of OTFTs [6]. These compact models are problematic in practical circuit simulations since it is very difficult to determine V_{th} for an OTFT, due to the large effect of trap sates on subthreshold swing (S) (Fig.1). S is degraded in the presence of trap states and this drastically affects the subthreshold behaviour. Also, above previous models consider an improper silicon resistance model and are overloaded with many modelling parameters which are difficult to determine and often irrelevant for OTFTs.

In this paper, we report on the development of a surface-potential based compact model for OTFTs including the both

tail and deep trap states across the band gap [7-9]. The drain current vs. gate voltage (I_{ds} vs. V_{gs}) fitting results are benchmarked with measured data from dinaphtho thieno thiophene (C_{10}-DNTT) based high-performance p-OTFT test devices [10]. The OTFT structure is different from the conventional MOSFET. OTFT features caused by the specific structure are investigated additionally.

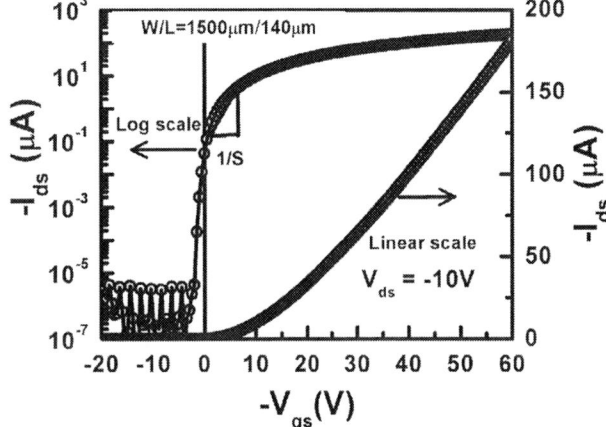

Figure 1. Measured drain current (I_{ds}) vs. gate voltage (V_{gs}) characteristics for a p-OTFT. The subthreshold swing (S) is the inverse of the slope in the subthreshold region.

II. TFT STRUCTURES AND MODEL DESCRIPTION

Two types of these OTFT architectures have been considered here as shown in Fig.2. One is the Top-Gate, Top-Contact (TGTC) OTFT structure, in which the gate is farthest away from the substrate and source and drain are placed at the top of the dielectric-semiconductor interface (Fig2.(a)). This corresponds to the basic conventional MOSFET structure. The other one is the Bottom-Gate, Top-Contact (BGTC) OTFT structure, in which the gate is embedded within the dielectric directly above the substrate and source and drain are now applied on top of the organic semiconductor layer as shown in Fig.2 (b). For simplicity we started modeling work with the TGTC OTFT structure since it is just like a-Si MOSFET structure. Then we considered the BGTC OTFT structure because of an easier fabrication process, commercially available high quality Si/SiO$_2$ substrates [11], and the availability of experimental data for the I-V characteristics of fabricated devices. The formation of good quality organic semiconductor layers on dielectric material is much easier by ink-jet printing or a deposition process than the reverse and

978-1-4673-4845-4/13 $31.00 © 2013 IEEE

metal can be deposited for top contacts using shadow masks to enable low cost fabrication. So, from fabrication point of view bottom gate OTFTs are much better than top gate OTFTs and hence BGTC OTFTs are most commonly used.

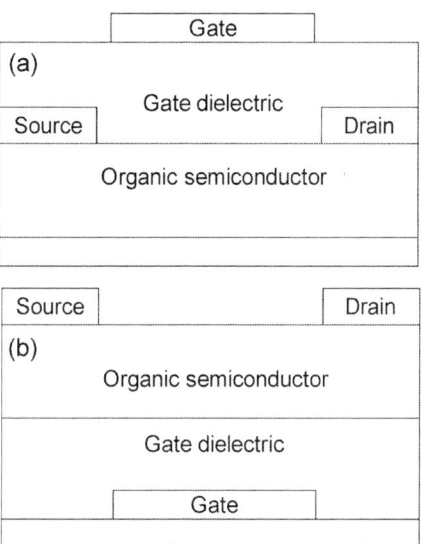

Figure 2. Schematic diagram of the benchmarked (a) Top-Gate, Top-Contact (TGTC) OTFT and (b) Bottom-Gate, Top-Contact (BGTC) OTFT.

A. Adopted Poisson's Equation for HiSIM-Org

Organic semiconductors are classified as either conducting polymers or small molecules (molecular crystals) and form grain boundaries as shown in Fig.3.

────── **Grain boundary**

Figure 3. Schematic diagram of grain boundaries in an organic semiconductor film fabricated by vacuum- deposition method

At these grain boundaries, molecular discontinuities locally cause a change in electronic structure, i.e. traps, resulting in band bending and barrier formation [12]. Consequently, current flow properties of carriers through OTFTs are dominated by the presence of trap states. We approximated the density of acceptor-type trap states $g_{A1}(E)$ and density of donor-type trap states $g_{D1}(E)$ as [13,14],

$$g_{A1}\left(E\right) = g_{C1}\exp\left(\frac{E - E_{LUMO}}{E_1}\right) \tag{1}$$

$$g_{D1}\left(E\right) = g_{C1}\exp\left(\frac{E_{HOMO} - E}{E_1}\right) \tag{2}$$

where, E_{LUMO} is the lowest unoccupied molecular orbital (LUMO) energy, E_{HOMO} is highest occupied molecular orbital (HOMO) energy, E_1 and g_{C1} are the inverse slope of the trap states and the trap states density at E_{LUMO} and E_{HOMO}, respectively. Integrating the product of the Fermi-Dirac distribution function and each of (1) and (2) across the band gap, the equation to describe the density of ionized acceptor-type traps and donor-type traps are obtained as,

$$N_{TA1}^{-} = g_{C1}E_1\frac{\left(\pi kT / E_1\right)}{\sin\left(\pi kT / E_1\right)}\exp\left(-\frac{E_{fn} - E_{LUMO}}{E_1}\right) \tag{3}$$

$$N_{TD1}^{+} = g_{C1}E_1\frac{\left(\pi kT / E_1\right)}{\sin\left(\pi kT / E_1\right)}\exp\left(-\frac{E_{fp} - E_{HOMO}}{E_1}\right) \tag{4}$$

where, E_{fn} and E_{fp} are the quasi-Fermi energies for electrons and holes, respectively. From Eqns. (3) and (4), total ionized trap density N_{t0} at the surface of the organic semiconductor layer surface can be obtained as,

$$N_{t0} = g_{C1}E_1\frac{\left(\pi kT / E_1\right)}{\sin\left(\pi kT / E_1\right)}\left\{\exp\left(\frac{E_{fn} - E_{LUMO}}{E_1}\right) - \exp\left(\frac{E_{fp} - E_{HOMO}}{E_1}\right)\right\} \tag{5}$$

The donor- and acceptor-like trap states are also approximated by the sum of the deep and tail trap states. Tail trap states are located near to both band edges, i.e. HOMO level edge and LUMO level edge. The suffix 1 used in Eqns. 1 to 5 indicates deep level trap states and for shallow states we have the same equations with suffix 2. To solve the Poisson's equation with the ionized trap profile in the organic semiconductor layer as a function of potential ϕ, we adopted the approximation of Eqns. 6 and 7 [7] for ionized trap densities as,

$$N_t\left(\phi\right) = N_{t0}\left(\frac{\phi - \phi_b}{\phi_s - \phi_b}\right) \tag{6}$$

$$N_t\left(x\right) = N_{t0}\exp\left(-\frac{K}{T_{org}}x\right) \tag{7}$$

where, ϕ_s is the surface potential, ϕ_b is the backside potential of the organic semiconductor layer, K is a fitting parameter, and T_{org} is the thickness of the organic semiconductor layer. In our investigation T_{org} is fixed to 50nm. Trapped charge in the organic semiconductor layer affects the space charge and that effect can be consider by adding trap charges in Poisson's equation as [15],

$$\nabla^2\phi = -\frac{q}{\varepsilon_{org}}\left(p - n + N_{TD}^{+} - N_{TA}^{-}\right) \tag{8}$$

where, ϕ, ε_{org}, and q are the potential, the organic semiconductor permittivity, and the electron charge, respectively. p and n are concentrations of holes and electrons, respectively. N_{TD}^{+} and N_{TA}^{-} are the ionized donor type and the ionized acceptor-type trap density, respectively. The carriers p and n are described as a function of ϕ.

B. Backside Potential Calculation

For compact modelling, device characteristics are described as a function of the surface potential ϕ_s [16]. To calculate ϕ_s, however, the backside potential as well as trap charge

978-1-4673-4845-4/13 $31.00 © 2013 IEEE

densities are required. The backside potential of the organic semiconductor layer is floating and varies with the applied voltages. Integrating depletion charges and trap charges in the organic semiconductor layer from the surface to the backside, the relation between the surface potential and the backside potential using Eqns. 6, 7, and 8 is derived as [7],

$$\phi_b = \phi_s - \phi_{sb0}\left[1 - \exp\left(-\frac{\phi_s}{\phi_{sb0}}\right)\right] \quad (9)$$

$$\phi_{sb0} = \frac{q}{\varepsilon_{org}} T_{org}^2 \left[\frac{N_A}{2} + N_{t0} \frac{e^{-K}(K-1)+1}{K^2}\right] \quad (10)$$

where, ϕ_s and ϕ_b are front and back surface potential at any point along y direction. ϕ_s and ϕ_b at the source side ($y = 0$) are denoted as ϕ_{s0} and ϕ_{b0}, respectively, and at the drain side ($y = L$) are denoted as ϕ_{sL} and ϕ_{bL}. The Poisson equation is solved independently at the source side and the drain side.

C. Surface Potential Calculation

Using Eqn. (11), the surface potential is calculated using Gauss's law as [7],

$$C_{org}\left(V_g - V_{FB} - \phi_S(y)\right)$$

$$= \sqrt{\frac{2q\varepsilon_{org}N_A}{\beta}}\left[\begin{array}{l}\left\{\left(e^{-\beta\phi_s(y)} - e^{-\beta\phi_b(v)}\right) - 1\right\} + \frac{n_i^2}{N_A^2}\left(e^{\beta\phi_s(y)} - e^{\beta\phi_b(v)}\right) \\ + \frac{\beta N_{t0}}{N_A}\left(\frac{\phi_S(y) - \phi_b(y)}{2}\right)\end{array}\right]^{\frac{1}{2}} \quad (11)$$

where, C_{org}, V_{FB}, β, and n_i are the gate insulator capacitance per unit area, the flat band voltage, the inverse thermal voltage, and the intrinsic carrier concentration, respectively. The developed model iteratively solves the surface potential at source and drain side after the HiSIM (Hiroshima University STARC IGFET Model) approach [16-18]. The model first calculates ϕ_{s0} at the source region end adjacent to the gate electrode in the organic semiconductor layer, then ϕ_{b0} at the backside of the source region end, after that ϕ_{sL} at the drain region end adjacent to the gate electrode in the organic semiconductor layer, and finally ϕ_{bL} at the backside of the drain region end. By consideration of both the charge sheet approximation and the gradual channel approximation all device characteristics are described analytically by the channel surface potential. The drain current is calculated with the derived surface-potential based device equation. We use the BGTC OTFT's measured data to verify the developed compact model.

Modeled drain current (I_{ds}) as a function of gate voltage (V_{gs}) based on our surface potential model are plotted in Figs.4 and 5 in log scale for n-OTFTs. We varied both tail-state and deep trap-state densities to study the trap effect on the *I-V* characteristics. Fig.4 shows the effect of deep-trap states on drain current (I_{ds}). Compared are three different trap densities (1e18 cm^{-3}/eV, 5e18 cm^{-3}/eV, and 1e19 cm^{-3}/eV) with trap-level energy E_{deep}= 0.2 eV at different drain voltages from

10V to 40V. It can be seen that deep-level-traps affect mainly the subthreshold current at low V_{gs}.

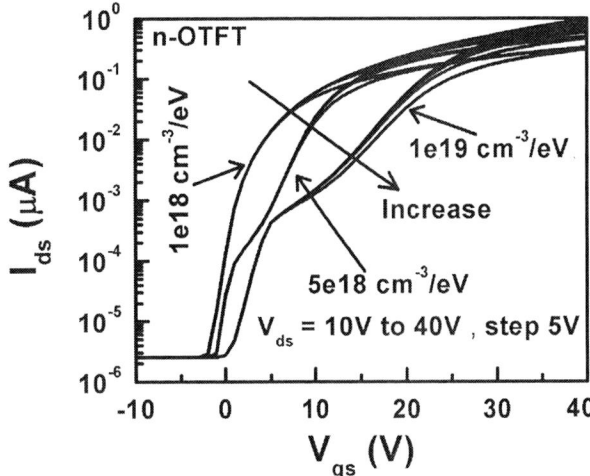

Figure 4. Effect of deep-trap states on drain current (I_{ds}), compared for three different trap densities with trap energy E_{deep}= 0.2 eV at different drain voltages from 10V to 40V.

Fig.5 shows the effect of tail-trap states on the drain current (I_{ds}). Compared are three different trap densities (1e19 cm^{-3}/eV, 5e19 cm^{-3}/eV, and 1e20 cm^{-3}/eV) with trap-level energy E_{low}= 0.085 eV at different drain voltages from 10V to 40V. Please notice that tail states also strongly affect the above subthreshold current at high V_{gs}.

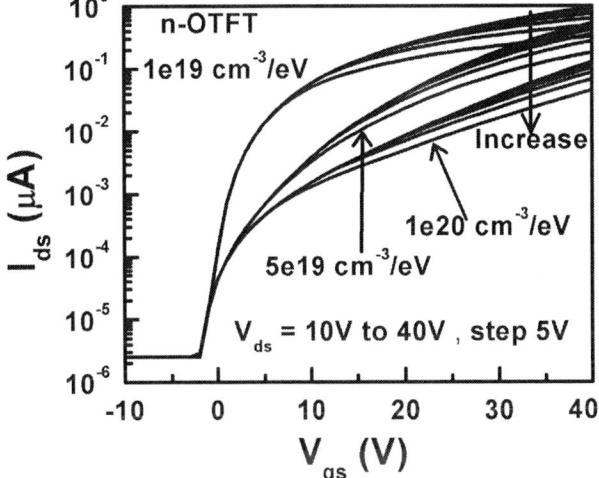

Figure 5. Effect of tail states on drain current (I_{ds}), compared for three different trap densities with trap energy E_{low}= 0.085 eV at different drain voltages from 10V to 40V.

III. MODELING OF STRUCTURAL FEATURES

To validate the model we have compared simulated drain current vs. gate voltage characteristics (I_{ds} vs. V_{gs}) and drain current vs. drain voltage characteristics (I_{ds} vs. V_{ds}) curves with measured C$_{10}$-DNTT based BGTC p-OTFTs [10]. A thin film (50nm thick) of C$_{10}$-DNTT was vacuum-deposited on bare Si/SiO$_2$ substrate. On top of the organic thin film, gold films (80nm) were deposited as source and drain electrodes through a shadow mask. Large I_{on}/I_{off} of 10^7 was extracted from

978-1-4673-4845-4/13 $31.00 © 2013 IEEE

measured *I-V* data. Figs. 6 and 7 compare the modeled (lines) and measured (symbols) I_{ds} vs. V_{gs} and I_{ds} vs. V_{ds} characteristics for a p-OTFT with L=140μm and W=1500μm, respectively.

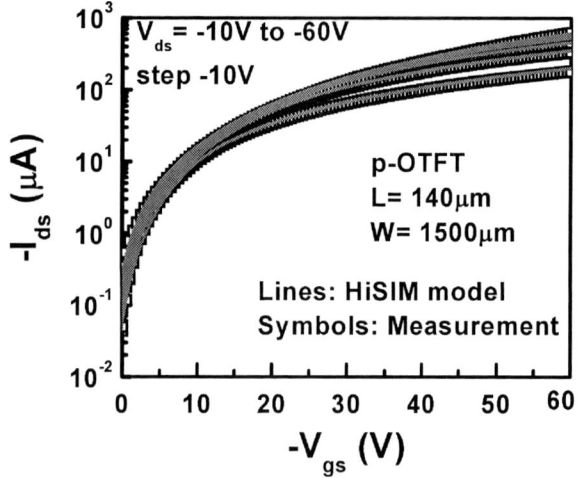

Figure 6. Comparison of measured (symbols) and modeled (lines) logarithmic drain current (I_{ds}) characteristics as a function of gate voltage (V_{gs}) for a p-OTFT

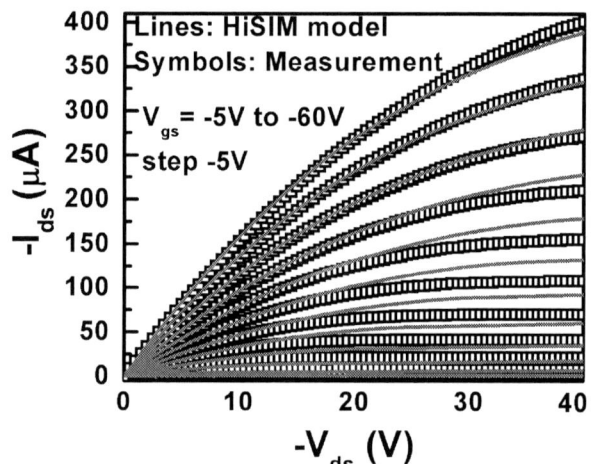

Figure 7. Comparison of measured (symbols) and modeled (lines) linear drain current (I_{ds}) characteristics as a function of drain voltage (V_{ds}) for a p-OTFT.

To extract the model parameters for measured data, we adjust first the flat-band voltage, then mobility and finally the trap parameters. The model considers the trap-state effects using model parameters as shown in Table I.

TABLE I. MODEL PARAMETERS TO CONSIDER TRAP-STATE EFFECTS

Parameters name	Description	Unit
TFTGC1	Deep states density	cm⁻³/eV
TFTE1	Inverse slope of deep states	eV
TFTGC2	Tail state density	cm⁻³/eV
TFTE2	Inverse slope of tail states	eV

To study the scaling effect, we also did a 2D device simulation of BGTC p-OTFTs (Fig.8) by varying channel length along horizontal direction and organic semiconductor layer thicknesses (T_{org}) along vertical direction using ATLAS 2D device simulator.

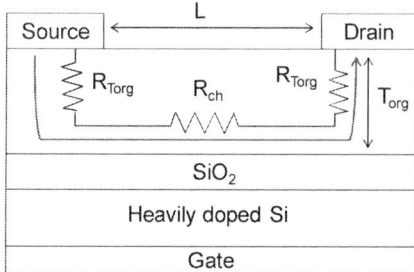

Figure 8. Schematic view of an OTFT shows the conducting path between source and drain. Channel length (L) and organic semiconductor layer thickness (T_{org}) affect the current conduction.

The simulated transfer characteristics (I_{ds} vs. V_{ds}) with W=1500μm and T_{org}=50nm of a p-OTFT are shown in Fig.9. Saturated drain current (I_{dsat}) variation at $V_{gs}=V_{ds}$=-60V with respect to T_{org} is shown in Fig.10. I_{dsat} decreases with increased T_{org} due to larger channel-to-source/drain resistances (R_{Torg}), which add to the channel resistance (R_{ch}). The variation of I_{dsat} vs. 1/gate-length ($1/L$) from measured and modeled transfer characteristics is shown in Fig. 11. For an ideal case, I_{dsat} should increase linearly with $1/L$. But in case of OTFT, the increase of I_{dsat} with respect to $1/L$ becomes smaller than expected and saturates already at relatively large L. This is due to the two dimensional (2D) effects of charge and electric-field distribution throughout the device, i.e., short-channel like effects in OTFT. Consequently, the channel to source/drain resistances lead to two additional effects in BGTC OTFTs when compared to conventional MOSFETs. Those are 1) strong organic semiconductor-layer thickness dependence and 2) short-channel like effects already at relatively large L.

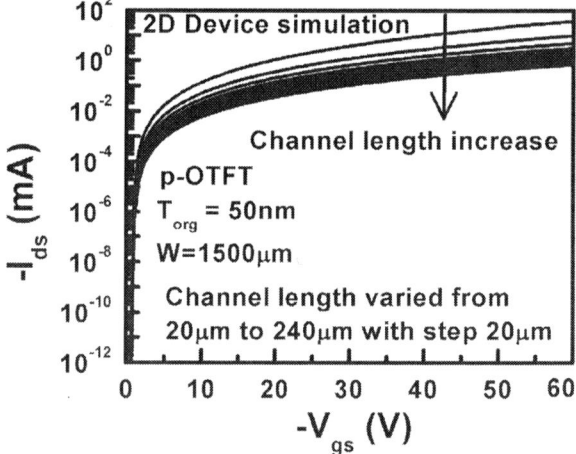

Figure 9. 2D device simulated drain current (I_{ds}) vs. gate voltage (V_{gs}) characteristics for the channel length variation from 20μm to 240μm with step 20μm of p-OTFTs of W=1500μm at V_{ds}=-60V.

978-1-4673-4845-4/13 $31.00 © 2013 IEEE   160

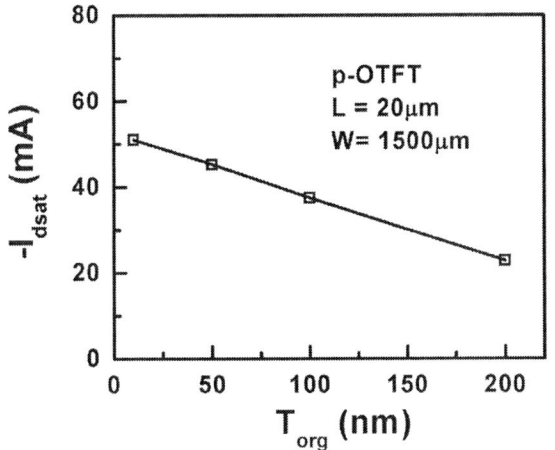

Figure 10. 2D device simulated drain current (I_{ds}) vs. gate voltage (V_{gs}) characteristics with different organic semiconductor (T_{org}) layer thickness of p-OTFTs.

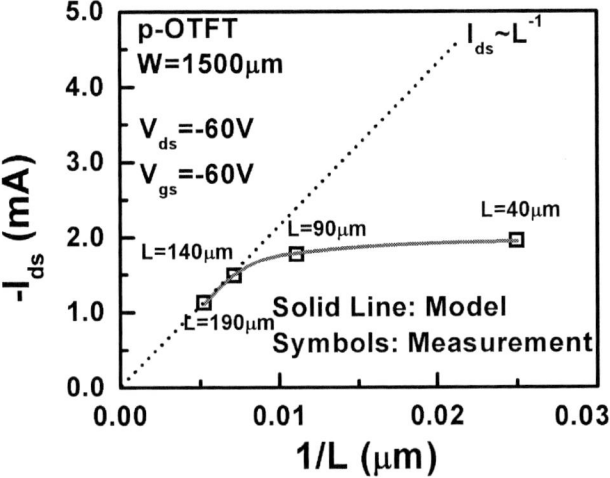

Figure 11. Saturated drain current (I_{dsat}) variation with respect to inverse of channel length (L) of p-OTFTs.

Further improvements of compact models for OTFTs will require additional developments of more precise descriptions of also other physical effects such as short-channel effects, resistance effects, overlap capacitance effects or noise effects.

IV. CONCLUSION

We presented a surface-potential based compact model for organic thin-film transistors (OTFTs) including trap effects. The model is developed on the basis of a complete surface-potential based description. This model can accurately describe the current in the subthreshold, linear, and saturation regions. The simulated I_{ds} vs. V_{gs} and I_{ds} vs. V_{ds} results are compared with measured C_{10}-DNTT field-effect transistor data [10]. Since the model is physically based, device parameters can be easily extracted accurately within a short time. It is also possible to simulate and fit the transistor characteristics of different OTFTs made from different organic materials.

ACKNOWLEDGMENT

The authors would like to thank Hiroshima University, Japan for the financial support and would also like to thank all members of the HiSIM team, Hiroshima University, Japan for their valuable inputs and discussions.

REFERENCES

[1] G. Horowitz, "Organic Field-Effect Transistors," *Adv. Mater.*, vol.10, no.5, pp.365-377, 1998.

[2] R. M. Meixner, H. H. Göbel, H. Qiu, C. Ucurum, W. Klix, R. Stenzel, F. A. Yildirim, W. Bauhofer, and W. H. Krautschneider, "A Physical-Based PSPICE Compact Model for Poly(3-hexylthiophene) Organic Field-Effect Transistors ," *IEEE Trans. Electron Devices*, vol. 55, no.7, pp.1776-1781, July, 2008.

[3] O. Yaghmazadeh, Y. Bonnassieux, A. Saboundji, B. Geffroy, D. Tondelier, and G. Horowitz, "A SPICE-like DC Model for Organic Thin-Film Transistors," *J. Korean Phy. Soc.*, vol. 54, no. 1, pp. 523-526, 2009.

[4] O. Marinov, M. J. Deen, U. Zschieschang, and H. Klauk, "Organic Thin-Film Transistors: Part I-Compact DC Modeling," *IEEE Trans. Electron Devices*, vol. 56, no. 12, pp. 2952-2961 , Dec. 2009.

[5] P. V. Necliudov, M. S. Shur, D. J. Gundlach, and T. N. Jackson, "Modeling of organic thin film transistors of different designs," *J. Appl. Phys.*, vol. 88, no. 11, pp.6594-6597, 2000.

[6] L. Li, H. Marien, J. Genoe, M. Steyaert, and P. Heremans, "Compact Model for Organic Thin-Film Transistor," *IEEE Electron Devices Lett*, vol. 31, no. 3, pp. 210-212, March. 2010.

[7] S. Miyano, Y. Shimizu, T. Murakami, and M. Miura-Mattausch, "A Surface Potential Based Poly-Si TFT Model for Circuit Simulation," *Int. Conf. on Simulation of Semiconductor Processes and Devices (SISPAD)*, pp.373-376, Sept. 2008.

[8] T. Nakahagi, D. Sugiyama, S. Yukuta, M. Miyake, M. Miura-Mattausch, and S. Miyano, "Modeling of enhanced 1/f noise in TFT with trap charges," *Int. Conf. on Simulation of Semiconductor Processes and Devices (SISPAD)* , pp.171-174, Sept., 2011.

[9] Y. Oodate, H. Tanoue, M. Miyake, A. Tanaka, Y. Shintaku, T. Nakahagi, A. Toda, T. Iizuka, H. Kikuchihara, H. J. Mattausch and M. Miura-Mattausch, "Characterization of Time Dependent Carrier Trapping in Poly-Crystalline TFTs and Its Accurate Modeling for Circuit Simulation," *Int. Conf. on Simulation of Semiconductor Processes and Devices (SISPAD)*, pp.71-74, Sept., 2012.

[10] M. J. Kang, I. Doi, H. Mori, E. Miyazaki, K. Takimiya, M. Ikeda, and H. Kuwabara, "Alkylated Dinaphtho [2,3-b:2',3'-f] Thieno[3,2-b] Thiophenes (C_n-DNTTs): Organic Semiconductors for High-Performance Thin-Film Transistors," *Adv. Mater.*, vol.23, pp.1222-1225, 2011.

[11] L.-L. Chua, J. Zaumseil, J.-F. Chang, E. C.-W. Ou, P. K.-H. Ho, H. Sirringhaus, and R. H. Friend, "General observation of n-type field-effect behaviour in organic semiconductors," *Nature*, vol.434, pp.194-199, 2005.

[12] G. Horowitz, "Tunnel current in organic field-effect transistors," *Synth. Met.*, vol.138, pp.101-105, 2003.

[13] W. Brütting, Physics of Organic Semiconductors, John Wiley & Sons, 2006.

[14] T. Leroux, "Static and dynamic analysis of amorphous-silicon field-effect transistors," *Solid-State Electron.*, vol. 29, Iss. 1, pp.47-58, Jan. 1986.

[15] T. Hayashi, Master thesis, Hiroshima University, 2011.

[16] M. Miura-Mattausch, H. J. Mattausch, and T. Ezaki, The Physics and Modeling of MOSFETS: Surface-Potential and Modeling HiSIM, World Scientific Pub Co Inc. 2008.

[17] M. Miura-Mattausch, N. Sadachika, D. Navarro, G. Suzuki, Y. Takeda, M. Miyake, T. Warabino, Y. Mizukane, K. Machida, R. Inagaki, T. Ezaki, H.J. Mattausch, T. Ohguro, T. Iizuka, M. Taguchi, S. Kumashiro and S. Miyamoto, "HiSIM2: Advanced MOSFET Model Valid for RF-Circuit Simulation," *IEEE Trans. Electron Devices*, vol.53, no. 9, pp.1994-2007, 2006.

[18] HiSIM2.7.0 Users Manual, Hiroshima University, STARC, Hiroshima, Japan, 2012.

Electrical and mechanical characterizations of a large-area, printed organic transistor active matrix with floating-gate-based nonuniformity compensator

Tsuyoshi Sekitani*, Tomoyuki Yokota, Takeyoshi Tokuhara, and Takao Someya

Department of Electrical and Electronic Engineering and Department of Applied Physics,
The University of Tokyo, 7-3-1 Hongo, Bunkyo-ku, Tokyo 113-8656, Japan
Exploratory Research for Advanced Technology (ERATO), Japan Science and Technology Agency (JST),
2-11-16, Yayoi, Bunkyo-ku, Tokyo 113-0032, Japan

Abstract—In this paper, we report the testing of the performance variations in a large-scale, printed, ultraflexible organic transistor active matrix on a 10-μm thin-film plastic substrate. A printed active matrix comprising printed floating gate organic transistors has been manufactured using high-definition screen-printing and inkjet-printing. Furthermore, by applying feedback control to the threshold voltages of the floating gate organic transistors, the circuit can be made to compensate for the device-to-device nonuniformity, which is less than 5%. The mechanical characteristics of the printed transistors are also evaluated. As a 10-μm thin film is used as the substrate, critical bending radii of less than 0.5 mm are achieved.

Keywords: variation characterizations, Printed electronics, Flexible electronics.

I. INTRODUCTION

Realization of a sustainable society will require the development of industrial manufacturing processes that have a minimal impact on the environment. From this viewpoint, the emerging field of printable electronics [1–8] should attract considerable attention: it has the potential to drastically reduce the ecological footprint, or the energy consumed, in manufacturing electronic products. In addition, developments in printing are allowing ever greater economization in the use of inks, thus minimizing material wastage in comparison to conventional manufacturing techniques.

However, a major obstacle to the development of large-area printed electronics such as large-area displays and sensors has been the fundamental compromise among manufacturing efficiency, transistor performance, and power consumption. Past improvements in manufacturing efficiency achieved through the use of printing techniques have come at the expense of inevitable low device performance and large power consumption, while attempts to improve performance or reduce power have led to higher process temperatures and increased manufacturing cost. Furthermore, crystal defects, grain boundaries, ion-based impurities, atmospheric air-associated degradations of organic semiconductors, and

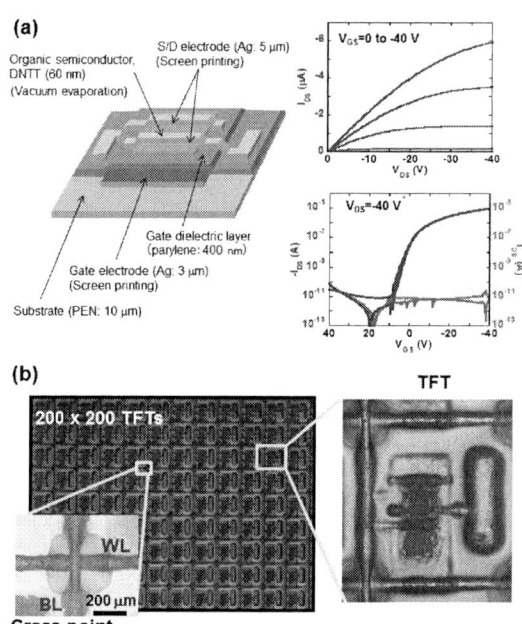

Fig. 1. (a) Schematic cross-sectional illustration of DNTT transistors. Output and transfer characteristics of the transistor are shown. (b) Pictures of test structure for performance variation on DNTT transistor active matrix with 200 × 200 transistors. Magnified picture of one transistor (TFT) and cross-point between word-line (WL) and bit-line (BL).

geometrical fluctuations of organic LEDs and printed organic transistors on plastics may result in large variations in electrical performance.

Here, we describe the testing of the performance variations (device-to-device nonuniformity) in a large-scale, printed, ultraflexible organic transistor active matrix on a 10-μm thin-film plastic substrate. Furthermore, we demonstrate how intrinsic large nonuniformity on printed transistors can be avoided with state-of-the-art screen printing technologies, printable organic materials, and circuit design, allowing the creation of large-area, ultraflexible, high-performance organic transistors and integrated circuits on ultrathin plastic substrates.

978-1-4673-4845-4/13 $31.00 © 2013 IEEE

The simultaneous characterization of mechanical and electrical performances for the printed organic circuits is also evaluated to demonstrate the feasibility of large-area printed circuits.

II. EXPERIMENTS

A. Printed organic transistors

1) Manufacturing process: Organic transistors were fabricated using source/drain electrodes formed by screen printing. Fig. 1 (a) shows a cross-sectional view of the organic transistors with top-contact geometry. A 10-μm-thick polyethylene naphthalate (PEN) film is used for flexible substrates. A Ag paste is formed using screen printing and dried at 120 °C in air to form 3-μm-thick gate electrodes. A parylene (diX-SR, Daisan-kasei) was formed using chemical vapor deposition for a 400-nm-thick gate dielectric layer.

A masking resist (#503B-SH, Asahi Chemical) is patterned by screen printing to form a shadow mask for the vacuum evaporation of organic semiconductors. The patterned resist is baked at 150 °C for 90 s in ambient air to remove solvents and to stiffen it. The base film with the printed shadow mask is rolled on a cylindrical bar with a diameter of 10 cm. For the organic semiconducting channel, 50-nm-thick dinaphtho[2,3-b:2′,3′-f]thieno[3,2-b]thiophene (DNTT) [9] was formed using vacuum evaporation with rotation mechanics, which is quite compatible with large-area material forming [10]. After the evaporation, the resist shadow mask is peeled off to pattern the organic channel layer. Ag pastes are again printed by screen printing to form printed source/drain electrodes. Finally, the whole device is coated by parylene as an encapsulation. The channel width and length of the transistors are 500 μm and 20 μm, respectively. For the remainder of this paper, these are referred to as DNTT transistors.

For comparison, a solution-processable polycrystalline organic semiconductor (Lisicon OSC) [11] with bottom-contact geometry was formed using dip-coating for the p-type channel of all-printed organic transistors. For the remainder of this paper, these are referred to as all-printed transistors.

2) Electrical performance: We have evaluated the electrical and mechanical characteristics of the printed transistors using a semiconductor parameter analyzer (B1500A, Agilent) in a light-shielding ambient atmosphere.

We measured the characteristics of the printed transistors manufactured by the above processes. The source-drain current (I_{DS}) of the transistors was measured as a function of the source-drain voltage (V_{DS}), as shown in Fig. 1 (a). The gate voltage (V_{GS}) was varied from 0 V to -40 V in steps of -10 V. The graphs in the figure show the corresponding transfer

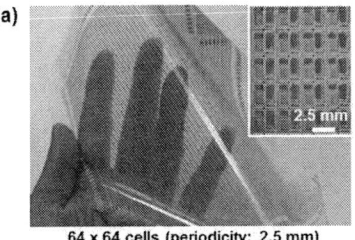

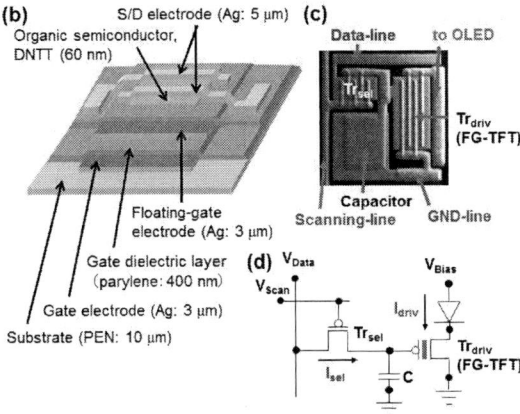

Fig. 2. (a) Pictures of printed active matrix LED pixel array comprising floating gate transistor. Inset shows magnified picture. (b) Schematic cross-sectional illustration of printed floating gate transistors. (c) and (d) Picture and circuit diagram of one pixel comprising printed floating gate transistor, normal DNTT transistor, and capacitor.

curves of the same device when V_{GS} was varied from 40 V to -40 V and V_{DS} was set at -40 V. The hysteresis was very small. In Fig. 1 (a), the on/off ratio exceeds 10^6, showing that the leakage current through the parylene layer is sufficiently low. The mobility in the saturation regime was 0.2 cm^2/Vs on average, indicating excellent transistor characteristics. This electrical performance is not degraded, even after 6 months of being stored in air, because of the good gas barrier characteristics of the parylene layer.

On the other hand, the all-printed transistors exhibit a mobility of 0.1 cm^2/Vs and on/off ratio of 10^6 at -40-V operation.

B. Printed organic floating gate transistors

The floating gate transistor is a field-effect transistor with two gate electrodes. In addition to the control gate, similar to that in a regular transistor, it has a floating gate embedded in the gate dielectric. When the dielectric is sufficiently thin, electronic charge can be brought into the floating gate by quantum tunneling or thermal emission when a sufficiently large program voltage is applied between the control gate and the source contact. Charging the floating gate changes the transistor's threshold voltage because the charge on the floating gate partially screens the electric field between the control gate and

978-1-4673-4845-4/13 $31.00 © 2013 IEEE   163

the semiconductor. This threshold voltage shift can be detected by measuring the drain current at a certain gate-source voltage. Because the floating gate is completely isolated by the dielectric, charge stored on the floating gate remains there without the need for any applied voltage (nonvolatile memory). To erase the memory, a voltage of opposite polarity is applied, discharging the floating gate through the dielectric.

Furthermore, the threshold voltage can be controlled by a high program voltage, which is typically two or three times higher than the normal operational (read-out) voltages. More details regarding organic floating gate transistors can be found in our previous reports [12,13].

1) Manufacturing process: Organic floating gate transistors were also fabricated using screen printing. Fig. 2 (a) shows a printed active matrix light-emitting diode (LED) driving a pixel circuit. Fig. 2 (b) shows a cross-sectional view of the organic floating gate transistors with top contact geometry. The manufacturing process is almost the same as that of the organic transistors mentioned in the previous session, and the channel width and length of the transistors are 500 μm and 20 μm, respectively.

2) Electrical performance: Fig. 2 (c) shows a photograph of the printed floating gate memory transistor, and Fig. 3 (a) shows a schematic illustration of the mechanism for controlling the threshold voltage (V_{th}) of the floating gate transistor. Initially, V_{th} is between +9 and +10 V (see Fig. 3 (b)). For programming, a voltage of -80 V is applied between the control gate and the source contact. This creates a gate current of more than 1 μA that charges the floating gate and shifts V_{th} to approximately -20 V. To erase, a voltage of +80 V is applied to discharge the floating gate and recover the initial threshold voltage. Cycling the gate-source voltage between +80 V and -80 V produces the hysteresis. The exact V_{th} shift (threshold voltage window) depends on the voltage and duration of the program pulse. A higher amplitude and longer duration on a program pulse lead to a larger shift in V_{th}. To read the stored information, V_{th} is determined, e.g., by measuring the drain current as a function of gate-source voltage. Figs. 3 (c) and (d) show the compensation of the device-to-device uniformity, which is sufficiently small after controlling V_{th} (nonuniformity compensation).

Taking full advantage of the state-of-the-art high-definition screen printing, low-resistivity Ag paste with a low drying temperature, and printed organic transistors, we have demonstrated ultraflexible active matrix organic LED pixel circuits on 10-μm thin-film PEN (Fig. 2 (a)). A circuit that utilizes printed floating gate organic transistors can compensate for the LED brightness variations and degradation for more than 6

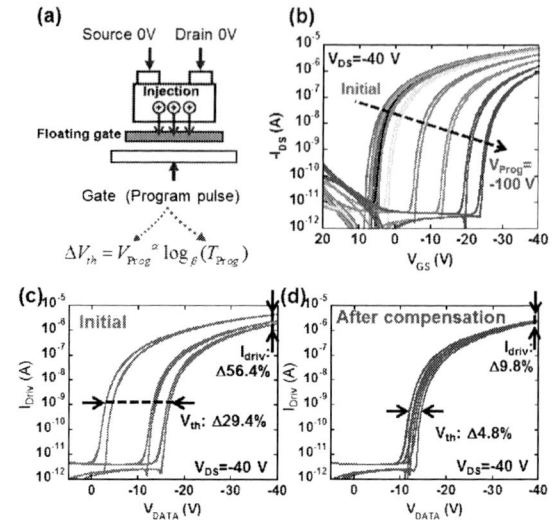

Fig. 3. (a) Schematic illustration of floating gate charging mechanism. (b) Transfer characteristics of the floating gate transistors after various program voltage pulses (100-ms duration) changing from -10 to -100 V, which were applied to the bottom (control) gate electrode. When reading out the data after programming, the voltage for the control gate (V_{GS}) is varied from +20 to -20 V with the application of a source/drain voltage (V_{DS}) of -40 V. (c and d) Compensation of performance variation. Pixel driving current (I_{Driv}) of three different floating gate transistors (FG-TFTs) are modified using dynamic control of V_{th} shift to reduce performance variation.

months. The 230 × 230 mm² printed active matrix circuit is comprised of 64 × 64 screen-printed organic 2-transistors-1-capacitor (2T1C) cells, and the periodicity is 2.5 mm (Fig. 2 (a)). Because 10-μm thin-film is used for the circuit's substrate, critical bending radii less than 1 mm are achieved.

By feedback controlling the threshold-voltages of the floating gate organic transistors and the organic LED driving currents, we can realize less than 5% nonuniformity with greater than 200 cd/m² luminance. Although the circuit design and system for feedback controlling have been reported in our previous SID [12,13], this paper has demonstrated, for the first time, the development of large-area, flexible active matrix circuits using high-definition screen printing. Furthermore, we have clearly demonstrated the feasibility of dynamic threshold voltage control as a feedback protocol for reducing the electrical performance variations in organic LEDs and printed organic transistors, taking advantage of the new device configuration of the organic floating gate transistor.

C. Mechanical characterization

1) Measurement system: The TFTs are stressed using a stress apparatus consisting of a precision mechanical stage that allows the varying of the bending radius down to 0.3 mm (Fig. 4 (a)). A capacitor is manufactured simultaneously on the same substrate in proximity to the TFTs. Its capacitance is measured for each bending radius to obtain the capacitance of the gate dielectric layers under strain. In addition, the capacitor functions as a strain gauge.

Bending experiments for flexibility tests have been performed in ambient air. The devices were stressed using a stress apparatus with a precision mechanical stage, and we evaluated the bending radii by precisely fitting the bending curvatures of the base film in the photographs taken from the side with a digital camera (Fig. 4 (a)) [14, 15].

2) Electrical and mechanical performance: The mechanical and electrical characteristics were evaluated simultaneously using a high-precision mechanical stage. Fig. 3 shows the transfer characteristics of the DNTT transistors manufactured using the printing electrodes mentioned above, and three different substrates were used. Saturation currents were decreased when the bending radius was 0.5 mm, and the threshold voltages were shifted in the negative direction. The mobility was decreased by 50% at 0.5 mm. However, these changes in performance were reversible at a bending radius of 0.5 mm, and the observed changes with bending stresses can be well explained by intergrain distance among polycrystalline DNTT semiconductors [14,15]. Further decreases in the bending radii result in irreversible degradations in performance, which were mainly due to the delamination of the source and drain electrodes from the DNTT layers. Similar bending experiments on amorphous Si transistors have been reported in other studies [16, 17].

D. Performance variation

1) Measurement system: Performance variations of printed transistors were evaluated using a semi-automatic probe and semiconductor parameter analyzer (B1500A, Agilent). Transfer curves with V_{GS} varying from 40 to -40 V and V_{DS} = -40 V were obtained in this system. Test structure of the performance variation was carried out using printed transistor active matrix, as shown in Fig. 1(b).

2) Nonuniformity characterization: Fig. 4(c) shows histograms of mobility with changing thickness of DNTT layers. The mobility systematically changed with changes in the thickness, and printed transistors with 66-nm-thick DNTTs show the highest mobility, which was 0.2 cm^2/Vs on average. Further increases in thickness in the DNTT results caused a decrease in mobility. For example, printed transistors with 89-nm-thick DNTTs showed mobilities less than 0.1 cm^2/Vs. This is mainly due to the large contact resistance between the printed source/drain electrodes and the channel layer.

Finally, we compare the performance variations between DNTT transistors and the all-printed transistors. The DNTT transistors show a device-to-device nonuniformity of 37%, while all-printed transistors show a value of 94%. Such a large

nonuniformity originates from a variation in the thickness of the solution-processed semiconductor layer and thickness of the gate dielectric layers.

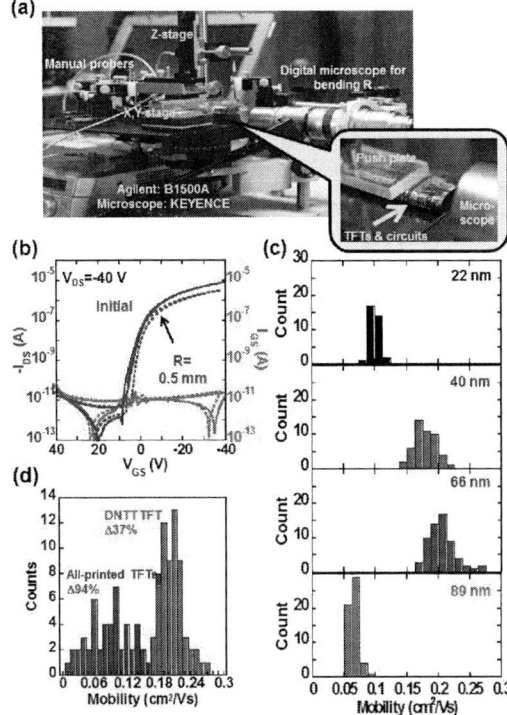

Fig. 4. (a) Pictures of measurement system for bending experiment. (b) Transfer characteristics of DNTT transistor measured before (blue solid line) and during bending stress (bending radius of 0.5 mm) (blue dashed line). Leakage currents are also shown (red line). (c) Histogram of mobility of DNTT transistor where thickness of semiconducting layer DNTT was systematically varied from 22 nm to 89 nm. (d) Histogram of mobility obtained from DNTT transistors (semiconducting layer is vacuum evaporated DNTT) (red) and all-printed transistors (semiconducting layer is solution-processed Lisicon OSC). Except for the semiconducting layer and gate dielectric layer, all electrodes are identical that are formed using screen-printing.

ACKNOWLEDGEMENTS

This study was partially supported by a Grant-in-Aid for Scientific Research (KAKENHI, WAKATE S & WAKATE A) and the Special Coordination Funds for Promoting Science and Technology. We thank Yasushi Sano (SP solutions), Takayasu Sakurai, and Makoto Takamiya (Univ. of Tokyo) for their technical support, sample preparation, and valuable discussions. We also thank Daisankasei Co., Ltd. for providing us with high-purity parylene (diX-SR).

REFERENCES

[1] R. F. Service, "Printable electronics that stick around," Science, vol. 304, pp. 675, 2004.

[2] H. E. Katz, "Recent advances in semiconductor performance and printing processes for organic transistor-based electronics," Chem. Mater., vol. 16, pp. 4748, 2004.

[3] J. A. Rogers, "Toward paperlike displays," Science, vol. 291, pp. 1502, 2001.

[4] M. Hamedi et al., "Towards woven logic from organic electronic fibres," Nature Mater., vol. 6, pp. 357, 2007.

[5] F. Garnier et al., "All-polymer field-effect transistor realized by printing techniques," Science, vol. 265, pp. 1684, 1994.

[6] J. A. Rogers et al., "Paper-like electronic displays: large-area rubber-stamped plastic sheets of electronics and microencapsulated electrophoretic inks," Proc. Natl. Acad. Sci. USA, vol. 98, pp. 4835, 2001.

[7] T. Sekitani et al., "A large-area wireless power-transmission sheet using printed organic transistors and plastic MEMS switches," Nature Mater., vol. 6, pp. 413–417, 2007.

[8] A. C. Siegel, S. T. Phillips, M. D. Dickey, N. Lu, Z. Suo, and G. M. Whitesides, "Printable electronics: foldable printed circuit boards on paper substrates," Adv. Funct. Mater., vol. 20, pp. 28–35, 2010.

[9] T. Yamamoto and K. Takimiya, "Facile synthesis of highly π-extended heteroarenes, dinaphtho[2,3-b:2',3'-f]chalcogenopheno[3,2-b]chalcogenophenes, and their application to field-effect transistors," J. Am. Chem. Soc., vol. 129, pp. 2224, 2007.

[10] Y. Noguchi, T. Sekitani, and T. Someya, "Organic-transistor-based flexible pressure sensors using ink-jet-printed electrodes and gate dielectric layers," Appl. Phys. Lett., vol. 89, pp. 253507, 2006.

[11] G. Lloyd et al., Tech. Dig. of IDW '10, 2010.

[12] T. C. Huang et al., Tech. Dig. of Society for Information Displays, vol. 13.2, 2011.

[13] T. Sekitani et al., Tech. Dig. of Society for Information Displays, 2012.

[14] T. Sekitani et al., "Ultraflexible organic field-effect transistors embedded at a neutral strain position," Appl. Phys. Lett., vol. 87, pp. 173502, 2005.

[15] T. Sekitani, U. Zschieschang, H. Klauk, and T. Someya, "Flexible organic transistors and circuits with extreme bending stability," Nature Mater., vol. 9, pp. 1015–1022, 2010.

[16] H. Gleskova, S. Wagner, W. Soboyejo, and Z. Suo, "Electrical response of amorphous silicon thin-film transistors under mechanical strain," J. Appl. Phys., vol. 92, pp. 6224–6229, 2002.

[17] L. Han, K. Song, P. Mandlik, and S. Wagner, "Ultraflexible amorphous silicon transistors made with a resilient insulator," Appl. Phys. Lett., vol. 96, pp. 042111, 2010.

Greek Cross Test Structure for Inkjet Printed Thin Films

Elkin Díaz, E. Ramon, Jordi Carrabina
CAIAC, Microelectronics and Electronic Systems Department
Universitat Autònoma de Barcelona
Barcelona, Spain
elkingonzalo.diaz@uab.cat

Abstract— **This paper reports on usage of Greek cross test structure to characterize geometry of inkjet printed electronics circuits. Geometric characteristics extracted from optical characterization can be correlated with electric measurements for square resistance in order to speed up the characterization processes. Design of inkjet printed Greek cross test structure should consider the ink coalescence and coffee ring effects.**

Keywords— *Inkjet Printed Electronics, Greek Cross Test Structure, Geometrical Characterization*

I. Introduction

Inkjet printing manufacturing process allows direct micro pattern of materials by printing from a computer generated graphics file reducing the need of photolithographic steps. This represents the main advantage of this technique [1]. The restriction is the lower quality of the resulting geometrical shapes [2] [3]. The key interesting point is that it is possible to produce micro and optoelectronic printed circuits using all polymer and metallic inks with a significant reduction of materials waste. Printed Electronics is also an attractive technology because of its advantages in low cost and flexibility of substrates.

Printing microelectronic devices built with an inkjet printer requires comprehensive knowledge of all aspects of the printing process. A set of materials and substrate must be compatible and appropriate according to electrical requirements. Opposite to traditional printing, electric characterization is required in order to check functional properties of printed layers [4]. Printed electronics manufacturers should be able to reuse or redefine characterization methods in order to develop their technologies. The usage of test structures, for a characterization methodology oriented to inkjet printed electronics, must include the following critical elements: the geometric characteristics of printed films, the impact of printing order over geometry of printed layers and the interaction between substrate and printed film of material.

The Greek cross is a standard test structure commonly used in electronics technology characterization [5] [6]. This article evaluates and analyzes the usage of this test structure for characterization of inkjet printed films. It also includes experiments with electrical and optical measurements.

II. Inkjet Printing of Materials

The process of printing material inks using a drop on demand printer will be summarized in this section. The beginning for printing process is with a rectangular matrix is used to represent each design layer; it is passed to the printer as design file in a black and white and bitmap. Next step is to jet droplets with certain distance between already ejected drops. Overlapped discrete spots of deposited fluid create physical geometries. Spot size and spot spacing determines drop overlap.

Printers cartridge will use a number of parallel nozzles (from 1 up to 16 for the DMP2831, industrial printers can go up 256 nozzles) to cover all the printing area. The cartridge is moved row by row (or raster scan for parallel jets), thus the printer jets small drops of material suspension ink (~pL) according to the design layout.

Liquid ink drops stay over the substrate, which is a process of reallocation of liquid over a solid in a wetting state. Partial wetting of the ink is expected at the end of the process, and then ink spreading parameter over the substrate must be negative. Printing leads a three mass states interphase system: liquid with liquid (drops of ink with other drops), liquid and solid (drops of ink with the substrate), liquid and air [7].

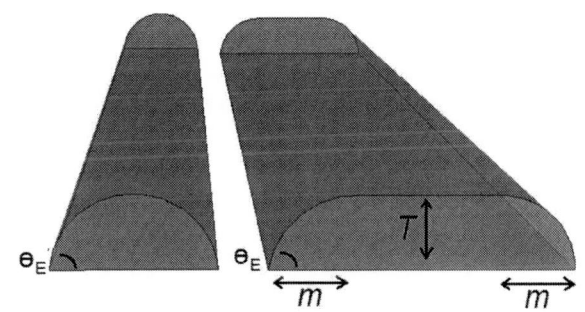

Fig. 1.   Semi-idealized conductive tracks.

Printed tracks of liquid ink have a shape similar to a parted cylindrical shape just before the solvents of inks are evaporated. Width of conductive tracks is adjusted by printing many rows of drops, or printing high volume drops. When the liquid tracks are wider then capillarity force loss dominance and the gravity force tend to planarize the liquid with a thickness T for the center of tracks.

These planarized printed tracks have a closer shape to rectangular tracks which is the most idealized model of tracks. Border of printed liquid ink tracks have a cylindrical contact angle θ_E as shown in Figure 1, which is similar to contact angle determined by Young's Law for a drop with a diameter equal to track width. Distance m is analogue to capillarity length of the same drop. Wider printed tracks of liquid ink will lead the drops to behave as heavy drops and thus to have printed tracks with a margin m and a regular thickness T.

During printing process drops are jetted close to each other previously deposited, and then drops in contact drive liquid ink to interact. Ink location changes due liquid coalesce. According to Laplace's theorem [8] for the increase in hydrostatic pressure that occurs traversing the boundary between two fluids: Previous jetted liquid ink forms a surface that has high volume and low inner hydrostatic pressure than new jetted drops. Hence, the smaller drop of ink will be attracted and forced into the larger surface of ink; it will occur if distance between them is smaller enough. At the end, the location of printed liquid ink is determined by equilibrium between gravity, capillarity and Van der Walls forces.

The following step in printing process is curing the printed pattern of ink. After that, the substrate holds an ink pattern which is cured for solvent evaporation until material layer is dry. Material inks must be formulated in order to be printable, which means that fluids used as inks must have the capability of being stably and accurately deposited as droplets by ink-jetting. Printability of inks is related with its fluid properties: viscosity, surface tension, and density of the fluid [9]. Thus, resulting shape of liquid ink is cured in an oven in order to have a film of "solid" material. The solvent is evaporated and the material particles collide according to the temperature profile performed by the oven.

The inkjet printing setup used for this work is a Fujifilm Dimatix DMP2831 printer. A specific ink for conductive layers was used, the Silver nanoparticle ink from Sunchemical (Suntronic EMD5603 with 20% of silver content). The substrate used is a DuPont Teijin PEN foil.

III. GEOMETRICAL CHARACTERIZATION

In standard semiconductor technology it is usual to perform 2D characterization for geometrical characteristics. This is a valid approach in the sense that this process allows the manufacturer to suppose that thickness is regular (though in aggressive nanometric processes this is not that true). As well, the profilometer is used to measure the thickness of layers and then the layer of material is supposed to have a regular thickness. Conversely, in the inkjet printed case it is possible to assume a mean thickness for every layer. This approach ignores coffee ring and coalescence effects [10], which are two main effects that distort thickness and width of printed layers.

The nature of liquid ink dynamics and curing processes make the thickness of material films irregular. Consequently, the instrumentation used for geometrical characterization of inkjet printed layers should consider thickness irregularities covering all the area of layers. Thicknesses are in nanometer range, width and length for layers of material devices are in the tenths of microns to millimeters range. Interferometer microscopy can perform this kind of measurements for 3D morphology fast extraction at this scale.

Comparing the thickness irregularity (~nm) with the irregularities in the planar dimensions of the layer (~mm) the thickness appear to be negligible. However, for designs with "wires" or long rectangles that interconnect different parts of the printed circuit, the minimum width is used and current flows in the planar direction (parallel to the substrate). Then cross sectional area of wires is one of the critical dimensions, and also the width-thickness ratio of layers. Therefore the role of thickness is important because it is directly associated with the conductivity of printed wires.

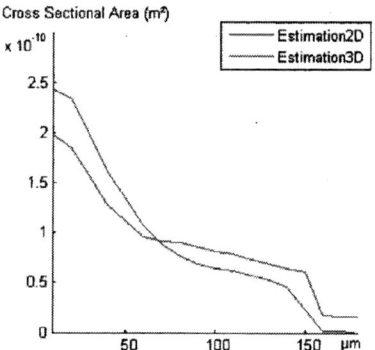

Fig. 2.   Estimation for Cross-sectional area microscopic and 3D profiling.

An example of the influence of thickness over cross sectional area is the inkjet printed Greek cross structure. Figure 2 shows the cross sectional area of the cross arm over distance from the center up to the end of the arm, width and thickness are bigger for regions close to the cross center. This irregularity in the arm cross sectional area over distance is caused by a center attraction of ink by coalescence.

The 2D estimation (blue line in Figure 2) of cross sectional area that supposes that thickness is regular over the cross. Cross-sectional area is proportional to the resistivity of the track, then in the case of assuming regular thickness it is just proportional to the track width. The 3D estimation (green line in Figure 3) is the result of integrating thickness data in the cross section of the arm, then for each point from the center to the arm end it shows the cross sectional area. Thickness dimension is important, as it allows quantifying the coalescence effects over resistivity of tracks.

In a first approach, we printed Greek crosses and extracted multiple profiles with stylus profiling. Geometrical characteristics were estimated from the superposition of the different sectional profiles for each sample. The coffee ring effect was detected and varying its intensity over the arms of the cross.

Measurements with stylus profiling are slow, speed of stylus is limited by the adhesion of printed tracks to substrate, the substrate has to be fixed to the holder but sample can be misallocated or destroyed if the speed of stylus is high. Those mechanical problems hind the characterization of geometrical characteristics for inkjet printed layers of material.

978-1-4673-4845-4/13 $31.00 © 2013 IEEE

The requirement for the microscope field of view is at least of 1mm², the vertical resolution must cover nanometer sensitivity because the thickness of layers is 400nm~1600nm. The required 3D profile for analyzing this kind of morphologies can be done using interferometry techniques by phase shift profiling, because it provide nanometer vertical resolution to measure large fields of view.

A second approach for geometrical characterization was performed using interferometry microscope. We fixed the size of each sample to an area of 1.2mm x 1mm, according to the resolution required for this kind of morphologies, and the available lenses with the Sensofar Interferometry Microscope used for this experiment.

IV. EXPERIMENT DESIGN CONSIDERATIONS

The minimum line spacing for printing conductive tracks of silver ink can be determined as a design rule. As well, the digital printing technique delimits the minimum line spacing in a discrete range with discrete spacing steps of the drop spacing size. The drop spacing is the theoretical spacing between two adjacent printed droplets of ink, normally for every ink depending on its properties exist fixed drop spacing. Drop spacing is parameter that also limits the speed of the printing process.

It is always possible to use the minimum drop spacing and reorganize the design patterns into the equivalent grid of minimum drop spacing; this allows designs to be less staggered. Drop spacing modification is possible by leaving holes between droplets when larger drop spacing is needed and reducing the number of holes between droplets for smaller drop spacing. The limitation of this approach is the related variation in printing speed.

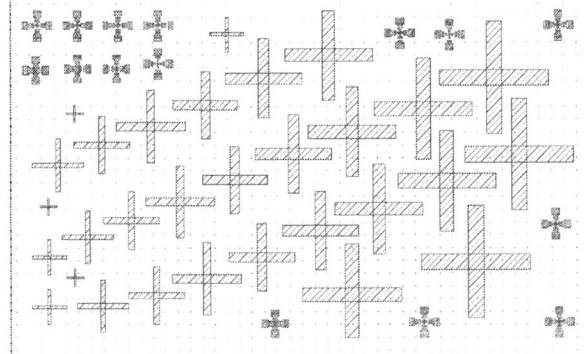

Fig. 3.   Experiment layout for cross regularity with different widths.

Printed samples were printed for different line widths ranging from 40μm up to 200μm conductive track with a cross structure. Design of these structures is shown in Figure 3. Those crosses were analyzed in terms of morphology and regularity.

The methodology proposed to compensate the Greek cross structure is shown in Figure 4. It involves three printing processes in order to design different patterns attempting to compensate regularity critical parts in designs. Many patterns

are proposed according to the location of irregularities in the critical parts of the design. The most regular pattern is selected and then the entire design is compensated according to compensation pattern.

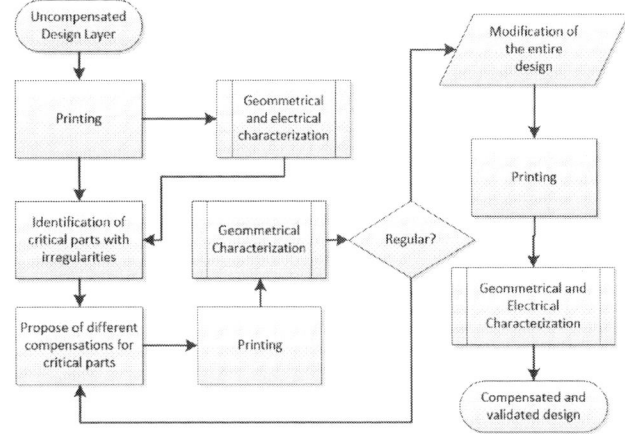

Fig. 4.   Methodology used for irregularity compensation.

The irregularity compensation patterns proposed allowed to select one pattern that improved the regularity of the cross structure. Examples of compensation patterns are shown in Figure 5, all of them were designed for a line width of 5 droplets. This compensation method is its limited by the number of parallel droplets, having no compensation possibilities for single droplet lines. But also it has overwhelming possibilities of compensation for wider lines.

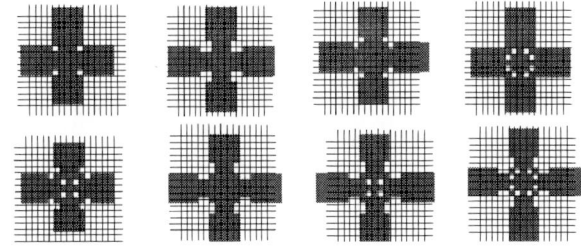

Fig. 5.   Compensation patterns example for inkjet printed cross structure.

Another experiment was intended to correlate optical measurements with electrical measurements. Two different sets of 90 test structures, same dimensions for compensated and uncompensated sets. We used only one fixed width for these samples, including pads for probe connection. Single width of droplets can vary, we used 40μm, but in order to print layers, droplets must overlap between them, this overlapping is determined by the drop space, we used a drop spacing of 20μm, then when we print 5 droplets overlapped by 20μm at the end we get 100μm tracks.

Electrical measurements for sheet resistance of Greek cross test structures with Ag-nano particles suspension ink samples. We used a B1500A Agilent Semiconductor Device Analyzer with 4 SMUs and a manual probe station for electric characterization. Retractable needles were used for the probe contacts.

978-1-4673-4845-4/13 $31.00 © 2013 IEEE   169

V. THEORY AND CALCULATIONS

The Greek cross test structure is based on the Van der Pauw measurement for resistivity. It is applied on a continuous surface of arbitrary shape. The assumptions that must be satisfied according to Van der Pauw technique formulation are the following: contacts must be in the circumference of the sample, contacts must be small enough, thickness of sample is homogeneous and there are no holes in the entire surface.

Geometry of inkjet printed crosses can be represented as the intersection of two printed tracks. According to the semi-idealized model of printed tracks previously used in Figure 1, the Greek cross center presents four regions with thickness less than *T*. In an ideal case the center should be a perfect square, in reality it is a shape close to a quadrangle or a four edges irregular polygon. But the corners of this irregular polygon are not directly seen in the structure, because the corners are presented in angles which result in the liquid morphology as rounded or curved angles.

Determination of cross center was done according the following criteria: The center of the structure is used to define X and Y axis according to the middle of each printed track. Once those axis are defined, then two 45° and two -45° lines are used to find the corner of the center by displacing until they are intersected with only one point in the rounded corner of the structure. Once the four corners are defined the center tetragon is defined and information can be extracted. This definition procedure is illustrated in Figure 6. The center area varies from structure to structure, and then an automated algorithm was used to determine each sample center.

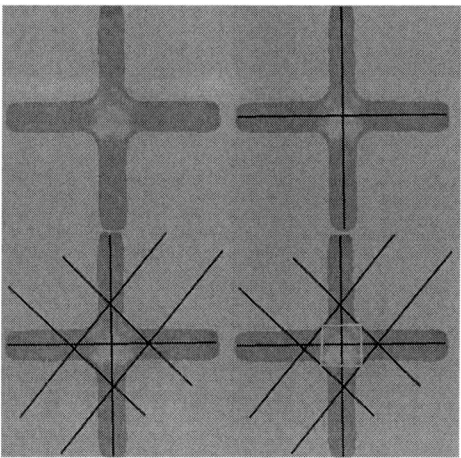

Fig. 6.   Center cross definition used in extraction of geommetrical data.

The center of the cross is the surface intended to have regular thickness, and arms are wires that connect the four contacts. Then irregularities in arms are non-influent over the measurement of sheet resistance, but the center parts with irregular thickness affect the estimation of sheet resistance.

There are many inherent capillarity effects in printed liquid ink expected to occur. Those effects appear when printing the cross. The start and end parts of the arms may form a softer shape or a semicircular shape instead of the rectangular cut.

The corners of the cross center may present also an edge softening due the proximity between two different liquid surfaces, and at the end the square of the center will have a bigger area than width², because of the additional spreading in the centers corners.

$$g[x,y] = \frac{h}{\sqrt{2\pi}\sigma} e^{-\frac{(x^2+y^2)}{2\sigma^2}} \qquad (1)$$

The shape of cross center hills are close to Gaussian hills, the variability of its geometry is analyzed and its height h and its width or smoothness determined by σ parameter. The analysis for samples requires the extraction of its characteristics according to a Gaussian hill model (1).

VI. RESULTS

The analysis of the different widths for cross structure revealed that irregularities are presented for all the cases, and demonstrated that irregularity get worst for low widths. However for all the sizes and samples analyzed center is critical and present peaks. Figure 7 shows results for six different widths and scales.

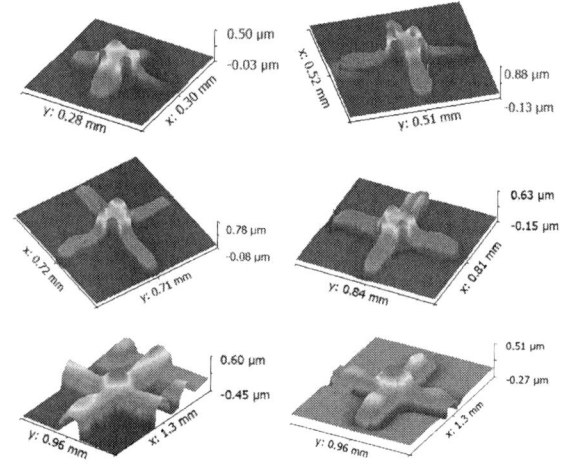

Fig. 7.   Center irregularity for different crosses with different widths.

Compensation methodology was applied in order to determine patterns for the Greek cross test structure. It is possible to reduce hills in the cross center for Greek cross test structures, which is desirable in order to use electrical measurements to estimate characteristics of conductive tracks. This compensation also reduces the area covered by the center of the cross.

The experiments with different widths of Greek cross test structure revealed the possibility to compensate the structure. Compensation was applied to the structure design, and geometrically irregularities and peaks were greatly reduced.

We printed two sets of samples, one using the compensated design, and other without compensation. All those samples have the same dimensions, with a width of tracks fixed to 100μm. Then we tried to correlate sheet resistance measurements with center geometrical characteristics.

The results for the uncompensated set of samples correlation between square resistance and mean thickness of the center did not present a correlation. A second set, with ninety modified printed test structures reflected improved geometry results. Compensated design for Greek cross test structure makes the center thickness more regular.

The analysis of data for compensated and uncompensated samples show no correlation between sheet resistance and mean height of the cross center. However, sheet resistance and volume for the uncompensated sample show power law correlation while the correlation between sheet resistance and volume for compensated samples was not feasible.

Irregular thickness prevents characterization procedures from assuming a linear correlation between square resistance and height of center's hill. Experimental data revealed an power law correlation (shown in Figure 8) between volume of cross center and square resistance estimation measured in Greek cross calculated with the Van der Pauw expression (using this calculation is not theoretically correct because in this case the thickness is irregular). Goodness of fit for this exponential model is 0.82.

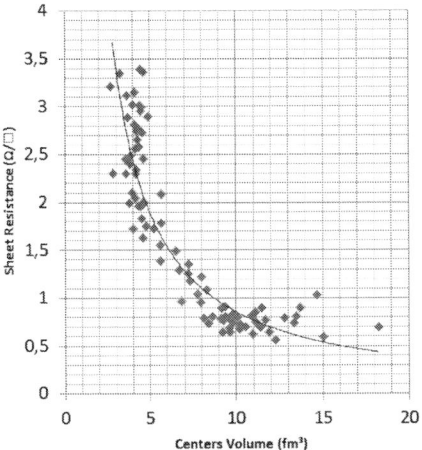

Fig. 8. Correlation between electrical and geometrical measurements, uncompensated Greek cross structure.

Uncompensated Greek cross structures all have the same dimensions, but according to the center definition, the corners made this center to vary too much, and also the variations in the process. This randomness scatters the centers volume as shown in Figure 8.

Uncompensated design shown in Figure 10, it has a smoothness (1) mean value of 40.36μm and smoothness standard deviation of 4.33μm. For the compensated design the smoothness mean value of 20.23μm and standard deviation is 0.947μm. For the irregularity of thickness the peak height h is reduced by the compensated design because the height mean value for uncompensated design is 254.73nm with a standard deviation of 30.125nm. Compensated designs shown in Figure 9, exhibits a reduced height mean of 63.93nm with standard deviation of 13.26nm.

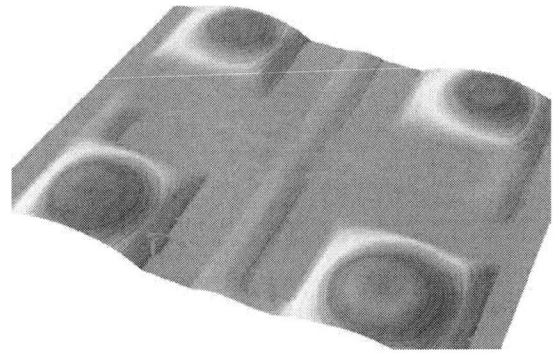

Fig. 9. Morphology of an inkjet printed Greek cross test compensated structure.

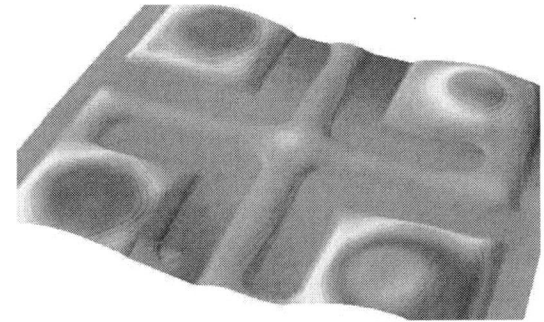

Fig. 10. Morphology of an inkjet printed Greek cross test uncompensated structure.

The correlation for uncompensated structures has higher dispersion of volume than compensated structure; volume variation is reduced by 4.5. However sheet resistance dispersion is multiplied by 3, which means that both designs have not enough correlation between Van der Pauw measurements and cross centers volume. Reducing coffee ring and coalescence effects over cross center thickness made a counter effect with irregularities and instabilities in the center borders, which is reflected in the increase of variability for sheet resistance.

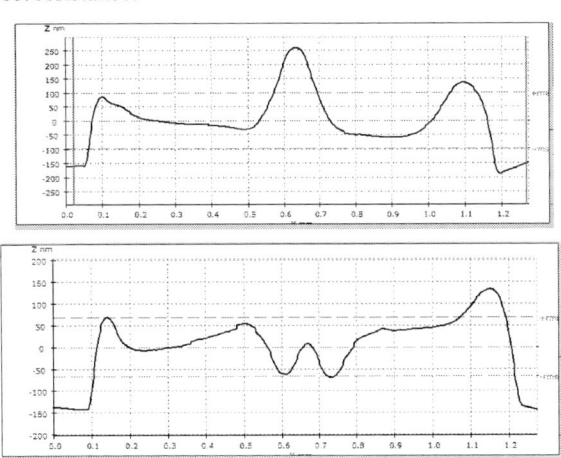

Fig. 11. Profile comparison between uncompensated and compensated structures.

Experiments show that it is possible to compensate the design of a Greek cross test structure. Compensation allows having center thickness closer to the thickness of interest; profiles of compensated and uncompensated structures are shown in Figure 11. Peak is reduced but in the counterpart valleys are created in the surroundings of the centers hill.

The thickness of interest is the mean thickness of printed conductive tracks. The technique used for compensation is an experimental patterning which reduces the volume of ink printed in the center of the cross. This compensation corrected the peaks in the center but it's unstable and the sheet resistance variability is increased.

In planar devices it is possible to assume that current flows only in planar direction, this is not the case for this inkjet printed electronics, where thickness is not regular then all morphology affects the direction of current. For example in the peaks of cross centers, where the current flow is distributed according to the morphology, resistivity cannot be simplified to rely inversely proportional to the thickness as thickness is not regular and current flow is distributed accordingly to surface morphology.

VII. CONCLUSIONS

The Greek cross test structure allows a fast characterization of the centers thickness, which is not directly equal to the thickness of printed tracks. It can be useful to characterize this liquid interaction effect and the ink-substrate relationship by contrasting printings for the same process, or comparing different inks or substrates with this kind of measurement.

A methodology for compensating the morphology irregularities in inkjet printed structures was proposed, and as an example the Greek cross test structure was compensated in order to improve its center thickness regularity and in order to reduce the peak to have center thickness similar to the thickness of printed tracks.

Optical characterization with thickness characteristics over wide areas is needed, in order to correlate geometrical and electrical characteristics. Inkjet printed test structures present irregularities due its process nature, those effects must be considered. It is possible to apply geometrical compensations while reducing irregularity effects but the stability and repeatability is critical in this kind of compensations.

ACKNOWLEDGMENT

This work was partly supported by the Spanish Ministry MITYC project TSI-020110-2009-360, the Spanish Ministry MICINN project TEC2011-29800-C03-03, the Grant GICSERV 2011 from ICTS-CNM (CSIC), the Catalan Government Grant Agency Ref. 2009SGR700 and the and the TDK4PE (FP7-ICT 2012-2014) "Technology & Design Kits Development for Printed-Electronics" under grant agreement n° 287682.

REFERENCES

[1] [1] C. Kim, M. Nogi, K. Suganuma, and Y. Yamato. Inkjet-printed lines with well-defined morphologies and low electrical resistance on repellent pore-structured polyimide films. ACS Applied Materials & Interfaces, 4(4):2168–2173, 2012.

[2] [2] R. Cauchois, M. Saadaoui, J. Legeleux, T. Malia, B. Dubois-Bonvalot, K. Inal, J.C. Fidalgo, et al. Chip integration using inkjet-printed silver conductive tracks reinforced by electroless plating for flexible board packages. In Acte de conférence IMAPS, 2012.

[3] [3] S.H. Ko, H. Pan, C.P. Grigoropoulos, C.K. Luscombe, J.M.J. Fréchet, and D. Poulikakos. All-inkjet-printed flexible electronics fabrication on a polymer substrate by low-temperature high-resolution selective laser sintering of metal nanoparticles. Nanotechnology, 18(34):345202, 2007.

[4] [4] G. Blanchet and J. Rogers. Printing techniques for plastic electronics. Journal of Imaging Science and Technology, 47(4):296–303, 2003.

[5] [5] A. Tsiamis, S. Smith, M. McCallum, A. Hourd, T. Stevenson, and A.J. Walton. Electrical test structures for the characterization of optical proximity correction. Semiconductor Manufacturing, IEEE Transactions on, 25(2):162–169, 2012.

[6] [6] S. Enderling, C.L. Brown III, S. Smith, M.H. Dicks, J.T.M. Stevenson, M. Mitkova, M.N. Kozicki, and A.J. Walton. Sheet resistance measurement of non-standard cleanroom materials using suspended greek cross test structures. Semiconductor Manufacturing, IEEE Transactions on, 19(1):2–9, 2006.

[7] [7] G.D. Martin, S.D. Hoath, and I.M. Hutchings. Inkjet printing-the physics of manipulating liquid jets and drops. In Journal of Physics: Conference Series, volume 105, page 012001. IOP Publishing, 2008.

[8] [8] P.G. De Gennes, F. Brochard-Wyart, and D. Quéré. Capillarity and wetting phenomena: drops, bubbles, pearls, waves. Springer Verlag, 2003.

[9] [9] D. Jang, D. Kim, and J. Moon. Influence of fluid physical properties on ink-jet printability. Langmuir, 25(5):2629–2635, 2009.

[10] [10] N. Jones, S.J. Sargeant, K. Sargeant, J.C. Briggs, and M.K. Tse. Characterizing and modeling coalescence in inkjet printing. Proc IS&T NIP14, 1998.

Process Control Monitors for Individual Single-walled Carbon Nanotube Transistor Fabrication Processes

Kiran Chikkadi, Miroslav Haluska, Christofer Hierold, Cosmin Roman

Micro and Nanosystems, Department of Mechanical and Process Engineering
ETH Zurich
Zurich, Switzerland
kiranc@ethz.ch

Abstract— The manufacturing yield of carbon nanotube transistors is very sensitive to changes in fabrication process parameters, while controlling length, density and orientation of nanotubes simultaneously is still proving elusive in batch fabrication processes. Here, we show an electrode design with a yield of up to 45% working transistors despite our batch fabrication process being based on randomly grown nanotubes. Transistor parameter distributions of 765 devices are shown, demonstrating the potential of our design for process monitoring and control.

Keywords—process monitoring; single-walled carbon nanotube; transistor; monte carlo; length and density distribution

I. INTRODUCTION

As the advantages of carbon nanotube-based electrical, mechanical and sensing elements become more apparent, processes to scale up the fabrication of single-walled carbon nanotube (SWNT) transistors from prototypes to wafer-scale are currently being developed [1-4]. Various large-scale-compatible processes such as top-down photolithography [5], dielectrophoresis [3], chemical self-assembly [2], spray and spin coating techniques are being explored for this purpose. However, the SWNT deposition or growth processes often result in SWNTs with varying length, density and diameter, which leads to devices with varying yields and having widely distributed electrical characteristics [6, 7]. As a result, process monitoring and feedback becomes essential for better control of carbon nanotube device fabrication processes. Currently, in situ or ex situ techniques such as optical (Raman spectroscopy), scanning probe microscopy (e.g. AFM) and scanning electron microscopy (SEM) are used for monitoring the SWNT deposition step. However, these techniques do not cover all relevant properties (e.g. contact resistance), they are slow and they might induce contamination (e.g. SEM-induced carbonaceous deposits).

In this work, we propose a set of test structures which can be electrically probed and implemented as process control monitors (PCM), allowing for feedback on device properties such as contact resistance, threshold voltage, ON/OFF ratio, and nanotube type (metallic or semiconducting). In particular, we focus on the problem of maximizing the yield of devices where an individual SWNT is bridging across electrodes in integration processes which lack SWNT orientation control. For obtaining statistics on transistor parameters, a main requirement for nanotube PCMs is to maximize the number of individually contacted SWNTs. Contacting multiple SWNTs complicates the interpretation of the results by rendering the extraction of individual SWNT property distributions difficult. Furthermore, the probability of incorporating a metallic SWNT into the device increases with the number of SWNTs in the device channel. Metallic SWNTs dominate charge transport in such channels, thus masking the properties of semiconducting nanotubes parallel to them [8]. Therefore, the main task was to design an electrode geometry aiming at maximizing individual-SWNT devices.

II. FABRICATION PROCESS

The fabrication process flow chosen to demonstrate our framework has been previously described in [9, 10]. This process is characterized by non-oriented SWNTs with a length following a Weibull distribution. In general, our methodology is also applicable to other batch-fabrication processes that lack control over nanotube length and orientation.

In short, our fabrication process involves the photolithographic definition of catalyst islands (ferritin is used as a nanotube growth catalyst precursor), followed by dip-coating and lift-off. A heavily p-doped Si wafer is used as the substrate (eventually used as the back-gate). 70 nm of dry thermal SiO_2 on the surface serves as the gate oxide. The density of catalyst particles (and thereby, the density distribution of the SWNTs) can be tuned by controlling the pH and concentration of the catalyst solution [11]. The size of the catalyst particles (and thus, the diameter of the SWNTs grown) can be adjusted by the amount of iron loaded into the ferritin cage [12]. Growth of SWNTs from the catalyst islands is done by a low-pressure chemical vapor deposition process with CH_4 and H_2 at 850 °C. The SWNTs are then contacted with Cr/Au metal electrodes using lift-off and passivated with atomic layer deposited Al_2O_3 for stability [13].

III. PCM DESIGN

The proposed electrode design consists of blocks of 20 electrodes as shown in Figure 1 (see also Figure 5). The design dimensions such as the electrode gap, width and the size of the catalyst islands were optimized using Monte Carlo simulations, with the objective of maximizing single-bridging SWNTs. The design requires about 0.5 mm^2 of chip area (excluding bond pads) for ~1000 test devices, providing sufficient device count to obtain statistics on the device characteristics. The small size also enables the devices to probe the uniformity across a wafer. For the Monte Carlo simulations, the distributions for length and density utilized here are obtained from the estimation procedure reported in [10], where we verified the length and density distributions using other monitoring structures.

Financial support from EU FP7 project Technotubes is gratefully acknowledged.

978-1-4673-4845-4/13 $31.00 © 2013 IEEE

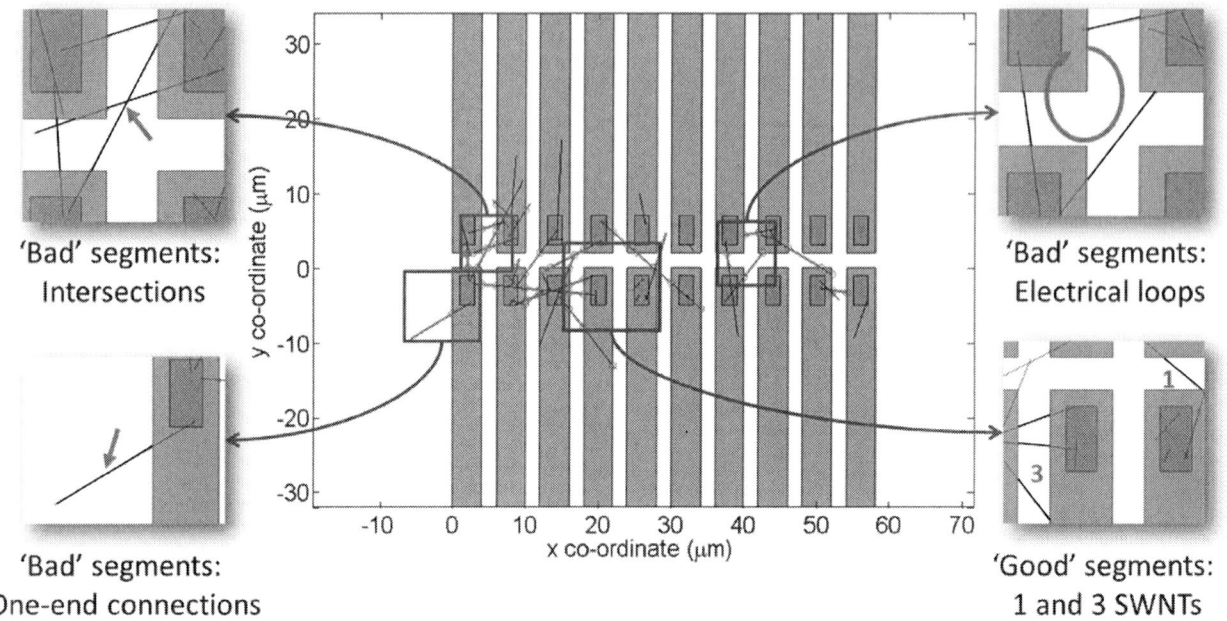

Figure 1. Design of the process control monitor structure. An instance of the Monte Carlo simulation for analyzing the connectivity is shown above. There are 20 electrodes of 4 μm width (light grey), with equal inter-electrode spacing. The catalyst island from which the SWNTs grow is located at the tip of the electrode, and is 2×4 μm^2 (dark grey). The total area of one such field of 20 electrodes is approximately 60×60 μm^2. The connectivity analysis distinguishes between intersections between two or more SWNTs, single-ended connections and electrical loops and excludes them from the analysis.

The length distribution for the SWNTs (assumed to be straight) is assumed to be a Weibull distribution with parameters a = 6.91 μm and b = 1.34. A Poisson distribution with parameter λ = 1.44 is assumed for the number of carbon nanotubes nucleating from a specific island. As the tube orientation is not controlled in this process, we assume a uniform angular probability distribution for it. After the tubes have been generated, a connectivity analysis is performed (figure 1). In addition to the simple bridging events between electrodes, the SWNTs can form connections leading to intersections between two or more SWNTs or single-end connections that are too short to bridge across. The connectivity analysis identifies these as 'bad' segments and parallel-connecting SWNTs as 'good' segments. Further, electrical loops formed by three or more mutually connected electrodes, are excluded (categorized as 'bad') since the devices part of a loop cannot be reliably characterized. It must also be noted that while the group contains 20 electrodes for computational simplicity, in principle a group may contain any number of electrodes as long as the geometric parameters are maintained.

Figure 2 shows the simulated bridging statistics for a 2 μm electrode gap size. To verify the simulation, we have gathered bridging statistics by AFM in the intermediate step after SWNT growth. These measurements show about 8 single-bridging tubes per 20 electrodes for an electrode gap of 2 μm, showing good agreement with simulated data. Furthermore, the SWNTs are straight as expected, although minor kinks and bends may be observed on some of the SWNTs.

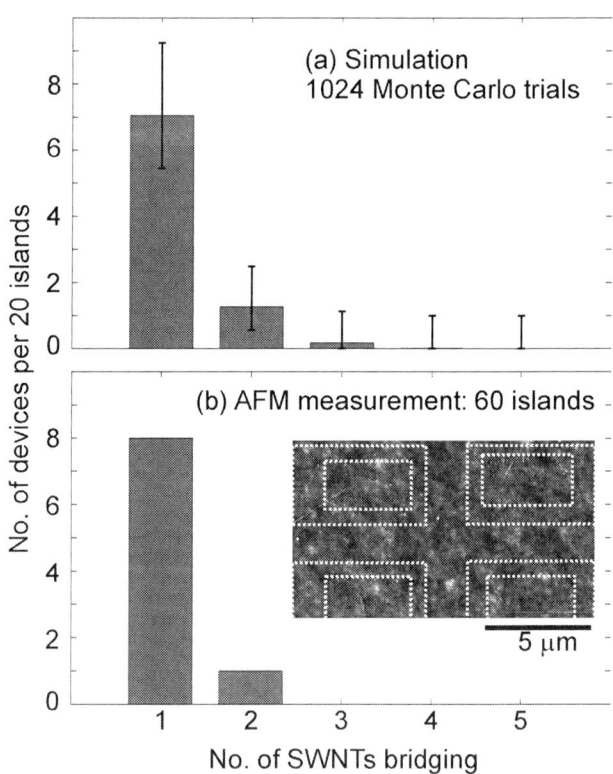

Figure 2. The number of single, double and multiple bridging SWNTs per set of 20 electrodes from (a) Monte Carlo simulations with an electrode gap of 2

μm and (b) Atomic Force Microscopy (AFM) measurements. Inset: an example AFM image used to obtain the bridging data is shown, with virtual electrodes overlaid to show the approximate position of the contacts after metal deposition. A good agreement between simulation (a) and experiment (b) is obtained for the number of single and double bridging SWNTs.

It is important to note that while it is possible to maximize the single-bridging nanotubes at the cost of the total device yield- for instance, placing the electrodes far apart reduces the probability of multiple bridging significantly, leading to mostly single bridging events. Therefore, it is necessary to optimize the yield of single-bridging devices with both the total yield as well as the proportion of single-bridging devices in mind. In this context, we define two quantities: (1) the single-bridging ratio, which is the abundance of the single-bridging events relative to the number all bridging events occurring; (2) the single-bridging yield, which is the number of single-bridging devices relative to the maximum number of devices possible (=19 for 20 electrodes).

Figure 3 shows the 'single-bridging ratio (SBR)' and the 'single bridging yield (SBY)' for varying electrode gap size. As expected, we observe that SBR monotonically increases while SBY monotonically decreases with electrode gap size. To obtain a reasonable trade-off between SBY (increasing) and SBR (decreasing) we have defined the function F = SBR×SBY as the one to maximize. It is seen that this optimum occurs at 2 μm, where SBY is about 39% and SBR is about 80%. This means that the final statistics are dominated by single-tube devices, which account for 80% of the devices obtained.

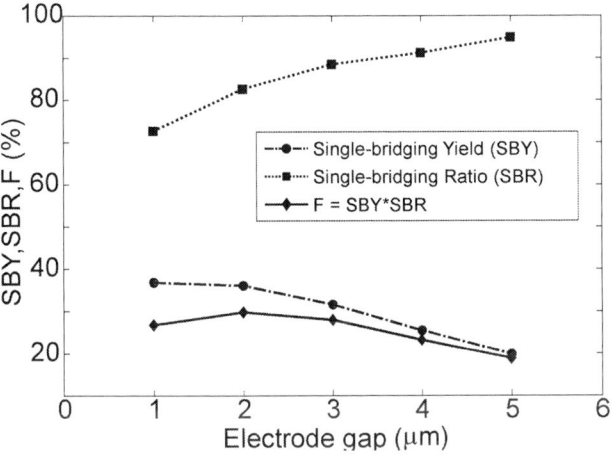

Figure 3. Optimization of the single-bridging device yield from the PCMs, obtained from 512 Monte Carlo trials. The single-bridging ratio (SBR) is the relative abundance of single-bridging events to the total number of devices obtained in the simulation. Single-bridging yield (SBY) is the ratio between the total number of single bridging events relative to the maximum number of possible devices (19 from 20 electrodes). It is seen that as the electrode gap increases, the total yield drops, while the relative single-bridging abundance increases. As a result, the compound function F = SBY*SBR is used to optimize the geometry, which peaks at 2 μm.

For comparison, we examine an electrode design involving a concentric inner and outer electrode (figure 4). Similar

designs have been previously investigated in [1] and [10]. Qualitatively, such a design can lead to much higher device yield, since SWNTs growing in nearly all orientations can be covered by the outer electrode, but it also simultaneously increases the probability of multiple bridging events. At a gap size of 1 μm the SBY for this device is observed to be only 22% while SBR is 25%. This means that multiple bridging events constitute a significant proportion of the final statistical distributions measured, making such a geometry less suitable for acquiring single-tube device statistics than the one proposed. As a comparison, at the same electrode gap, our design results in a SBY of about 39% and a SBR of 72%.

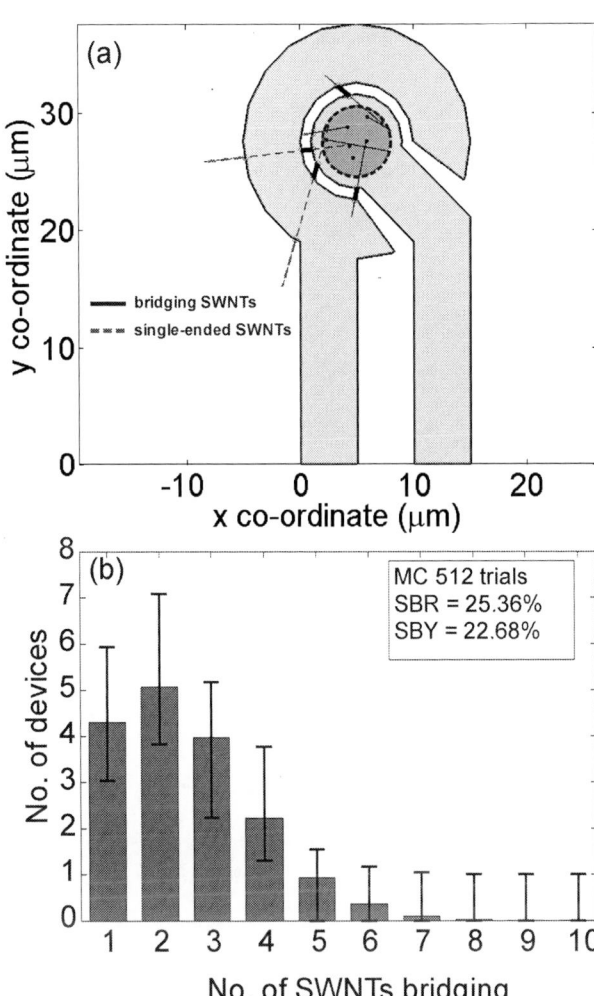

Figure 4. (a) An alternative design with a concentric electrode and gap size of 1 μm simulated with identical parameters. The ferritin island (dotted circle) has a similar area to the previous design, and identical parameters are used for the SWNT distributions. In this instance of the simulation, 4 SWNTs may be observed bridging between electrodes. (b) From the bridging yield, is it clear that both the single-bridging yield and the single-bridging ratio suffer due to multiple bridging events, rendering such a structure less suitable for acquiring single-tube statistics.

IV. RESULTS AND DISCUSSION

Figure 5 shows an SEM image of the completed PCM structures. To characterize these devices, transfer characteristics with pulsed gate voltage sweeps (with 1 ms pulsed width) were obtained with reduced hysteresis as described in [14]. A total of 765 devices (out of 1680 electrodes) were fabricated from two distinct runs, at a total device yield of 45.5%. This is comparable to the simulated yield of 44.7% (that accounts for multiple bridging events). From the fabricated devices, about 57% of the devices were semiconducting, which also agrees well with the predicted semiconducting tube proportion (50-60%).

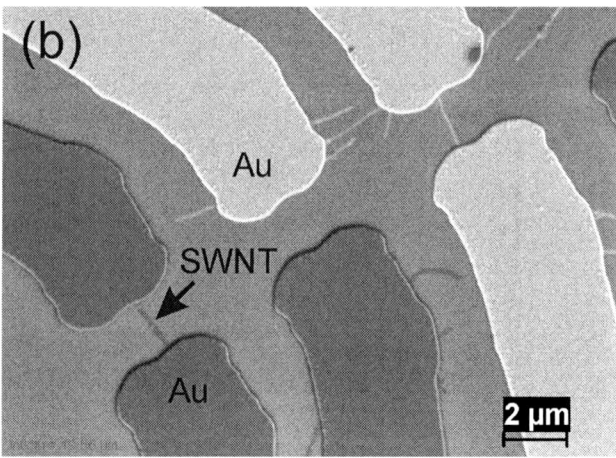

Figure 5. Scanning electron microscope images of the fabricated PCM device structures. (a) The 20 electrodes that form the source-drain contacts are seen (the difference in color is presumed to be from charging effects in the SEM; the material is Cr/Au for all electrodes). (b) close-up of the devices with the SWNT channel are seen. Three single-bridging devices and several single-ended connections are seen. The Si substrate is used as the back-gate for electrical characterization.

Statistics for the on-state resistance, threshold voltage, on-off ratio and hysteresis as measured from the semiconducting devices (~400) are shown in figure 6. Thus, the capability to extract such information rapidly from a distribution dominated by single-tube devices provides the possibility to measure and gain feedback on the fabrication process. Currently, the electrical measurements cannot distinguish single- from multi-bridging or tube-tube intersections due to the wide resistance distribution. With better control of the contact resistance distribution, we expect that these will become distinguishable as well.

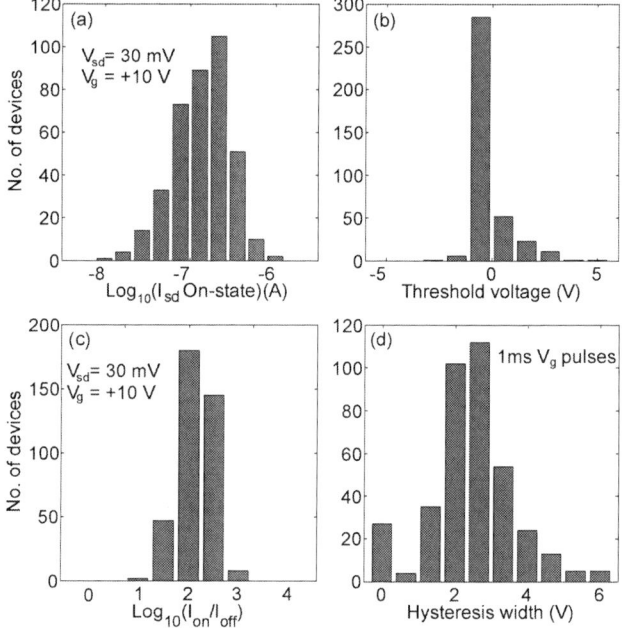

Figure 6. Histograms for (a) the maximum current (at V_g =10 V, V_{sd} =30 mV), (b) threshold voltage, (c) ON/OFF ratio (OFF-state limited by setup noise) and (d) gate hysteresis width at 1 ms gate pulse width. The mean on-state resistance is estimated to be 280 kΩ. Variations in these histograms can provide feedback on the process flow.

V. CONCLUSIONS

In conclusion, for carbon nanotube device fabrication processes where the length and the orientation are widely distributed, device yields of around 45% of which 80% are expected based on MC simulations to be single-bridging SWNTs can be obtained by our electrode design. We have demonstrated the possibility of extracting statistics for parameter distributions such as threshold voltage, on-state current, on/off ratio and hysteresis width, from small device areas, thus opening the way for these devices to be applied as PCMs in carbon nanotube device fabrication processes.

VI. ACKNOWLEDGMENT

The authors acknowledge fabrication support from the clean room staff at FIRST lab at ETH Zurich. M.H. acknowledges the support and exchange of ideas enabled through the COST NanoTP action MP0901. Helpful discussions with Lukas Durrer, Moritz Mattmann, Matthias Muoth, Shih-Wei Lee and Valentin Döring are gratefully acknowledged.

[1] I. Martin-Fernandez, M. Sansa, F. Perez-Murano, P. Godignon, and E. Lora-Tamayo, "A test vehicle and a two step procedure to evaluate a massive number of single-walled carbon nanotube field effect transistors," 2010, pp. 48-51.

[2] Hongsik Park, A. Afzali, S.-J. Han, G. S. Tulevski, A. D. Franklin, J. Tersoff, J. B. Hannon, and W. Haensch, "High-density integration of carbon nanotubes via chemical self-assembly," *Nature Nanotechnology,* vol. Advanced online publication, 2012.

[3] A. Vijayaraghavan, S. Blatt, D. Weissenberger, M. Oron-Carl, F. Hennrich, D. Gerthsen, H. Hahn, and R. Krupke, "Ultra-large-scale directed assembly of single-walled carbon nanotube devices," *Nano Letters,* vol. 7, pp. 1556-1560, 2007.

[4] J. Kim, "Large-scale integrated carbon nanotube gas sensors," *Journal of Nanomaterials,* 2012.

[5] I. Martin-Fernandez, M. Sansa, M. J. Esplandiu, E. Lora-Tamayo, F. Perez-Murano, and P. Godignon, "Massive manufacture and characterization of single-walled carbon nanotube field effect transistors," *Microelectronic Engineering,* vol. 87, pp. 1554-1556, 2010.

[6] S. R. Wang, Z. Y. Liang, B. Wang, and C. Zhang, "Statistical characterization of single-wall carbon nanotube length distribution," *Nanotechnology,* vol. 17, pp. 634-639, 2006.

[7] T. S. Cho, K. J. Lee, J. Kong, and A. P. Chandrakasan, "A 32-mu w 1.83-ks/s carbon nanotube chemical sensor system," *IEEE Journal of Solid-State Circuits,* vol. 44, pp. 659-669, 2009.

[8] P. G. Collins, M. S. Arnold, and P. Avouris, "Engineering carbon nanotubes and nanotube circuits using electrical breakdown," *Science,* vol. 292, pp. 706-709, 2001.

[9] K. Chikkadi, C. Roman, L. Durrer, T. Süss, R. Pohle, and C. Hierold, "Scalable fabrication of individual swnt chem-fets for gas sensing," *Procedia Engineering,* vol. 47, pp. 1374-1377, 2012.

[10] K. Chikkadi, C. Roman, and C. Hierold, "Process control monitors for single-walled carbon nanotube based sensor fabrication processes," to appear in *IEEE 26th International Conference on Microelectromechanical Systems (MEMS) 2013,* Taiwan, 2012.

[11] K. Chikkadi, M. Mattmann, M. Muoth, L. Durrer, and C. Hierold, "The role of ph in the density control of ferritin-based catalyst nanoparticles towards scalable single-walled carbon nanotube growth," *Microelectronic Engineering,* vol. 88, pp. 2478-2480, 2011.

[12] L. Durrer, J. Greenwald, T. Helbling, M. Muoth, R. Riek, and C. Hierold, "Narrowing swnt diameter distribution using size-separated ferritin-based fe catalysts," *Nanotechnology,* vol. 20, 2009.

[13] T. Helbling, C. Hierold, C. Roman, L. Durrer, M. Mattmann, and V. M. Bright, "Long term investigations of carbon nanotube transistors encapsulated by atomic-layer-deposited al2o3for sensor applications," *Nanotechnology,* vol. 20, pp. 434010, 2009.

[14] M. Mattmann, C. Roman, T. Helbling, D. Bechstein, L. Durrer, R. Pohle, M. Fleischer, and C. Hierold, "Pulsed gate sweep strategies for hysteresis reduction in carbon nanotube transistors for low concentration no2 gas detection," *Nanotechnology,* vol. 21, 2010.

978-1-4673-4845-4/13 $31.00 © 2013 IEEE

978-1-4673-4845-4/13 $31.00 © 2013 IEEE   178

SESSION 9: Memory

978-1-4673-4845-4/13 $31.00 © 2013 IEEE

Gap in pagination due to withheld paper.

Pages 181-186

A proper approach to characterize retention-after-cycling in 3D-Flash devices

Fengying Qiao[1,2,*], Antonio Arreghini[2], Pieter Blomme[2], Geert Van den bosch[2], Liyang Pan[1], Jun Xu[1] and Jan Van Houdt[2]

[1] Institute of Microelectronics, Tsinghua University, Beijing, China 100084

[2] Imec, Kapeldreef 75, B-3001 Leuven, Belgium

* qfy08@mails.tsinghua.edu.cn

Abstract— **We propose a procedure to evaluate retention-after-cycling in 3D-Flash devices. Proper comparison of retention transients requires the initial charging level to be as close as possible, but P/E cycling results in serious I_D-V_G degradation, preventing a consistent extraction of the threshold voltage. We introduce a test where a relaxation phase is added after cycling, consisting in baking samples for 24 hours at 200 °C. This relaxation appears to anneal interface traps and to remove locally accumulated charge, restoring similar shape of I_D-V_G curves before and after cycling, hence allowing a proper comparison of retention of fresh and stressed devices.**

Keywords— Retention-After-Cycling; 3D Flash; Interface Traps; BiCS

I. INTRODUCTION

BiCS-like 3D-Flash memories are promising solutions to overcome scaling limitations of traditional cells, adopting more relaxed memory pitch and achieving higher density by vertically stacking cells on multiple layers [1-3]. These devices are intrinsically different from classical planar transistors, as the channel is vertical and deposited after the gate stack; furthermore, proper capacitive coupling is achieved by exploiting the cylindrical geometry [4]. Such differences with conventional cells demand for in-depth investigation of reliability issues in simplified test structures. In [5] we proposed a single vertical cell test vehicle for BiCS-like memory evaluation: being fabricated with the same approach as the final device, it is relevant to study most of the performance and reliability aspects, such as program/erase (P/E) characteristics and retention [6]. Post cycling retention is a major concern for flash memory technology and is also used to characterize the tunnel oxide degradation [7].

In this paper we propose a procedure to measure retention after cycling in such structures and to compare results with uncycled devices in a relevant way, by clearing the measurement from artifacts related to device degradation and to the simplified vehicle.

II. DEVICE DISCRIPTION

Our simplified 3D-Flash test vehicle consists of a single vertical memory cell, as illustrated in the schematic diagram and the SEM cross-section in Fig. 1(a) and (b), respectively. The device is realized by first depositing a 200-nm-thick p^+ doped poly-silicon layer, representing the control gate of the device, sandwiched by two oxide layers. Cylindrical holes are etched in this stack, and the Oxide-Nitride-Oxide (ONO) gate dielectrics are deposited in the sidewall of the holes. ONO is etched from the hole bottom to enable connections with the silicon substrate. Finally Source is diffused from substrate and Drain is implanted. Then the hole is filled with undoped poly-silicon, to form the channel. The full procedure is described in detail in [5]. Available test structures include single cells (3-terminals transistors) and devices, further denoted as "capacitors", with 90,000 cells in parallel to increase the area and obtain a detectable capacitance. We investigate devices with a total diameter of ~70 nm (~45 nm poly-silicon channel diameter) and gate stack consisting in 4.5 nm blocking oxide, 3.7 nm nitride and ~4 nm tunnel oxide.

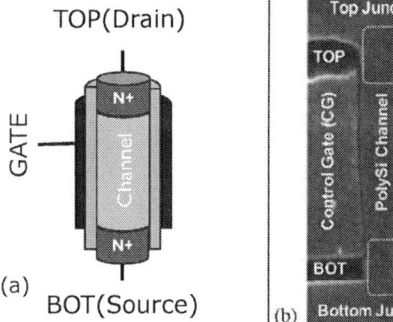

Figure 1. (a) Schematic diagram and (b) SEM cross-section of the adopted 3D-Flash vertical vehicle.

III. EXPERIMENT RESULTS

Long time charge loss in charge trapping devices is a combination of trap-to-band tunneling of stored charge toward the substrate (or the gate) and thermally driven Poole-Frenkel emission of charge out of the gate stack [8]. Degraded oxides are expected to have increased tunneling component because of more defects in the SiO_2 barriers, and therefore to analyze post-cycling tunnel oxide degradation, room temperature (RT) retention is a more relevant test than the one performed at high temperature.

To degrade the device we follow the typical endurance test procedure, consisting in providing a repetitive sequence of P/E cycles at fixed pulse amplitude and duration (without verify operation), to get a relevant window. As shown in Fig. 2, our cells withstand 100k cycles under typical P/E conditions, consisting in program pulses of 17 V for 100 µs and erase pulses of -16 V for 10 ms. The threshold voltage V_T is extracted using a fixed drain current criterion ($I_D = 11$ nA) and the plotted ΔV_T is relative to the fresh state.

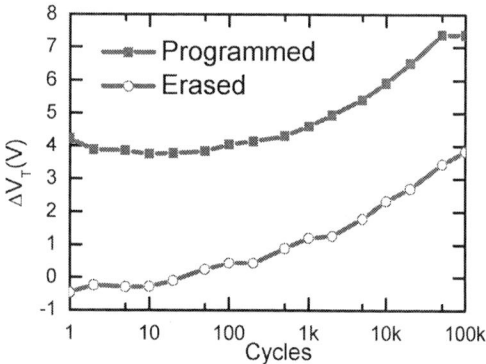

Figure 2. Memory window evolution during device degradation at constant P/E amplitude and duration. V_T is measured at fixed drain current of 11 nA.

In Fig. 2 a large positive window shift is observed, ~4.5 V for the eased state and ~3.5 V for the programmed state after 100k cycles. The origin of this shift can be explained by plotting the intermediate I_D-V_G curves of the cell transistor during 100k P/E cycling, as shown in Fig. 3(a). Cycling results in strong distortion of I_D-V_G curves, hampering a consistent extraction of V_T during the full test and leading to situation of Fig. 2. The shift is mostly related to sub-threshold slope degradation, and not necessarily to changes in charge injection into the nitride.

Fig. 3(b) shows the endurance test when V_T is extracted at different I_D (from 0.01 nA to 100 nA) from I_D-V_G curves shown in Fig. 3(a). Window shift is much smaller by extracting V_T at a lower current level, where I_D-V_G suffers less distortion induced ΔV_T. However, this is not sufficient to fully eliminate the effects, especially for large number of cycles. Furthermore, V_T measurements at very low current are more complicated, due to smaller signal-to-noise margin.

Long term charge loss is strongly dependent on the initial level the device are programmed to [9], hence the same program charge must be present in uncycled and cycled devices for a fair comparison. Degradation of I_D-V_G curves makes it virtually impossible to ensure that the same programmed state is achieved.

Therefore, we further analyzed the root cause of I_D-V_G curve degradation. The observed behavior is compatible with either interface trap generation (a phenomenon intrinsic to cycling degradation [10]) or with non-uniform charge accumulating along the channel during cycling. The latter may be caused by a limitation of our test structure, given by the presence of n^+ junctions in the device itself, as shown in Fig. 1(a), resulting in non-uniform electric field, especially during erase condition. This limitation would not be present on devices of a real NAND string [1]. Fig. 4 compares the shape of I_D-V_G of the programmed/erased state after different cycles with that of fresh devices. For better readability, I_D-V_G curves of fresh devices on the right graph are parallel shifted towards the programmed sate. Worse sub-threshold slope degradation is

typical in the erased state, therefore confirming the presence of non-uniform charge accumulation along the channel direction.

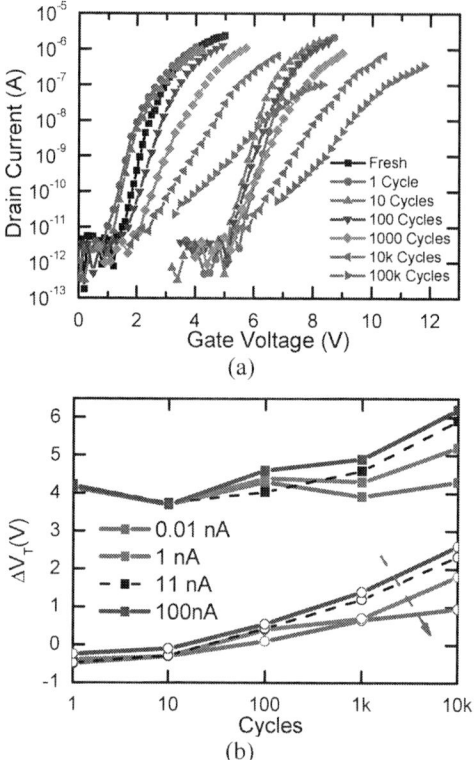

Figure 3. (a) I_D-V_G characteristics degradation during the cycling test of Fig. 2. (b) Memory window evolution during cycling obtained by extracting V_T at different drain currents.

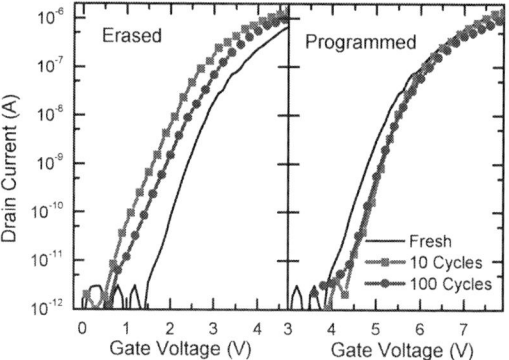

Figure 4. Comparison of I_D-V_G shape for the programmed and erased state after different cycles (I_D-V_G curve of fresh devices on the right graph parallel shifted).

To deeper investigate the role of interface state generation, we performed endurance experiments on capacitors. Fig. 5(a) shows the shape of C-V curves during cycling: almost no C-V deformation is observed after 1 cycle, while broader and shifted curves are observed after 3k cycles. This C-V shape degradation can be attributed to a combination of the non-uniformities along the channel and interface traps generation [11], similar as in the transistor. G-V curves in Fig. 5(b) reveal

978-1-4673-4845-4/13 $31.00 © 2013 IEEE

an increase of the conductance peak with cycling, up to ~50% after 3k cycles. The conductance peak increase clearly demonstrates the formation of a large number of interface states in the poly-silicon channel [12].

To obtain similar characteristics on uncycled and cycled devices, we propose a "recovery" step between cycling degradation and any subsequent operations: after 3k cycles, devices in the erased state are stored at RT for 4 days or at 200 °C (HT) for 24 hours. Fig. 6(a) and (b) show the C-V and

G-V characteristics of the cycled capacitors before and after the 4-days RT relaxation: no significant C-V shape recovery or conductance peak decrease is observed. Fig. 7(a) and (b) show instead the C-V and G-V characteristics of the cycled capacitors before and after the 24 hours 200 °C relaxation, the shape of both the C-V and G-V of the cycled devices is almost totally recovered to the one of uncycled devices. Therefore, the 200 °C bake recovery step is very effective both to anneal interface states [9] and to cancel non-uniformities generated during cycling, by promoting lateral charge redistribution [13].

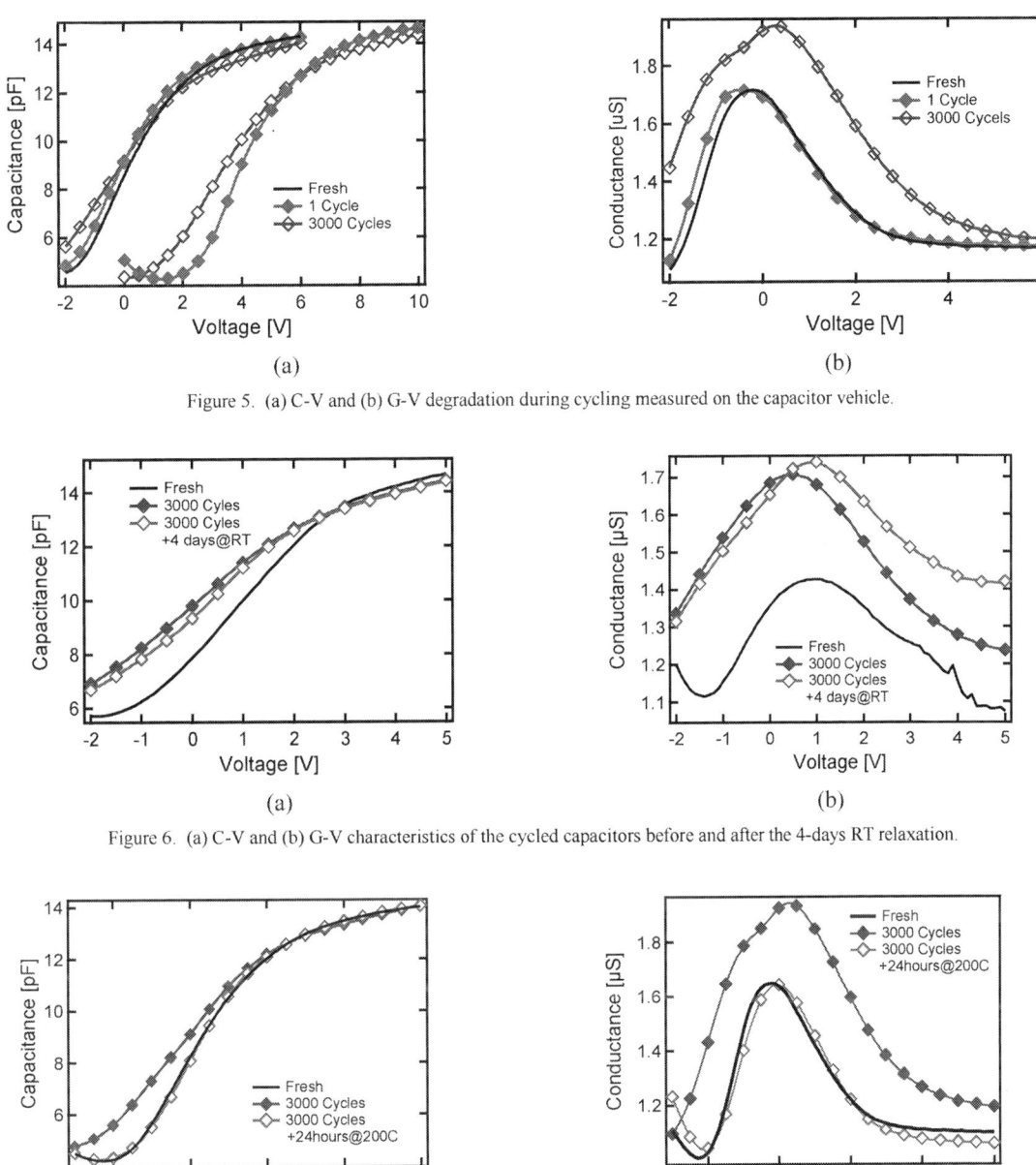

Figure 5. (a) C-V and (b) G-V degradation during cycling measured on the capacitor vehicle.

Figure 6. (a) C-V and (b) G-V characteristics of the cycled capacitors before and after the 4-days RT relaxation.

Figure 7. (a) C-V and (b) G-V characteristics of the cycled capacitors before and after the 24 hours 200 °C relaxation.

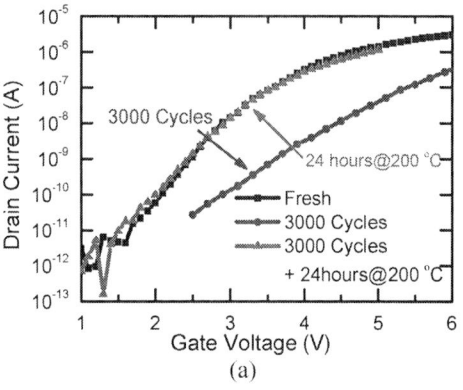 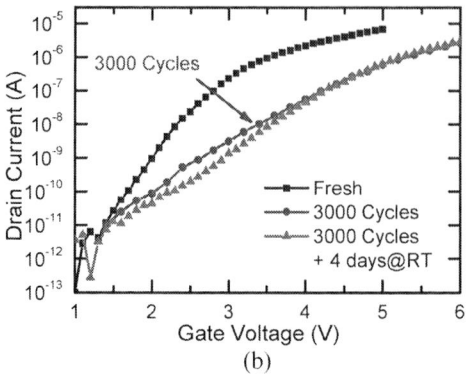

Figure 8. I_D-V_G characteristics of cycled cells (erased state) before and after the relaxation at (a) 24 hours 200 °C and (b) 4 days RT.

Similar experiments were also repeated on transistors: after 3k cycles, erased devices underwent the 4-days RT storage or the HT baking procedure. Also for this case HT baking recovery is effective for recovering the initial I_D-V_G shape, as shown in Fig. 8(a). On the contrary RT storage only induces slight shifts of the I_D-V_G curves, as shown in Fig. 8(b).

In addition, charge pumping measurements was also performed on cells after different cycles, and the result is shown in Fig. 9. Charge pumping current I_{CP}, which is proportional to interface trap density, increases by almost ~ 50% in stressed devices, further confirming the generation of interface traps [14].

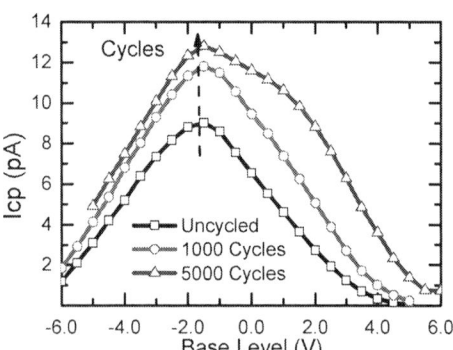

Figure 9. Comparison of I_{CP} for a erased cell after different cycles.

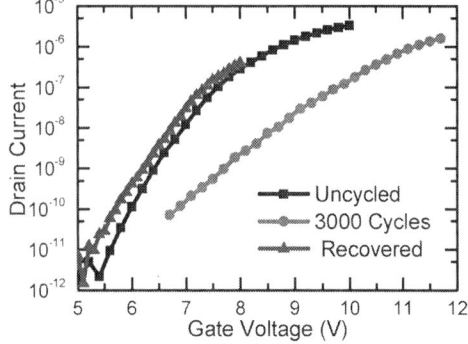

Figure 10. I_D-V_G curves of the programmed state for an uncycled, 3k-times cycled and recovered device.

After the baking operation, the recovered devices are ready for the retention-after-cycling test, consisting in re-programming them to the same level used in retention before cycling (4 V in our case). It is important to ensure that this re-program does not degrade I_D-V_G, frustrating the benefits of the recovery operation. Fig. 10 compares the I_D-V_G characteristics of programmed state for an uncycled, cycled and recovered device. It is obvious that a single program operation after recovery does not significantly degrade the I_D-V_G curve, hence the proposed procedure can be safely applied.

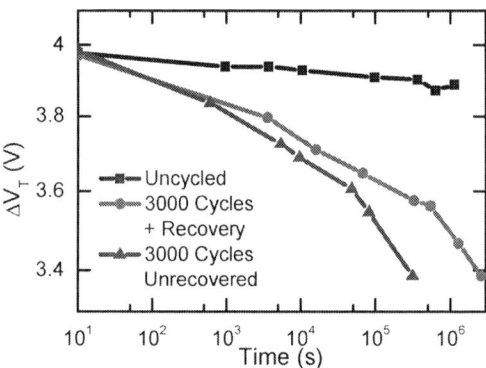

Figure 11. Comparison of RT retention of uncycled and cycled devices (with or without recovery)

Fig. 11 compares retention of the devices before and after 3k cycles by applying our newly developed recovery procedure (circle symbol) or not (up triangle symbol). The difference in retention loss with or without recovery is mostly attributable to a different programmed state, masked by I_D-V_G degradation. This difference is anyway less relevant than the behavior of uncycled devices. Retention-after-cycling is worsen by tunnel oxide degradation, which is not affected by the recovery operation.

For the curve of recovered devices in Fig. 11, the I_D-V_G characteristics during retention was also recorded and is shown in Fig. 12. Only rigid shifts of I_D-V_G are observed, indicating that V_T is extracted always in a consistent way, hence the test is trustable.

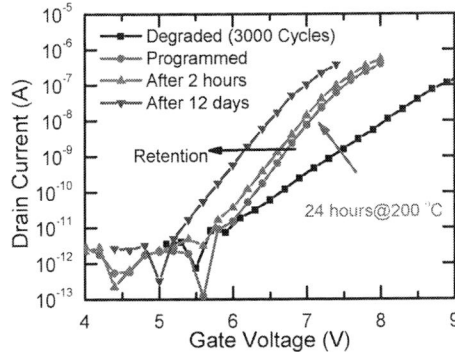

Figure 12. I_D-V_G characteristics recorded during retention test of Fig. 11, for cycled devices with recovery procedure.

IV. CONCLUSIONS

We propose a procedure to properly evaluate retention-after-cycling in simplified 3D-flash vehicles. During cycling we observe severe degradation of I_D-V_G curves that prevents proper comparison of V_T before and after stress. This degradation has been identified as the combined result of interface-state generation and vehicle artifacts (charge non-uniformities given by the presence of junctions). We propose a 24 hours 200 °C relaxation phase to anneal interface traps and promote lateral redistribution of charge, allowing proper extraction of device V_T and hence more relevant comparison of retention before and after cycling.

REFERENCES

[1] H. Tanaka, M. Kido, K. Yahashi, M. Oomura, R. Katsumata, M. Kito, Y. Fukuzumi, M. Sato, Y. Nagata, Y. Matsuoka, Y. Iwata, H. Aochi and A. Nitayama, "Bit cost scalable technology with punch and plug process for ultra high density flash memory," in VLSI Symp. Tech. Dig., 2007, pp. 14–15.

[2] R. Katsumata, M. Kito, Y. Fukuzumi, M. Kido, H. Tanaka, Y. Komori, M. Ishiduki, J. Matsunami, T. Fujiwara, Y. Nagata, L. Zhang, Y. Iwata, R. Kirisawa, H. Aochi and A. Nitayama, "Pipe-shaped BiCS flash memory with 16 stacked layers and multi-level-cell operation for ultra high density storage devices," in VLSI Symp. Tech. Dig., 2009, pp. 136–137.

[3] Y. Komori, M. Kido, R. Katsumata, Y. Fukuzumi, H. Tanaka, Y. Nagata, M. Ishiduki, H. Aochiand A. Nitayama, "Disturbless flash memory due to high boost efficiency on BiCS structure and optimal memory film stack for ultra high density storage," in IEDM Symp. Tech. Dig., 2008, pp. 851–854.

[4] E. Gnani, S. Reggiani, A. Gnudi, G. Baccarani, J. Fu, N. Singh, G. Q. Lo and D. L. Kwong, "Modeling of nonvolatile Gate-All-Around charge-trapping SONOS memory cells," in Proc. ESSDERC, 2009, pp. 280-283.

[5] G. S. Kar, G. Van den bosch, A. Cacciato, P. Blomme, A. Arreghini, L. Breuil, V. Paraschiv, C. Vrancken, B. Douhard, O. Richard, I. Debusschere, J. Van Houdt, S. Van Aerde and B. Tang, "Novel bi-layer Poly-Silicon channel vertical Flash cell for ultrahigh density 3D SONOS NAND Technology," Proc. IMW 2011, pp. 65-68.

[6] G. Van den bosch, G. S. Kar, P. Blomme, A. Arreghini, A. Cacciato, L. Breuil, A. De Keersgieter, V. Paraschiv and C. Vrancken, B. Douhard, O. Richard, S. Van Aerde, I. Debusschere, and J. Van Houdt, "Highly scaled vertical cylindrical SONOS cell with bilayer polysilicon channel for 3-D nand flash memory," IEEE Electron Device Lett., vol. 32, no. 11, pp. 1501–1503, Nov. 2011.

[7] W. H. Lee, C. Hur, H. Lee, H. Yoo, S. Lee. B. Lee, C. Park and K. Kim, "Post-cycling data retention failure in multilevel NOR flash memory with nitrided tunnel oxide," in Proc. IRPS 2009, pp. 907-908.

[8] A. Arreghini, N. Akil, F. Dirussi, D. Esseni, L. Selmi and M. J. van Duuren, "Long term charge retention dynamics of SONOS cells," Solid State Electron., 2008, vol. 52, pp. 1460-1466.

[9] J. D. Lee, J. Choi, D. Park, and K. Kim, "Effects of interface trap generation and annihilation on the data retention characteristics of flash memory cells," IEEE Transactions on Devices and Materials Riliability, vol. 4, no. 1, pp. 110-117, Mar. 2004.

[10] G. Van den bosch, L. Breuil, A. Cacciato, A. Rothschild, M. Jurczak and J. Van Houdt, "Investigation of window instability in program/erase cycling of TANOS NAND Flash memory," in Proc. IMW 2009, pp. 84-87.

[11] R. Castagne and A. Vapaille, "Apparent interface state density introduced by the spatial fluctuations of surface potential in an MOS structure," Electronics Letters, vol. 6, no. 22, pp. 691-694, Oct. 1970.

[12] A. Pacelli, A. L. Lacaita, S. Villa and L. Perron, "Reliable extraction of MOS interface traps from low-frequency CV measurements," IEEE Electron Device Lett., vol. 19, no. 5, pp. 148–150, May 1998.

[13] A. Maconi, A. Arreghini, C. Monzio Compagnoni, G. Van den boscha, A. S. Spinelli, J. Van Houdta and A. L. Lacaita."Impact of lateral charge migration on the retention performance of planar and 3D SONOS devices," In Proc. ESSDERC 2011, pp. 195-198.

[14] G. Groeseneken, H. E. Maes, N. Beltran, R.F. De Keersmaecker," A. reliable approach to charge pumping measurements and in MOS transistors," IEEE Trans. Electron Devices, vol. 31, no. 1, pp. 42-53, Jan. 1984.

978-1-4673-4845-4/13 $31.00 © 2013 IEEE

A Novel Test Structure to Implement a Programmable Logic Array Using Split-Gate Flash Memory Cells

Henry Om'mani, Mandana Tadayoni, Nitya Thota, Ian Yue, and Nhan Do

Silicon Storage Technology
A Subsidiary of Microchip Technology Inc.
450 Holger Way
San Jose, California 95134, USA
Email: nhan.do@microchip.com

Abstract— We developed a novel configurable logic array test structure using a highly scalable 3rd generation split-gate flash memory cell that features low power and fast configuration time. This split-gate SuperFlash® configuration element (SCE) has been demonstrated with a 90nm embedded Flash technology. The resulting SCE eliminates the need for esoteric fabrication process, and sensing, and SRAM circuits and reduces configuration time for programmable arrays (PA) such as FPGAs and CPLDs. Additionally, SCE inherently ports the advantages of SST's split-gate Flash memory technology with compact area, low-voltage read operation, low-power poly-to-poly erase and source-side channel hot electron (SSCHE) injection programming mechanisms, along with superior reliability.

Index CPLD; FPGA; Programmable Logic Array; Split-Gate Flash Memory

I. INTRODUCTION

Most programmable arrays use volatile memory SRAM as a configuration data storage element. Recently, there are efforts to replace SRAM with non-volatile memory (NVM). NVM-based FPGA are ideal for embedded IP applications with an architecture and many features enhanced to improve chip integration, IP usage, and test time. Robert Lipp et al [1] proposed a high-voltage triple-well EEPROM sense scheme with a shallow trench isolation (STI) process. For timely programming, the switching device using this approach requires voltages higher than ± 16V. Kyung Joon Han et al [2] improved on this scheme by replacing the high-voltage triple-well process with the deep trench isolation process. However, the deep-trench NVM device operation employs FN tunneling which requires a voltage split between gate and channel of ± 10V. This split voltage operation requires the triple-well process.

In this work, a novel test structure, for the first time, is proposed to create the configuration element by using twin split-gate Flash memory cells. SST's split-gate flash memory cell is well known as the leading NVM solution for embedded applications [3-5] thanks to its compatible integration with the baseline logic process offered by foundries, low power and highly reliable poly-to-poly erase with thick tunneling oxide, very efficient source-side-channel-hot-electron (SSCHE) injection programming, low-voltage read operation, and high-endurance capability. As a result, SCE technology eliminates the triple-well process requirement, and the

sensing and SRAM circuits by using SST's split-gate flash memory technology that is amenable to the standard CMOS process and employs the split-gate memory cell to store configuration data and to directly configure the logic array (SLA) switch element. In addition, SCE inherits the technological advantages offered by the split-gate flash memory aforementioned.

II. TEST STRUCTURE

Fig. 1 shows a cross section of the SST's 3rd generation split-gate flash memory cell. The operation of the memory cell is described in Table I.

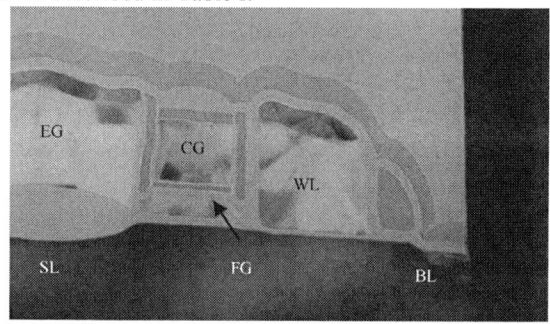

Fig.1. Cross section of 3rd generation SST's Split-gate memory cell.

TABLE I: ERASE, PROGRAM AND READ OPERATION OF SUPERFLASH® CELL

	Bit Line	Source Line	Word Line	Coupling Gate	Erase Gate
Erase	0V	0V	0V	0V	11V
Program	1μA	4.5V	~1V	10V	4.5V
Read	0.8V	0V	V_{CC}	V_{CC} / 0V	0V

The memory cell element is programmed when electrons are efficiently injected from the channel into the floating gate (FG). This is realized with a small current of ~1-2μA biasing the bitline (BL) while horizontal and vertical electrical fields are generated with medium and high voltages applied to the sourceline (SL)/erase gate (EG) and coupling gate (CG), respectively, as described in Table I. The horizontal field generates the energetic electrons near the gap between the wordline (WL) and FG and the vertical field helps to inject the electrons into FG. To erase the memory cell element, a high voltage (11V) is applied to the erase gate (EG) which removes the electrons from FG, and raises its potential, therefore turning on the FG channel. The cell states can be sensed with a small bias on the BL, Vcc on the WL, and either Vcc or 0V on the CG.

Fig. 2 shows the SCE schematic and layout which includes a pair of split-gate Flash memory cells, and CELL1 and CELL2, sharing a common BL. Two NMOS transistors, T1 and T2, connect the BL to two different nodes, flash BL (FBL) or data output (DOUT) during Flash or SLA modes, respectively. Flash mode (FM) is used to perform Flash operations on the cells. SLA mode (PAM) biases both cells for read by turning off T1 and connects BL to the switching element (SE) via T2. During Flash mode operation, CELL1 and CELL2 are configured to be in complementary states depending on whether SE is to be off or on during SLA mode. Programming CELL1 to the OFF state and erasing CELL2 to the ON state allows transmission of GND or SL GND (SLGND) to SE. VDD is transmitted to SE when the cells are configured in reverse. Table II shows SCE operation modes and their respective bias conditions.

TABLE II: SCE MODES OF OPERATION.
$V_{CGP/E/R}$ = CG VOLTAGE DURING PROGRAM / ERASE / READ OPERATIONS.

Signal	Flash Mode (FM)					Programmable Mode (PAM)
	Erase (V)	Program (V)		Read (V)		
		CELL 1	CELL 2	CELL 1	CELL 2	
SLVDD	0	V_{SLP}	0	V_{SLR}	0	V_{SLR}
EG1	V_{EGE}	V_{SLP}	0	V_{SLR}	0	V_{SLR}
CG1	0	V_{CGP}	0	V_{CGR}	0	V_{CGR}
WL1	0	V_{WLP}	0	V_{WLR}	0	V_{WLR}
BL	0	V_{BLP}	0	V_{BLR}	0	V_{BLR}
WL2	0	0	V_{WLP}	0	V_{WLR}	V_{WLR}
CG2	0	0	V_{CGP}	0	V_{CGR}	V_{CGR}
WL2	0	0	V_{WLP}	0	V_{WLR}	V_{WLR}
EG2	V_{EGE}	0	V_{WLP}	0	V_{WLR}	V_{WLR}
SLGND	0	0	V_{SLP}	0	V_{SLR}	V_{SLR}
FM	ON	ON	ON	ON	ON	0
PAM	0	0	0	0	0	ON
SE	OFF	OFF	OFF	OFF	OFF	ON

III. RESULTS AND DISCUSSION

Results of the SCE's characterizations in both the Flash memory cell and SLA modes are presented. Fig. 3 illustrates the flash cell operation mode while T2 is turned off. With independent accesses to different memory cell nodes, top and bottom memory cells (cell1 and cell2) can be programmed or erased separately. For instance, the top cell can be erased by applying 11V on the top erase gate (EG1) while holding other nodes at 0V. Likewise, the bottom cell can be erased similarly when bottom erase gate (EG2) is biased with 11V. Erase, programming, and read operating conditions described for the split-gate flash memory cell in the previous section can be used for SCE characterizations. As shown in Fig. 4 and Fig. 5, the cell current measured in the Flash mode for the erased cell is around 27 µA and for the programmed cell is < 10 pA. The difference in erased cell currents between top and bottom cells may be particularly due to the process mis-alignment of the WL poly. At higher temperature, the leakage for the programmed cell increases but, fortunately, can be suppressed by changing the bias voltage for CG during read from V_{CC} to 0V. Furthermore, endurance characteristics of the SCE can be characterized using this operating mode.

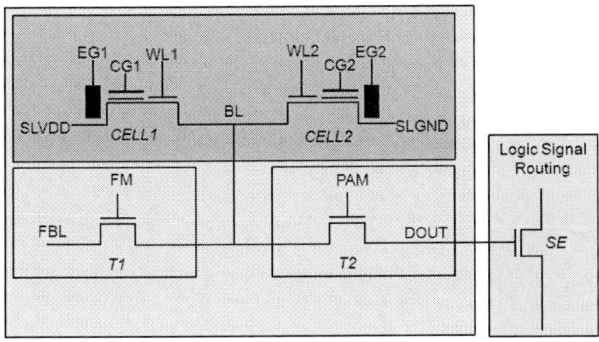

Fig. 2a. Schematics of split-gate Flash configuration element (SCE) and programmable array switch.

Zoomed Layout view

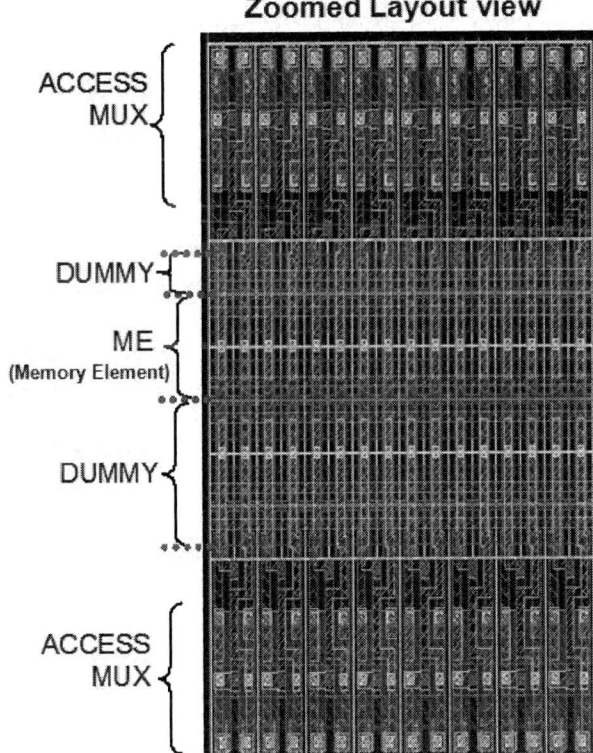

Fig. 2b. Layout SCE and programmable array switch.

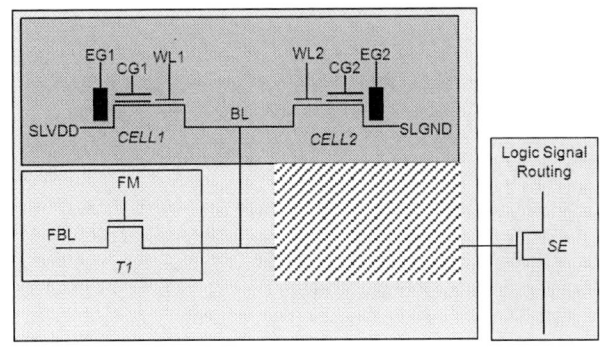

Fig. 3. Illustration of SCE flash cell operation mode.

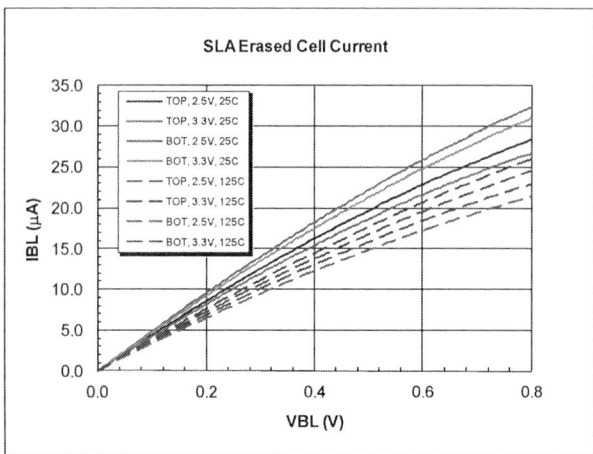

Fig. 4. Erased flash cell current as a function of BL voltage

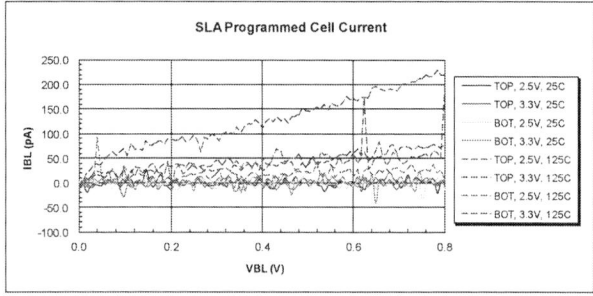

Fig. 5. Programmed flash cell current as a function of BL voltage.

Fig. 6 illustrates the SCE schematics during the SLA mode where T1 is turned off. SLA operation includes passing either VDD (SLVDD) or GND (SLGND) to DOUT. To characterize the performance of the device, the following procedure had been performed: (1) Simultaneously program all cells, (2) Erase top or bottom row which leaves the other row in the programmed state, and (3) Measure DOUT and leakage.

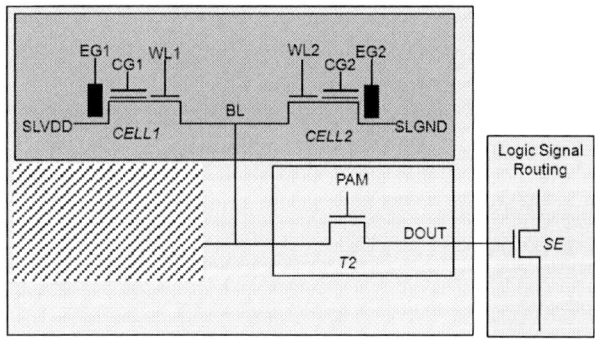

Fig. 6. Illustration of SCE SLA operation mode

Fig. 7 shows the transmission of the VDD signal to DOUT with negligible degradation. This means the flash cell is fully erased and able to transfer the data without loss. Once the cell is programmed, very low current (<10 pA) is expected. However, this leakage can be significant if the array of cells becomes large and is operated at high temperature. For example, an array of 10Kbits can induce a total of 100 μA standby leakage measured at 125°C, which may not be suitable for certain applications that require low power. Fortunately, this leakage can be reduced with longer programming time and lower CG voltage (Vcg) during read. Fig. 8 shows the leakage as a function programming time with

different Vcg. Both longer programming time and lower Vcg reduces leakage significantly.

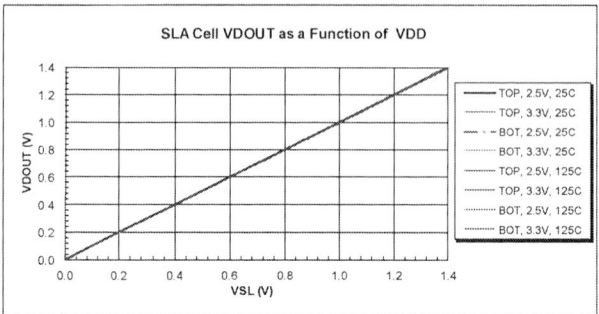

Fig. 7. SCE Output as a function of VDD

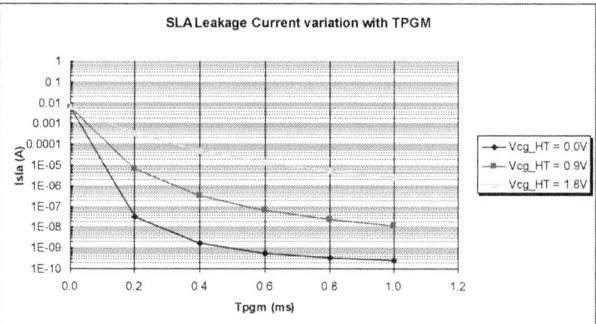

Fig. 8. SCE leakage as a function of programming time and Vcg.

IV. CONCLUSION

A Flash-based configuration element has been demonstrated by using a very novel test structure that employs a pair of SST's split-gate 3rd generation flash memory cells. With the same operation used in the memory cell, acceptable erased and programmed cell currents were achieved in the SCE Flash mode. The SLA configuration can fully transmit the VDD signal without degradation to the output DOUT through the memory element. A low leakage level can also be achieved with a longer programming time and lower Vcg during read.

ACKNOWLEDGMENT

The authors would like to thank Mr. Mark Reiten for providing management support and encouragement, Randy Yach and Alex Kotov for useful suggestions.

REFERENCES

[1] R. Lipp, R. Freeman, T. Saxe, IEEE 2000 Custom Integrated Circuits Conference, 2000.

[2] K. J. Han, N. Chan, S. Kim, B. Leung, V Hecht, B. Cronquist, IEEE Non-Volatile Semiconductor Memory Workshop, 2007, pp. 32-33.

[3] S. Kianian, A. Levi, D. Lee, and Y.-W. Hu, "A novel 3 Volts only, small sector erase, high density flash EEPROM," in VLSI Symp. Tech. Dig.,1994, v.6A, pp. 71-72.

[4] S. N. Keeney, M. Gill, and D. Sweetman, "NOR Flash Stacked and Split-Gate Technology," in Nonvolatile Memory Technologies with Emphasis on Flash. A Comprehensive Guide to Understand and Using NVM Devices, ed. J. E. Brewer and M. Gill, John Wiley & Sons, 2008. pp. 179-222.

[5] Kai Man Ian Yue, Bomy Chen, Geeng Chuan Michael Chern, Tsung-Lu Syu, "Storage Element for Controlling a Logic Circuit, and a Logic Device Having an Array of Such Storage Elements," US Patent 7,701,248.

On-wafer integrated system for fast characterization and parametric test of new-generation Non Volatile Memories

Erika Covi*, Alessandro Cabrini*, Loris Vendrame†, Luca Bortesi†, Roberto Gastaldi† and Guido Torelli*

*Dipartimento di Ingegneria Industriale e dell'Informazione
University of Pavia
Via Ferrata 1, 27100 Pavia - Italy
e-mail: erika.covi@unipv.it
†Micron Semiconductor Italia s.r.l.
R&D - Technology Development
Via C. Olivetti 1, 20864 Agrate Brianza (MB) - Italy

Abstract— In new and future generations of Non Volatile Memories such as Phase Change Memories (PCMs) and Resistive-RAMs (ReRAMs), having accurate and controllable program pulses is fundamental to adequately characterize the memory cell, since the obtained cell status is a function of the applied pulse parameters. In order to massively test new cells and enhance conventional instrumentation flexibility, an accurate on-chip pulse generator, which is able to provide pulses with different amplitude, falling time, and duration, has been designed, fabricated, and experimentally evaluated. The designed device can generate pulses with amplitude, fall time, and time duration programmable from 0.5 V to 4.5 V, from 10 ns to several μs, and from 50 ns to 350 ns, respectively.

I. INTRODUCTION

In new and future generations of Non Volatile Memories such as Phase Change Memory (PCM) and Resistive RAM (ReRAM) [1] [2], remarkable efforts must still be devoted to optimize materials, cell architecture, and/or programming algorithms, also aiming at production yield improvement. In particular, accurate and controllable pulses are required to experimentally investigate the programming performance of such memories [3]. A massive characterization at the wafer level is highly desirable so as to collect a sufficient amount of data and monitor the cell performance during the production phase.

The information in a PCM cell is stored by exploiting two different structural solid-state phases (i.e., the crystalline and the amorphous phase) of a small portion of a chalcogenide alloy. The typical PCM cell resistance varies between a few kΩ and a few MΩ [4], [5]. The phase transition in PCMs is controlled by temperature and, hence, adequately controlling the current flowing through the alloy is essential for optimal programming operation [6].

The operating principle of ReRAMs is based on a dielectric which may be made conductive by generating a conductive filament. The conductive filament can be broken or re-formed by applying adequate voltage pulses [2], [7].

As the programming pulse shape determines the final state of the memory cell [8] [9] [10], programming operation during the research phase demands high accuracy and high flexibility in varying the amplitude, the duration, and the rise and fall time of program pulses, so as to allow investigating the cell behaviour under a variety of programming conditions. In addition, overshoots and any other disturbance affecting the program pulse could casually alter the cell state and, consequently, the program operation could lead to different results than expected or even fail. The applied pulses must therefore be carefully controlled across their whole voltage swing, down to safe voltage levels in order to perform successful programming operations. This is even more true when multilevel-cell programming is considered [11] [12] [13], as in this case the value of the programmed resistance must be very tightly controlled.

On the one hand, the conventional instrumentation available for on-wafer automatic parametric testing features high accuracy, but is mainly conceived for DC measurements of elementary devices. The speed of this instrumentation is therefore not sufficient to allow the fast parametric tests required to perform massive statistical studies of the I-V characteristic of new cells, with the aim of investigating the cell behaviour and improving both the cell geometry and the materials used. On the other hand, long cables are necessary to connect the selected cell(s) on the wafer to commercial pulse generators. Non negligible noise can therefore be added to the generated waveforms, thus degrading the signal and, hence, limiting both the accuracy and the controllability of program pulses. Moreover, the usual hybrid configuration (an Automatic Parametric Wafer Testing System, APWTS, and a Pulse Generator, PG, that cooperate to perform tests) uses a switch matrix to switch the connections of the pads of the memory chip under characterization to the different signals provided by the test equipment, thus decreasing testing speed due to switching and settling times. As a consequence, the need exists for designing a flexible integrated system able to generate accurate and controllable

978-1-4673-4845-4/13 $31.00 © 2013 IEEE

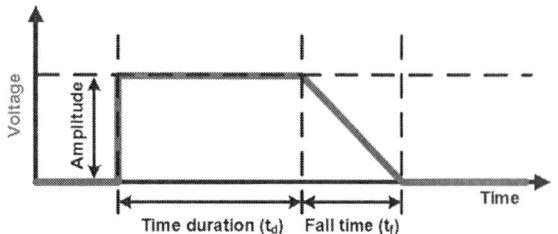

Fig. 1.   Programmable parameters of the program pulse.

program pulses. For easier use, the system should be driven by the existing commercial instrumentation.

The on-chip pulse generator should be integrated in the wafer scribe lanes, which implies low area occupation and a particular layout aspect ratio, and should include a mini-array of the cells to be characterized. The test chip must be provided with the capability of selecting cells in different array positions so as to allow exploring possible behaviour differences related to the cell location.

The last requirement limits the number of pads available for controlling pulse generation because most pads are used to address the mini-array. When using a digital system with the above constraint, the digital control data must therefore be fed to the test chip through serial communication, thus compromising testing speed. The above issues led us to the choice of developing a fully analog system rather than a digital one.

This paper is organized as follows. In Section II, the basic operating principle of the designed system will be shown. In Section III, the system architecture will be explained in detail. Section IV will focus on the experimental results from a fabricated prototype. Eventually, some conclusions will be drawn in Section V.

II. BASIC OPERATING PRINCIPLE

In the presented implementation, the program pulse to be applied to the selected memory cell has a trapezoidal shape (Fig. 1) with pulse amplitude, pulse fall time, and time duration programmable from 0.5 V to 4.5 V, from 10 ns to several μs, and from 50 ns to 350 ns, respectively. The pulse rise time was set to about 10 ns. The programmable pulse parameters are controlled by the amplitude of three external analog voltage signals. More specifically, the required value of pulse amplitude is directly fed to the test chip by the APWTS as a DC level, V_{ampl}, whereas the pulse fall time and the pulse time duration are encoded by means of the amplitude of pulses V_{tf} and V_{td}, respectively, which are provided by the PG (Fig. 2) to best optimize the usual testing configuration interfacing.

The time distance, Δt, between the rising edges of these two pulses acts as a time reference for the system. The value of Δt is externally programmable and must take transients of external signals into account. This leads to a lower bound for the value of Δt. Capacitance values and charging currents were designed so as to ensure that capacitors can be charged

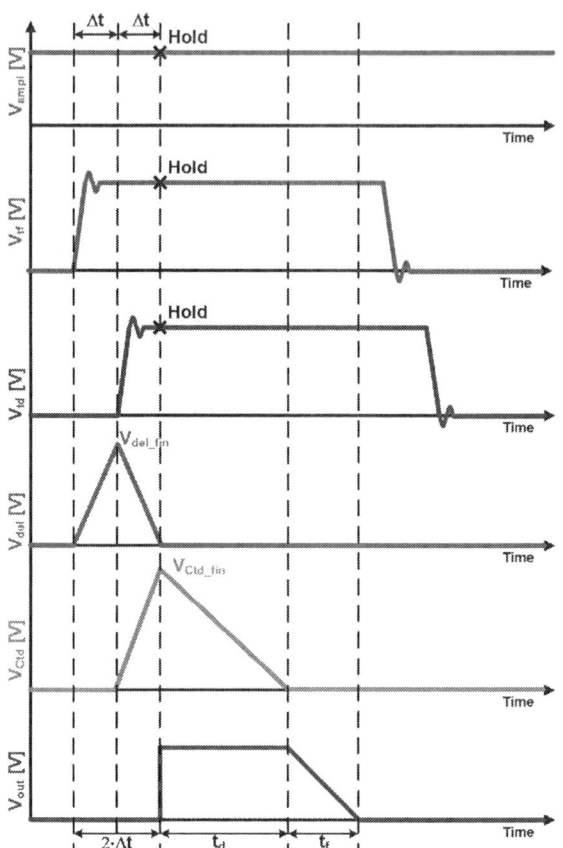

Fig. 2.   External (V_{ampl}, V_{tf}, V_{td}) and internal (V_{del}, V_{Ctd}, V_{out}) waveforms.

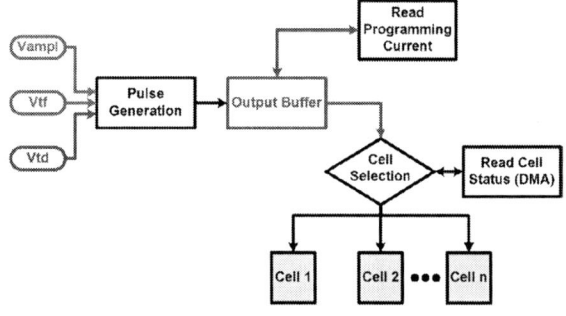

Fig. 3.   High-level scheme of the proposed system for memory cell characterization.

to the desired values with no risk of saturation. Moreover, as will be shown in next Section, Δt can act as an additional control parameter in generating the programming pulse.

A high-level scheme of the developed test chip is illustrated in Fig. 3. The analog signals from the external instrumentation (V_{ampl}, V_{tf}, and V_{td}) are recombined in the test chip in order to generate the desired pulse, which is then applied to the cell through a buffer able to provide the selected cell with the amount of current needed for program operation. Cell selection is performed at the beginning of a test sequence

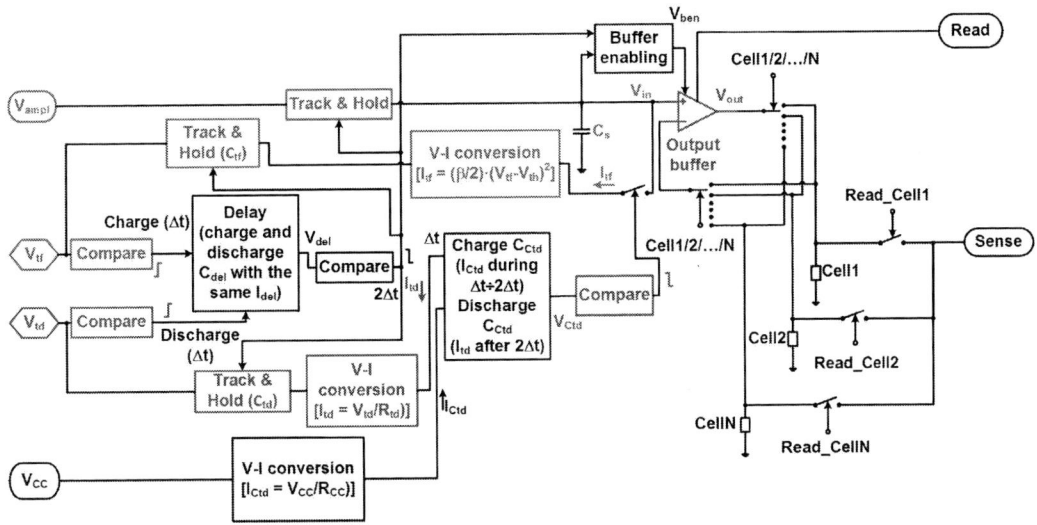

Fig. 4. Detailed block diagram of the proposed integrated system.

by suitable digital circuits driving few pass-gates under the control of external signals. Once cell selection is performed, the test sequence starts and evaluates the selected cell with negligible time overhead between the program and the read sequence. The program pulse is made available a time interval $2\Delta t$ after the starting instant, which corresponds to the rising edge of V_{tf}. A current mirror connected to the output allows the programming current to be measured externally by the APWTS, which can also read the state of the selected cell in Direct Memory Access (DMA) mode when required.

III. SYSTEM ARCHITECTURE

The proposed system will be described referring to the waveforms in Fig. 2 and the block diagram in Fig. 4, where the presence of some capacitors is indicated in their respective blocks rather than including their circuit symbols.

All the external signals (V_{ampl}, V_{tf}, and V_{td}) are tracked after the rising edge of V_{tf} is detected and are then held across three respective capacitors (C_s, C_{tf}, and C_{td}) after $2\Delta t$. As shown in the figure, the delay Δt allows getting rid of external signal transients.

Using Track-and-Hold circuits assures that a stable voltage value is fed to the cascaded circuits for pulse generation. An adequately large size of the capacitors reduces charge injection effects, thus improving system accuracy.

As voltage V_{ampl} needs no conversion to set the amplitude of the generated pulse, it is held across a storage capacitor, C_s, which is initially discharged, and directly fed at buffer input, V_{in}.

The buffer is initially disabled and its output voltage V_{out} is forced to 0 V by an internally generated signal until the system is ready to deliver the output pulse (i.e., until $t = 2\Delta t$). V_{out} is thus held at 0 V even though the buffer input rises from 0 V to V_{ampl}.

Time delay $2\Delta t$ is obtained by first charging and then discharging a capacitor (C_{del}): after the rising edge of V_{tf}

is detected, capacitor C_{del}, which is initially discharged ($V_{del} = 0$ V), is charged with a constant current (I_{del}) until the rising edge of V_{td} is detected, after a time interval Δt. At this instant ($t = \Delta t$), the system begins to discharge C_{del} with the same current I_{del}. Voltage V_{del} will then approach 0 V after an additional time delay Δt, i.e., at $t = 2\Delta t$. A simple comparator is then able to detect this time instant. The value of I_{del} does not need to be very accurate: the key factor for this conversion is an adequate matching between the charging and the discharging current, which is easily achieved on-chip by means of current mirrors.

Program pulse duration t_d is obtained by means of voltage-to-time conversion. A capacitor (C_{Ctd}) is charged, during the time interval from Δt to $2\Delta t$, at a constant rate by an internally generated current

$$I_{Ctd} = \frac{V_{CC}}{R_{CC}} \qquad (1)$$

where V_{CC} is the supply voltage and R_{CC} is an n-well resistor, thereby reaching a final voltage level $V_{Ctd-fin}$. At $t = 2\Delta t$, the output buffer is enabled and makes V_{out} rise from 0 V to V_{ampl}. At this instant, C_{Ctd} begins to be discharged with an internally generated current

$$I_{td} = \frac{V_{td}}{R_{td}} \qquad (2)$$

R_{td} being an n-well resistor matched to R_{CC}. The voltage, V_{Ctd}, across this capacitor will approach 0 V after a time interval

$$t_d = \frac{I_{Ctd}}{I_{td}}\Delta t = \frac{R_{td}}{R_{CC}}\frac{V_{CC}}{V_{td}}\Delta t \qquad (3)$$

A simple comparator will therefore be able to detect the time instant in the above equation, thus starting the discharge of capacitor C_s and, hence, setting the length of the generated pulse. Pulse duration t_d is therefore dependent on ratio $\frac{R_{td}}{R_{CC}}$,

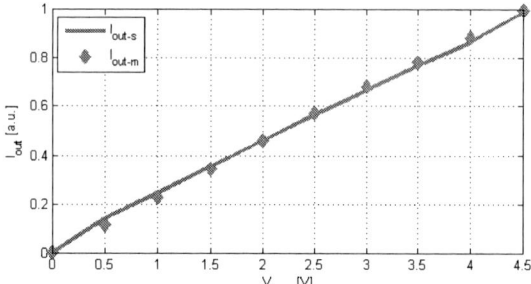

Fig. 5. Comparison between the measured (I_{out-m}) and the simulated (I_{out-s}) output current at pulse plateau.

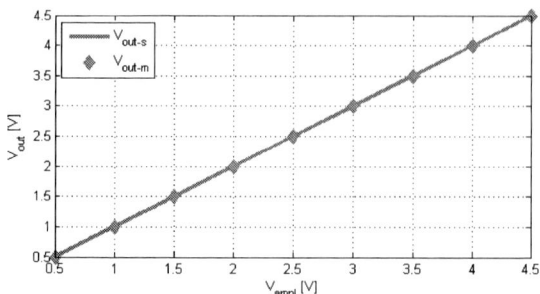

Fig. 6. Comparison between the measured (V_{out-m}) and the simulated (V_{out-s}) output current at pulse plateau.

which allows overcoming any problem due to fabrication process spreads.

The desired program pulse fall time is obtained by means of a voltage-to-time conversion, which is performed by discharging C_s with a current, I_{tf}, generated by an MOS transistor operated in the pinch-off region under the control of voltage V_{tf}:

$$I_{tf} = \frac{\beta}{2}(V_{tf} - V_{th})^2 \qquad (4)$$

with obvious meaning of symbols. It is worth to point out that the non-linear voltage-to-current conversion in (4) allows a wide current range of the discharge current and, hence, a wide range of the fall time in spite of the limited range of V_{tf}, which allowed us to meet target specifications whilst still minimizing silicon area.

Even though the designed test chip reduces the effects due to component non-idealities and process spreads, still some impact of these effects limits system accuracy. An automatic procedure to calibrate the system, compatible with the instrumentation characteristics (such as analog variable ranges and resolution), has been conceived in order to enhance the overall accuracy of the entire measurement chain, although it will not be shown in this paper.

IV. EXPERIMENTAL RESULTS

The test chip described was designed in 180-nm CMOS technology, which corresponds to a typical fabrication process for high-voltage peripheral circuitry of memory chips. High-voltage devices were used to provide the system with adequate reliability even in the presence of the used supply voltage (6 V). The resulting total area of the pulse generator is 14,310 μm^2, not including pads.

The test chip was provided with the possibility to vary an integrated load resistor (which emulates the memory cell) so as to allow its performance to be evaluated under different load conditions. The resistor value can be directly measured in DMA mode to achieve high accuracy. In the following, experimental results obtained with a supply voltage of 6 V and an equivalent load resistance of 14.5 kΩ, which represents worst case condition, will be illustrated.

Fig. 5 shows the measured (I_{out-m}) and the simulated (I_{out-s}) current at pulse plateau for different values of V_{ampl}.

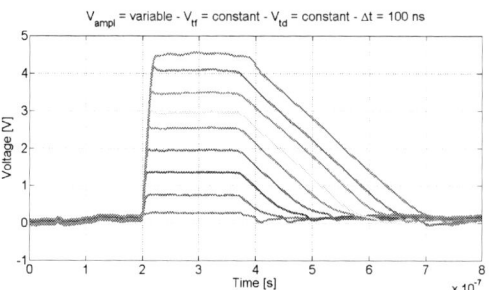

Fig. 7. Measured voltage program pulse generated by varying the amplitude of V_{ampl} (500 mV step) while keeping V_{tf} and V_{td} constant ($\Delta t = 100$ ns).

A comparison between the experimental (V_{out-m}) and the simulated (V_{out-s}) output voltage at pulse plateau is provided in Fig. 6. Very good agreement between experimental data and simulated results is apparent in both figures.

Further experimental analysis of the test chip was carried out by means of an active microprobe. Each of the program pulse control parameters (V_{ampl}, V_{tf}, V_{td}, and Δt) was separately varied in order to explore program pulse flexibility.

At first, the pulse voltage amplitude capability was investigated by setting $\Delta t = 100$ ns, keeping V_{tf} and V_{td} at a constant value, and varying V_{ampl} with steps of 500 mV. It is apparent from Fig. 7 that design specifications are fully met, since the generated program pulse amplitude ranges from less than 0.5 V up to 4.5 V.

Then, pulse fall time variability was investigated by setting $\Delta t = 100$ ns, $V_{ampl} = 3$ V, keeping V_{td} at a constant value, and varying V_{tf}. Since V_{tf}-to-t_f conversion is non-linear, a variable step of V_{tf} was chosen. According to the target, measurements showed a minimum and a maximum fall time of about 9 ns and 5 μs, respectively (Fig. 8). The minimum fall time is limited by the chosen maximum value of V_{tf}, whereas the maximum fall time is limited by the minimum value of V_{tf} which ensures correct operation of the NMOS transistor performing the voltage-to-current conversion, according to (4).

The pulse time duration performance was finally investigated by setting $V_{ampl} = 3$ V, keeping V_{tf} at a constant value, and varying the two variables that control t_d according to (3). In detail, first V_{td} was varied while keeping $\Delta t = 100$ ns,

978-1-4673-4845-4/13 $31.00 © 2013 IEEE

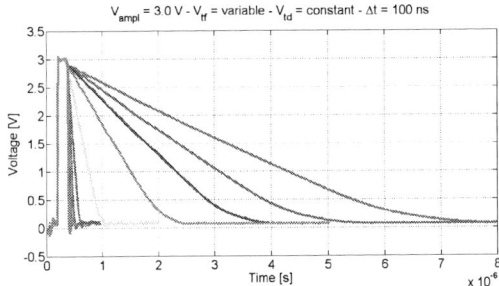

Fig. 8. Measured voltage program pulse generated with different values of voltage V_{tf} (variable step) while keeping V_{ampl} and V_{td} constant ($\Delta t = 100$ ns).

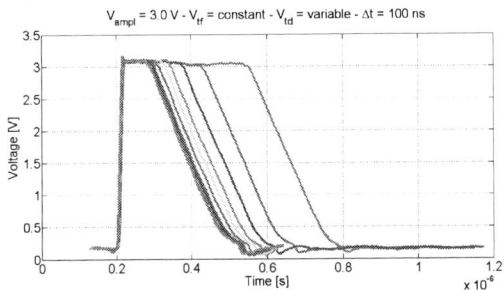

Fig. 9. Measured voltage program pulse generated by varying V_{td} (500 mV step) while keeping V_{ampl} and V_{tf} constant ($\Delta t = 100$ ns).

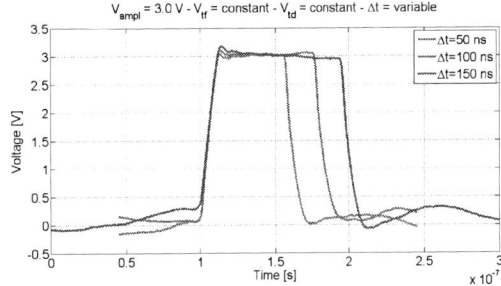

Fig. 10. Measured voltage program pulse generated by varying Δt (50 ns step) while keeping V_{ampl}, V_{tf} and V_{td} constant (the trigger occurs in different time instants for the three pulses).

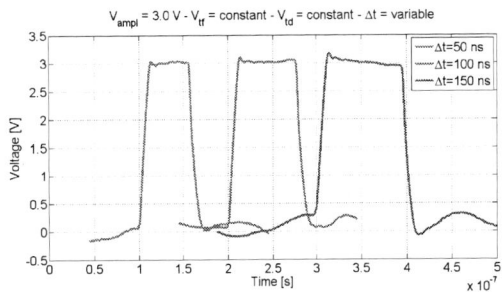

Fig. 11. Measured voltage program pulse generated by varying Δt (50 ns step) while keeping V_{ampl}, V_{tf} and V_{td} constant (the trigger occurs in the same time instant for all pulses).

then Δt was varied from 50 ns to 150 ns with 50 ns steps, while keeping V_{td} at a constant value. The first measurements (Fig. 9) showed that the specified minimum (50 ns) and maximum (350 ns) time duration values are fully achieved. The second set of measurements showed two effects: *i*) the variation of t_d is proportional to Δt (Fig. 10), as illustrated in (3), thus enhancing the programmable range of t_d, and *ii*) the time instant in which the program pulse is generated varies with Δt (Fig. 11).

V. CONCLUSION

An accurate on-chip pulse generator has been designed in 180-nm CMOS technology, fabricated, and experimentally characterized. The presented system allows generating pulses with different amplitude, duration, and fall time in order to meet the flexibility, controllability, and accuracy specifications required to massively characterize new-generation Non Volatile Memories, thus overcoming the limits of commercial APWTS and PG equipments.

ACKNOWLEDGMENT

The authors wish to thank A. Calderoni, A. Calloni, A. Redaelli, and D. Ventrice for many useful discussions and for their advices in test chip architecture and equipment interfacing.

REFERENCES

[1] K. Prall *et al.*, "An update in emerging memory: progress to 2xnm," in *Proc. of IEEE Int. Memory Workshop (IMW)*, 2012, pp. 1–5.

[2] H. Akinaga and H. Shima, "Resistive random access memory (ReRAM) based on metal oxides," *Proc. of the IEEE*, vol. 98, no. 12, pp. 2237–2251, Dec. 2010.

[3] A. L. Lacaita *et al.*, "Electrothermal and phase-change dynamics in chalcogenide-based memories," in *IEEE Int. Electron Device Meeting (IEDM), Tech. Dig.*, 2004, pp. 911–914.

[4] F. Wang and X. Wu, "Non-volatile memory devices based on chalcogenide materials," in *Proc. of Int. Conf. on Information Technology: New Generations*, 2009, pp. 5–9.

[5] K. Byeungchul *et al.*, "Current status and future prospect of phase change memory," in *Proc. of IEEE Int. Conf. on ASIC (ASICON)*, 2011, pp. 279–282.

[6] A. Redaelli *et al.*, "Electronic switching effect and phase-change transition in chalcogenide materials," *IEEE Electron Device Lett.*, 2004.

[7] D. Lee *et al.*, "Resistance switching of copper doped MoOx films for nonvolatile memory applications," *Applied Physics Letters*, vol. 90, no. 12, pp. 2237–2251, Mar. 2007.

[8] G. De Sandre *et al.*, "Program circuit for a phase change memory array with 2 MB/s write throughput for embedded applications," in *European Solid-State Circuits Conference (ESSCIRC)*, 2008, pp. 198–201.

[9] J.-T. Lin *et al.*, "Operation of multi-level phase change memory using various programming techniques," in *Int. Conf. on IC Design and Technology (ICICDT)*, 2009, pp. 199–202.

[10] C. Ahn *et al.*, "Crystallization properties and their drift dependence in phase-change memory studied with a micro-thermal stage," *J. of Applied Physics*, vol. 110, no. 11, pp. 114 520–114 520–6, 2011.

[11] T. Nirschl *et al.*, "Write strategies for 2- and 4-bit multi-level phase change memory," in *Proc. of IEEE Int. Electron Devices Meeting (IEDM)*, 2007, pp. 461–464.

[12] F. Bedeschi *et al.*, "A bipolar-selected phase change memory featuring multi-level cell storage," *IEEE J. of Solid-State Circuits*, vol. 44, no. 1, pp. 217–227, Jan. 2009.

[13] S. Braga *et al.*, "Voltage-driven partial-reset multilevel programming in phase-change memories," *IEEE Trans. on Electron Devices*, vol. 57, no. 10, pp. 2556–2563, Oct. 2010.

978-1-4673-4845-4/13 $31.00 © 2013 IEEE

978-1-4673-4845-4/13 $31.00 © 2013 IEEE   200

SESSION 10: Arrays and Ring Oscillators

978-1-4673-4845-4/13 $31.00 © 2013 IEEE   202

Tr variance evaluation induced by probing pressure and its stress extraction methodology in 28nm High-K and Metal Gate process

T. Okagaki, T. Hasegawa, H. Takashino, M. Fujii, A. Tsuda, K. Shibutani,
Y. Deguchi, M. Yokota, K. Onozawa

Renesas Electronics Corp., 4-1 Mizuhara Itami, Hyogo 664-0005, Japan

Fax: +81-72-789-3004 **Tel:** +81-72-787-2406
e-mail: takeshi.okagaki.xg@renesas.com

I Abstract

We discuss characteristics variance in detail, caused by probing stress in 28nm High-K and Metal Gate process. The V_{th} variation of nch large size transistor increases by 20% comparnig with weak probing pressure($\simeq 0$). Regarding small size transistors, probing stress impact both on V_{th} fluctuation and on T_{pd} fluctuation is small.

Moreover, we extracted the space distribution of probing stress quantitatively. It is useful to calibrate a stress simulation methodology and to facilitate evaluation of the mechanical strength of the material.

II Introduction

Because of the demand for smaller chip size, active circuits under bonding/probing pads are becoming a general technology. Several reports have described the fluctuation of characteristics such as propagation delay and drain current caused by the bonding and probing stress[1][2]. As described in this paper, we discuss transistor and propagation delay characteristic variance caused by probing stress in 28nm High-K and Metal Gate processes. The characteristic variance is compared directly with the local variation, which originates from dopant fluctuation. Results indicate whether probing stress should be examined at the time of circuit design stage.

Moreover, the stress which reaches the silicon surface during probing test is extracted using an accurate physical model. The quantitative space distribution of stress is useful to calibrate a stress simulation methodology and to facilitate evaluation of the mechanical strength of the material.

III Test Structure

Figure 1 shows a schematic of the test structure layout for transistor characteristics. For detailed evaluation of the fluctuation of characteristics, an arrayed transistor circuit is placed under the probing pad. These transistors can be measured selectively and independently using a decoder circuit. The probing pad size is 50 μm $\times$ 50 μm, and transistors are arrayed in 30 μm $\times$ 30 μm area. The total transistor number is 128 in an array. The characteristic fluctuation, induced by probing stress, is extracted by changing the probing pressure mechanically. Local variation is extracted with weak probing pressure ($\simeq$0) conditions. Two types of gate length, minimum length, and 0.9 μm are included in an array. The number of 0.9

μm gate length transistors is nine. These are located at the center and corner uniformly in the array. The remaining number is the minimum gate length transistor. Three types of gate widths of minimum width, 0.135 μm and 0.27 μm are arranged at a separate array.

Figure 2 presents a schematic of the test structure layout for propagation delay(T_{pd}). Four inverter ring oscillators are placed under the probing pad. These are measured independently. The gate length and gate width are, respectively, minimum, and 0.135 μm.

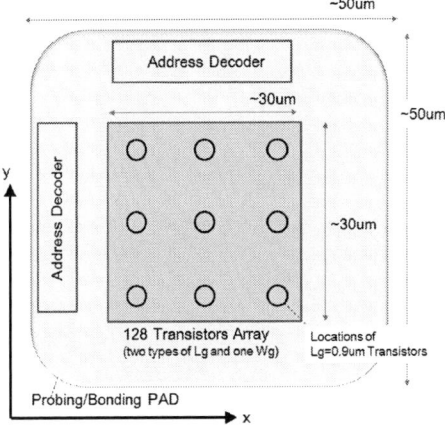

Fig. 1: Layout schematic of transistors. These are located under the probing pad.

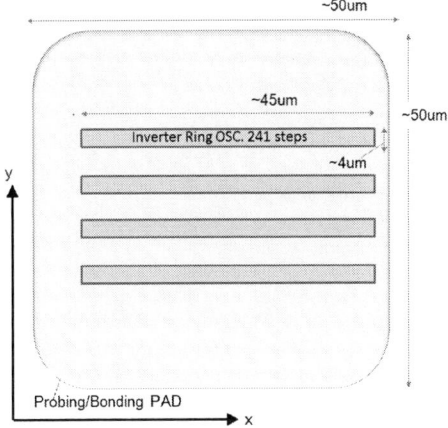

Fig. 2: Layout schematic of ring oscillators. Four ring oscillators are located under the probing pad.

978-1-4673-4845-4/13 $31.00 © 2013 IEEE   203

IV Results and Discussion

Figure 3 shows the spatial distribution of nch and pch transistor ΔV_{th}, which is defined as follows.

$$\Delta V_{th} = V_{th}(\text{strong probing pressure} >> 0)$$
$$- V_{th}(\text{weak probing pressure} \simeq 0).$$

The probing trace is shown in Fig. 4. V_{th} clearly changes under the probing position. Its change occurred between 10 μm area. ΔV_{th} cumulative distributions of small area nch and pch transistors are presented respectively in Figs. 5– 6. The probing stress induces V_{th} lowering in the nch transistor because of compressive stress perpendicular to the channel [3]. In contrast, V_{th} of the pch transistor rises. The V_{th} fluctuation range of the nch transistor is larger than that of the pch transistor.

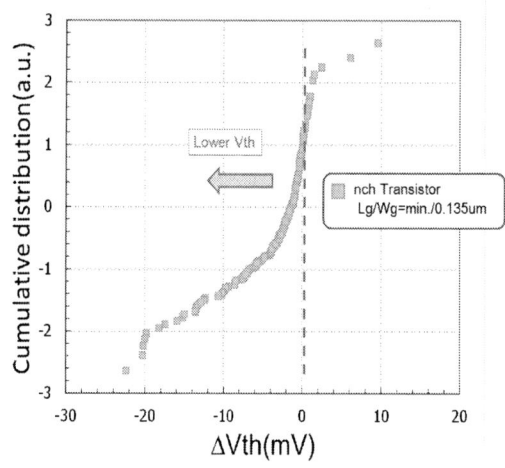

Fig. 5: Cumulative distribution of ΔV_{th} for nch transistors. Each point is shown by summarizing the result of 2 arrays.

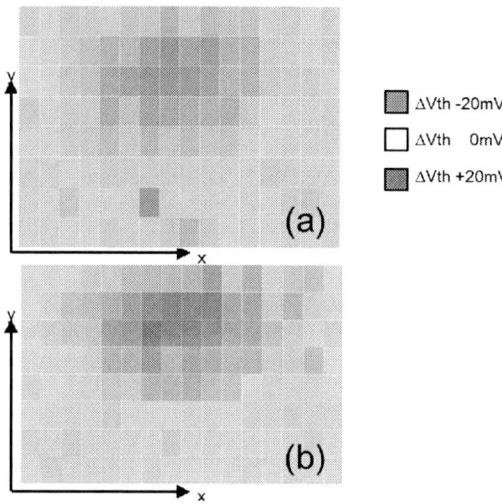

Fig. 3: ΔV_{th} spatial distribution under probing pad: (a) nch transistor, (b) pch transistor.

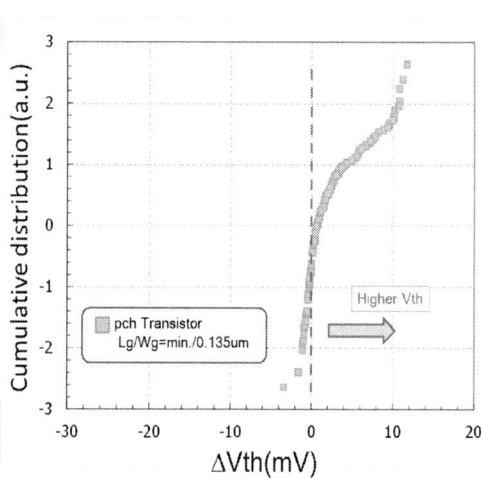

Fig. 6: Cumulative distribution of ΔV_{th} for pch transistors. Each point is shown by summarizing the result of 2 arrays.

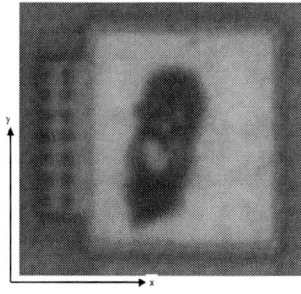

Fig. 4: Photograph of probing trace. The probe edge touches the upper part of the pad.

Figure 7 shows a Pelgrom plot of nch transistor V_{th} variation(σV_{th}) with two kinds of probing pressure (weak and strong). σV_{th} is extracted in the 30 μm $\times$ 30 μm arrayed area. It is confirmed that V_{th} variation is independent from the probing stress for small area transistors (Lg=minimum). In contrast, for the large area transistor (Lg=0.9 μm), V_{th} variation under strong probing stress is 1.2 times greater than that which occurs under weak probing stress.

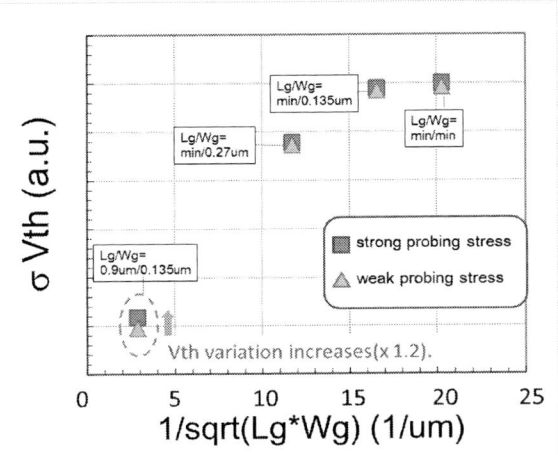

Fig. 7: V_{th} variations of nch transistors with weak/strong probing conditions.

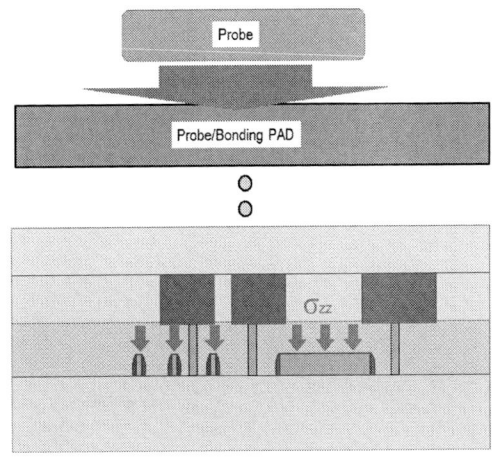

Fig. 9: Schematic view of probing stress. The probing test induces uniform compressive stress to the transistors.

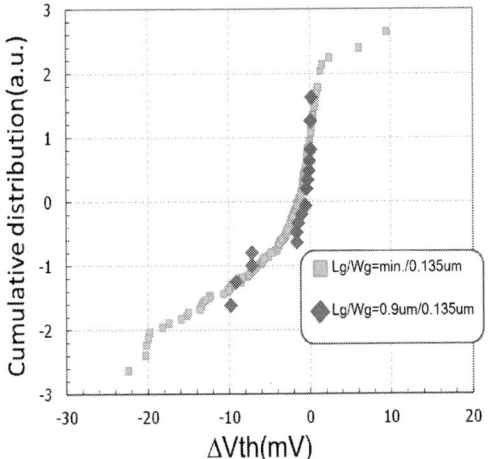

Fig. 8: Cumulative distribution of ΔV_{th} for both small and large area nch transistors. Each point is shown by summarizing the results of two arrays. The results of small area transistors are the same as that shown in Fig. 5.

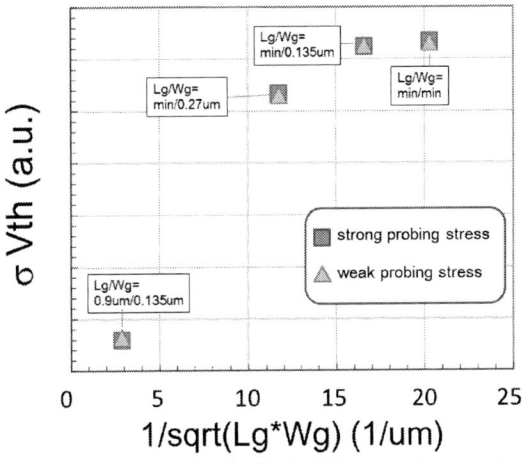

Fig. 10: V_{th} variations of pch transistors with weak/strong probing conditions. No noticeable difference is observed under strong probing stress.

Next we examine the relation between transistor area and probing stress impact. ΔV_{th} cumulative distributions of the small area transistor and large area transistor are shown in Fig. 8. Results reveal that V_{th} fluctuation under strong probing stress is independent of the transistor size. We consider that this is because the probing stress is not localized in a channel, but is applied about equally to the whole channel (Fig. 9). The V_{th} variation under probing test condition can be expressed as

$$\sigma V_{\text{th}} = \sqrt{\Delta V_{\text{th}}^2 + \sigma V_{\text{th}}(\text{local})^2},$$

where $\sigma V_{\text{th}}(\text{local})$ represents the local variation, which is originated by dopant fluctuation. Therefore, in a large area transistor with small $\sigma V_{\text{th}}(\text{local})$, the σV_{th} variation increases by 20% under strong probing stress.

Figure 10 shows a Pelgrom plot of pch transistor V_{th}

variation with probing stress of two kinds. No noticeable difference is observed under strong probing stress because ΔV_{th} is sufficiently smaller than $\sigma V_{\text{th}}(\text{local})$.

The propagation delay fluctuation(ΔT_{pd}) is evaluated in Fig. 11. ΔT_{pd} is defined as follows.

$$\Delta T_{\text{pd}} = T_{\text{pd}}(\text{strong probing pressure} >> 0)$$
$$/ \quad T_{\text{pd}}(\text{weak probing pressure} \simeq 0)$$

There is little influence of the probing stress to propagation delay fluctuation. It is considered that localized V_{th} fluctuation, which is induced by probing stress, is averaged, and that the influence on the ring oscillator thereby becomes small.

Next, we strive for extraction of applied stress, which causes V_{th} fluctuation. A physically based calculation result of the energy band edge shift under diverse stress conditions has been reported in [3]. With the relation be-

978-1-4673-4845-4/13 $31.00 © 2013 IEEE

tween band edge shift and ΔV_{th} of nch transistor, which has been reported in [4], an induced stress to the silicon surface can be extracted. It is shown in Fig. 12. The compressive stress is maximized under the probing position. Up to 250 MPa is observed.

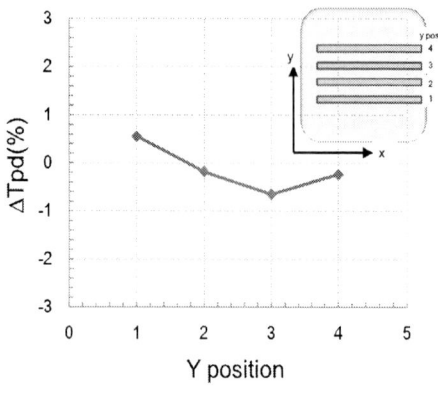

Fig. 11: Propagation delay fluctuation induced by probing stress. A range of ΔT_{pd} is within $\pm 1\%$.

V   Conclusion

Detailed analysis of characteristic fluctuation induced by probing stress shows that V_{th} variation of the nch large area transistor increased by 20%. Regarding small size transistors, probing stress impact both on V_{th} fluctuation and on T_{pd} fluctuation is small.

When using a large area nch transistor under a pad, we must be more careful about an increase of the local variation.

Moreover, we extracted the channel stress quantitatively, which was induced by probing tests. It was up to 250 MPa. The results of spatial stress distribution at the silicon surface are applied to improvement of the simulation accuracy.

References

[1] K. Chou et al., *EDL*, p.466 2001.
[2] K. Takemura et al., *SSDM*, p.324 2006.
[3] H. Takashino et al., *SISPAD*, p.107 2011.
[4] J. Lim et al., *EDL*, p.731 2004.

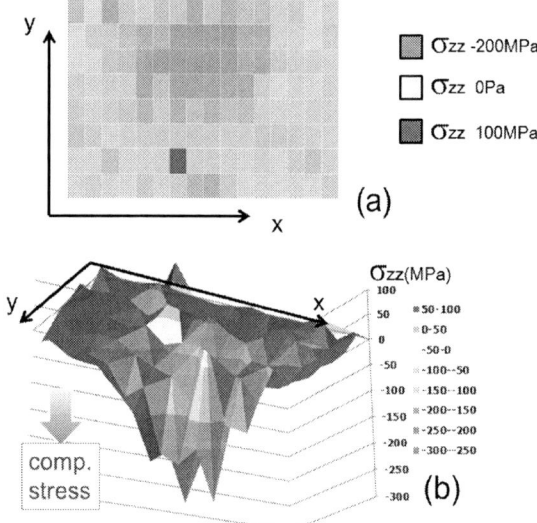

Fig. 12: Extracted stress distribution under probing pad. The 2-D plot is shown in (a). The 3-D plot is shown in (b).

Efficient Technique for Si Validation of Level Shifters

Puneet Sharma[1], Brad Smith[2], Donald Hall[2], Mike Nelson[2] and Umesh Lohani[1]

[1]Freescale Semiconductor, Plot No 2&3, Sector-16A, NOIDA INDIA
[2]Freescale Semiconductor, 7700 W. Parmer Lane, Austin, TX USA
Contact: puneets@freescale.com

Abstract— This paper presents a new structure that uses an addressable parametric array to validate level shifter cells. This structure is very area efficient and allows direct measurement of input and output voltages. Being a parametric structure enabled direct measurement of the output voltages, a critical parameter for level shifters. Experimental data confirmed the utility of this approach, validating level shifters in three different power domains including source biasing on the same 22-pad design. The simulation results show good correlation with the measured data.

Keywords—Level shifters; Si validation; addressable array

I. INTRODUCTION

In SOC's, level shifter cells are essential elements as multiple power domains are used along with various low power techniques. Today, level shifters are not just restricted to voltage translation but are also required to support various low power techniques, such as source biasing where the source node of a NMOS is biased to a voltage higher than 0 V. With these requirements it is critical to validate level shifter cells early in the design cycle to ensure first-time Si success. Testing level shifters in SOC form is complex and may not allow easy access for direct measurement of input and output voltages. Wiring individual cells to their own probe pads is an easy solution but is not very area-efficient because it consumes lot of pads. In this paper a new structure is presented which uses an addressable parametric array to test level shifter cells. This new structure is very efficient in terms of area and at the same time allows direct measurement of output voltages. Depending upon the configuration of array and size of the probe pads, the area could be a factor of 10 to 15 smaller than discrete test structures.

The level shifters cells to be validated in this work were functionally all buffer cells with the input and output pins are driven to different voltage levels (VDDin, VSSin, VDDout and VSSout). See Fig. 1. Here, these cells support three voltage domains, as shown in Table I. In this paper, the term "core" shall refer to the 1.20 V voltage domain and the term "IO" shall refer to the 3.30 V voltage domain. The level shifter library to be validated contains more than 100 cells to translate signals from one voltage domain to another and back again, as demonstrated in Fig. 2. For digital cells, measurement of functionality and speed is sufficient for validation [1-6] but for level shifters it is important to also measure the actual output voltage levels. This is especially important for the domain that supports source biasing, in which a logical "0" is actually a voltage measured in mV (i.e., not 0 V). The work reported here concerns such DC measurements and ring oscillator results are not included.

TABLE I:
VOLTAGE DOMAINS TO BE SUPPORTED BY LEVEL SHIFTER CELLS

Voltage Domain	VDD	VSS
Core voltage	1.20 V	0.00 V
Core voltage with source biasing	1.20 V	0.20 V
IO voltage	3.30 V	0.00 V

II. STRUCTURE DESIGN

A. Direct-Connect

The simplest structure for measuring a standard cell in a parametric tester is to connect each of its pins to its own probe pad. The generic direct-connect arrangement used in this work is shown in Fig. 3. Each device under test (DUT) had dedicated input and output pads while its power supply pads were shared across all DUTs. With this structure, the number of pads required is a direct function of the number of DUTs to be tested. The direct-connect structure presented in this paper contained seven DUTs and were tested with a 22-pad probe card.

B. Parametric Array

The array configuration is shown in Fig. 4. Each DUT cell was connected to shared input & output bus lines through CMOS passgates. These passgates were built from devices that can tolerate the highest voltages that will be applied to the DUTs. The addressing scheme was such that only one DUT at a time was selected; all other DUTs were disconnected from the input and output buses by use of the passgates. This sort of passgate muxing has been used before to enable measurements of single devices [7, 8], but is now being used to connect standard cell signal pins to shared test buses. Each DUT was connected directly to the appropriate power domain buses (e.g., no power gating was used). Given the isolation provided by the passgates, it was possible to place IO-to-core and core-to-IO level shifter cells in the same array. Even when an IO-level voltage was present on the input or output buses, the unselected core-level DUTs never experienced that high voltage. The inputs of the unselected DUTs were pulled down with NFET devices to ensure no switching when the shared input voltage changed. The addressable array structure presented in this paper contained 96 DUTs, broken into three power groupings and was tested with a 22-pad probe card.

III. EXPERIMENTAL RESULTS

Structures as described above were designed and built in a 90 nm low-power technology. The final structure was 2860 um wide x 80 um high for both the array and direct-connect structures. The four cells listed in Table II were placed into

978-1-4673-4845-4/13 $31.00 © 2013 IEEE

array structures. Because of area limitations, only one type of cell was placed into direct-connect structures — those that output a core voltage with source bias (the last row in Table II). An Agilent 4071 parametric tester and a 300 mm wafer prober were used to measure the test structures. Use of a parametric tester enabled the high-precision voltage and current measurements report below. Table II also shows the voltage levels that were applied during simulations and during testing. Where source bias was required, 0.20 V was used. Process matrix wafers with a variety of process conditions (fast vs. typical vs. slow) were run. SPICE simulations of the individual DUTs were performed and compared to the electrical measurements. Unless otherwise stated, the term "measured" in this paper shall refer to data measured in the addressable array structures.

TABLE II:
VOLTAGE LEVELS APPLIED DURING SIMULATIONS AND TESTING

Cell Type	VDD of Input	VSS of Input	VDD of Output	VSS of Output
C2I: Core to IO	1.20 V	0.00 V	3.30 V	0.00 V
I2C: IO to Core	3.30 V	0.00 V	1.20 V	0.00 V
B2U: Core With Source Bias to Core Without Source Bias	1.20 V	0.20 V	1.20 V	0.00 V
U2B: Core Without Source Bias to Core With Source Bias	1.20 V	0.00 V	1.20 V	0.20 V

IV. ANALYSIS

A. V_{OUT} vs. V_{IN}

Fig. 5 shows the measured output voltage of each of the four cells as a function of the input voltage. The input voltage was swept from VSS to VDD and then back to VSS (per the values in Table II). For the two cells that use source bias, both 0.00 V and 0.20 V were used and overlaid on their corresponding plots. In these figures, the high degree of correlation between the measured and simulated voltages is clear. In fact, some hysteresis can be seen in several of the cells, generally as predicted by simulations.

The parameters VIH, VIL (input voltages that are detected as "1" or "0", respectively), VOH and VOL (output voltages corresponding to "1" and "0" outputs, respectively) can be measured directly from these curves. The "1" output voltages did indeed match VDD, as expected (e.g., 1.20 V for core voltages, 3.30 V for IO voltages). For the outputs without source biasing, the "0" output voltages were 0.00 V. For the source-biased outputs, the "0" voltage level was 0.20 V, matching expectation. Such direct voltage measurements of outputs are critical for full validation of cells such as level shifters, especially when source biasing is to be used.

The results from the array structures were also compared to those from the corresponding direct-connect structures. Fig. 6 shows those results, again with a source bias of both 0.00 V

and 0.20 V. The direct-connect results match well those from the array structures.

Fig. 7 shows five process splits overlaid in one plot for the core-to-DGO level shifter cell with input voltage swept in only one direction. The response of the cell to changes in the devices is clearly visible.

B. Switchpoint

"Switchpoint" was defined in this work as the input voltage at which the output voltage changed by ~25% as the input was swept. Fig. 8 compares the switchpoint values measured in the array vs. the values that simulation predicted. These plots show all process splits, both 0.00 V and 0.20 V source bias (where appropriate) and all cell types; the core voltage and IO voltage plots were separated into (a) and (b), respectively, for clarity. While the correlation is fairly good, the simulated values tended to be smaller than the measured values. This is exactly the sort of small deviations from ideal that can be seen when using a high-precision parametric tester to make these measurements. It should be noted that these differences are below 100 mV in magnitude, well within the required margin for predicting the behavior of a digital standard cell.

The switchpoint values measured in the array structures matched those measured in the direct-connect structures very well, as seen in Fig. 9. The high degree of correlation observed between the two structures confirms the effectiveness of the array technique. Fig. 9 contains only one cell (the one labeled "U2B" in Table II), so has a fraction of the data points that Fig. 8 has. However, re-plotting Fig. 8 with just the U2B cell (Fig. 10) produces the same result, indicating that the differences between the model and measured values are real and not due to the structure.

C. Current Profile

Fig. 11 shows the supply currents for all the level shifter DUTs as a function of input voltage. The complex response of the supply currents predicted by the simulations is clearly visible in the measured data. Note that the measured currents hit a minimum value around 0.1-1 nA for the I2C and C2I cells and at about 1 uA for the core-voltage cells. This is due to the fact that all DUTs in a power grouping share power pins. While the strength of the array structure is the ability to measure the voltages of individual DUT cells, it is still capable of measuring supply current for those same cells.

This background leakage floor increases with temperature as can be seen in Fig. 12. Such background levels are acceptable for these types of measurements.

V. SUMMARY

A new and efficient structure using an addressable parametric array to validate level shifter cells has been demonstrated. This technique allows many more DUTs to be placed in the same area with direct measurement of input and output voltages. Measured voltages and currents correlated well with simulations and corresponding simple, direct-connect measurements.

REFERENCES

[1] J.-S. Goo, *et al*, "SPICE parameter extraction and RO validation of a 65nm SOI technology", IEEE International SOI Conference, Oct 2008

[2] W. Agatstein, K. McFaul, P. Themins, "Validating an ASIC Standard Cell Library", Proceedings of the Third Annual IEEE ASIC Seminar and Exhibit, Sep 1990

[3] R.-B. Lin, I. Chou, C.-M. Tsai, "Benchmark Circuits Improve the Quality of a Standard Cell Library", Proceedings of the Asia and South Pacific Design Automation Conference (ASPDAC), Jan 1999

[4] A. Gupta, "A Robust Level-Shifter Design for Adaptive Voltage Scaling", 21st Conference on VLSI Design, Jan 2008

[5] S. Monga, "High speed stress tolerant 1.6 V – 3.6 V low to high voltage CMOS level shift architecture in 40 nm", IEEE International Symposium on Circuits and Systems (ISCAS), May 2012

[6] V. Kolagunta, *et al*, "Calibration of library element optimization to improve static power", ICMTS, Mar 2012

[7] K. Doong, *et al*, "Field-Configurable Test Structure Array (FC_TSA): Enabling Design for Monitor, Model and Manufacturability", ICMTS, Mar 2006

[8] B. Smith, *et al*, "A Novel Biasing technique for Addressable Parametric Arrays", ICMTS, Mar 2008

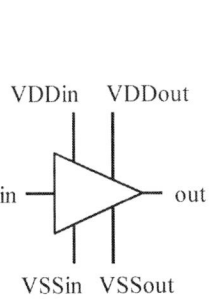

Fig. 1: Schematic of a generic level shifter DUT cell.

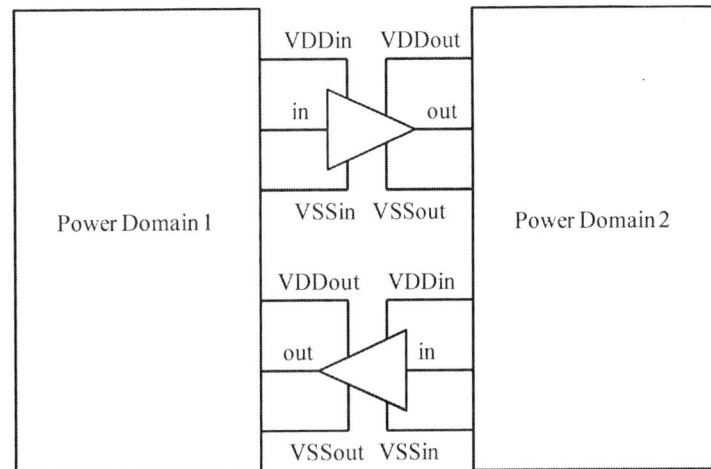

Fig. 2: Power Level shifter voltage translation across power domains.

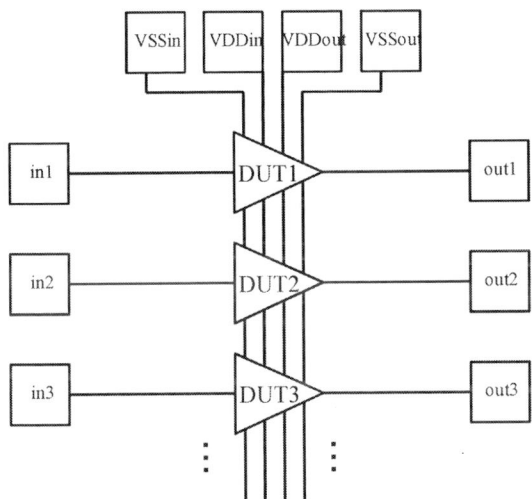

Fig. 3: Schematic of the direct-connect structures.

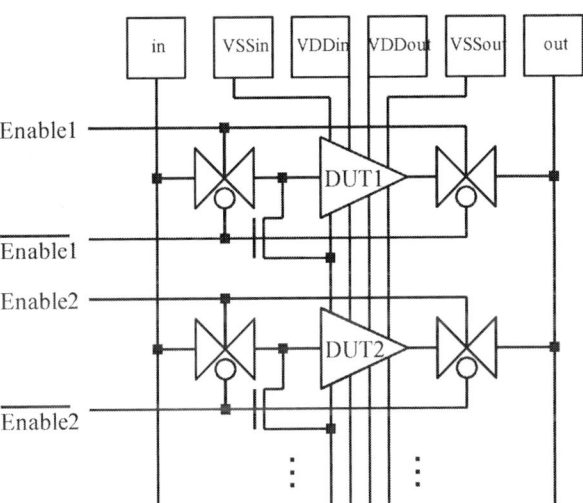

Fig. 4: Schematic of the addressable array structures.

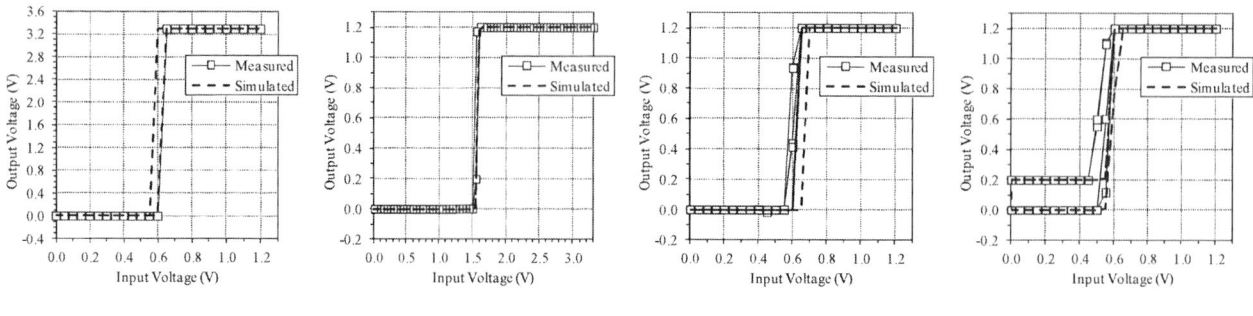

(a) C2I – Core to IO voltage.
(b) I2C – IO to core voltage.
(c) B2U – Core voltage with source bias (0.00 V and 0.20 V) to core voltage without source bias.
(d) U2B – Core voltage without source bias to core voltage with source bias (0.00 V and 0.20 V).

Fig. 5: Measured and simulated output voltage as a function of input voltage. Input voltage was swept from 0.00 V to VDD (per the values in Table II), then back again. Simulated = dashed lines, measured = symbols.

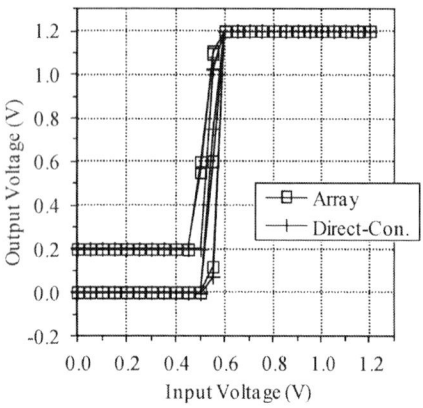

Fig. 6: Measured output voltage for the U2B cell (source bias = 0.00 V and 0.20 V) in array and direct-connect format. Input voltage was swept from 0.00 V to VDD (per the values in Table II), then back again.

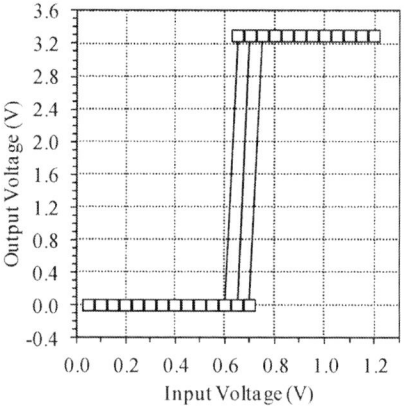

Fig. 7: Output voltage for the C2I cell, five process splits. Input voltage was swept from 0.00 V to 1.20 V.

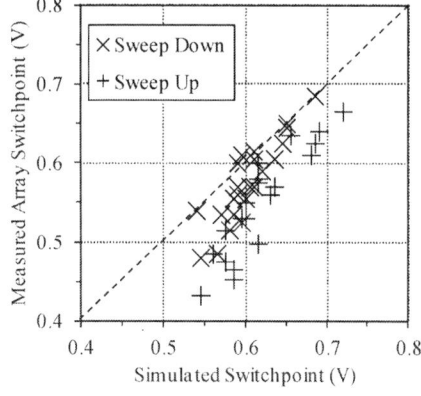

(a) C2I, U2B and B2U cells

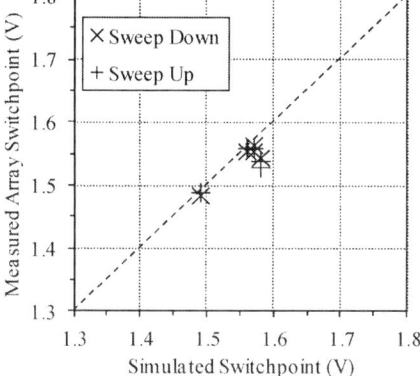

(b) I2C cell

Fig. 8: Measured switchpoint voltages vs. simulated for all process splits, source bias = 0.00 V and 0.20 V.

978-1-4673-4845-4/13 $31.00 © 2013 IEEE

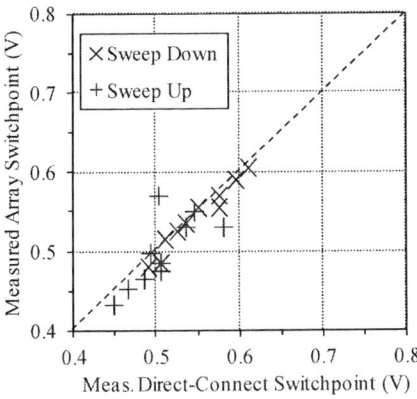

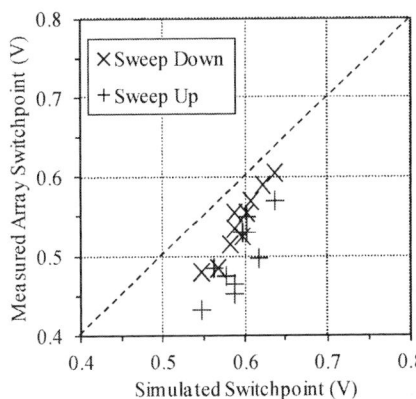

Fig. 9: Switchpoint measured in array format vs. measured in direct-connect format for all process splits, source bias = 0.00 V and 0.20 V.

Fig. 10: Measured switchpoint voltages vs. simulated for the U2B cell only (a subset of Fig. 8) for all process splits, source bias = 0.00 V and 0.20 V.

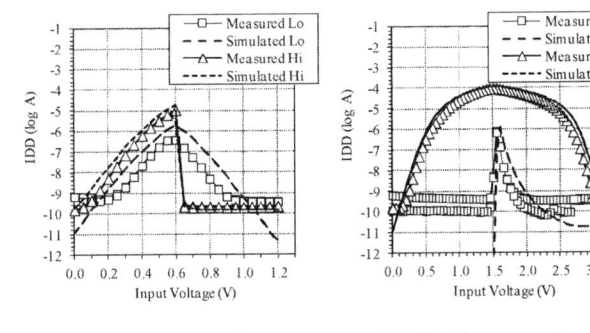

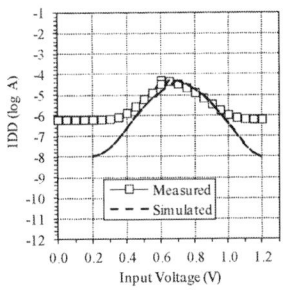

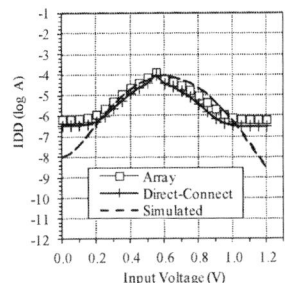

(a) C2I – Core to IO voltage.

(b) I2C – IO to core voltage.

(c) B2U – Core voltage with source bias (0.20 V) to core voltage without source bias.

(d) U2B – Core voltage without source bias to core voltage with source bias (0.20 V) in both array and direct-connect format.

Fig. 11: Measured and simulated power supply currents as a function of input voltage. Input voltage was swept from 0.00 V to VDD (per the values in Table II), then back again. Simulated = dashed lines, measured = symbols. "Hi" = IO voltage, "Lo" = core voltage. In (c) and (d), both input and output power supplies were shorted together.

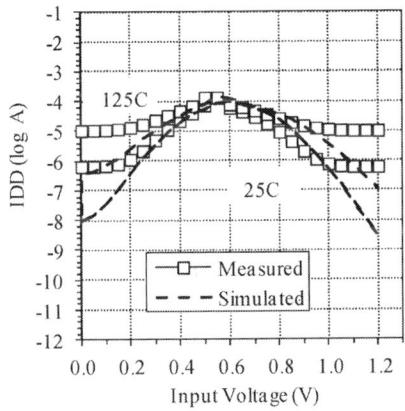

Fig. 12: Measured and simulated power supply current as a function of input voltage for 25 C and 125 C. Input voltage was swept from 0.00 V to VDD (per the values in Table II), then back again.

Mosaic SRAM Cell TEGs with Intentionally-added Device Variability for Confirming the Ratio-less SRAM Operation

Hitoshi Okamura, Takahiko Saito, Hiroaki Goto, Masahiro Yamamoto and Kazuyuki Nakamura

Center for Microelectronic Systems, Kyushu Institute of Technology
680-4 Kawazu, Iizuka, Fukuoka 820-8502, Japan
Tel: +81-948-29-7584, Fax: +81-948-29-7586, Email: hitoshi_okamura@cms.kyutech.ac.jp

Abstract—MOSAIC SRAM Cell TEGs consisting of memory cells having all combinations of gate sizes of transistors differing by two orders of magnitude were developed with 0.18um CMOS process to verify the operation margins for SRAM circuits. The measured results show the operation of the ratio-less SRAM is completely independent of the size of transistors in the memory cell.

Keywords—*SRAM; Variability; Ratio-less; Static Noise Margin;*

1. Introduction

In the development of SRAMs for the state-of-the-art technology generation, it is difficult to maintain a sufficient operation margin in the SRAM, because of supply voltage scaling and device variability [1]. Figure 1(a) shows a conventional 6-transistor CMOS SRAM cell which has been widely used in the past and up to the present as the basic SRAM memory cell circuit. In this circuit, the write (flip/flop inversion) operation is performed from the outside through the impedance of the transfer transistors (SN1, SN2), while the read operation must be performed by outputting the stored information to the bit-lines through the same transfer transistors without destroying the information in the flip/flop. In the design of the transfer gate transistors, therefore, there exist the upper and lower limit values that are permissible from considerations of impedances, and the ratio design is thus required for stable operation. As a method of quantitatively evaluating this design margin in SRAM cells, there are the Read and Write Static Noise Margin (RSNM and WSNM) indexes as shown in Fig. 1(b) [2]. In recent years, with drastically decreasing the device dimensions, it is becoming difficult to secure these SNMs simultaneously, because they suffer direct effects of the lowering of the power supply voltage and the increase in process variation among devices [3].

To overcome this problem, a circuit named 8T-SRAM has been proposed (Fig. 2(a)) [4]. In this circuit, the RSNM restriction can be avoided since it has a dedicated read buffer, however, the WSNM restriction still exists. In order to avoid both RSNM and WSNM restriction simultaneously, we had proposed the Ratio-less 10T-SRAM [5]. The operation of the Ratio-less SRAM is completely independent of the size for transistors in the memory cell. In order to confirm the Ratio-less operation, we developed the MOSAIC SRAM Cell TEGs.

2. Structure of MOSAIC SRAM Cell TEG

We developed MOSAIC SRAM Cell TEGs using 0.18μm CMOS in which the gate width (W) value of each set of transistors in a memory cell was varied by a range of two orders of magnitude. The device matrix array (DMA) TEGs to evaluate the device variability of MOSFETs or SNMs in SRAM cells were developed in the past [6]. However, a SRAM cell TEG with intentionally-added device variability to confirm the wide operation margin has not been developed. Figure 3 shows the design example of MOSAIC SRAM cell array for conventional 6-transistor SRAM. Our MOSAIC SRAM TEG includes the complete set of size combinations (the number of different designs are two to the sixth power, or 64) for 1-bit memory cells. The cell array consists of 64 memory cells in which the gate widths are designed with the design allocation table in Fig. 3(a). The size of each transistor composing a memory cell is designed to be either 0.3μm or 30μm. The cell design examples are shown in Fig. 3(b)-(d). Cell_0 in Fig. 3(b) is the case in which all transistors have the minimum size of 0.3μm, Cell_63 in Fig. 3(c) is the case in which all transistors have the maximum size of 30μm, and Cell_28 in Fig. 3(d) is an example of a mix of 0.3 and 30μm transistors.

3. Layout Method for MOSAIC SRAM Cell Array

Figure 4 shows the layout method we developed to form the MOSAIC cell array. In order to reduce the layout complexity, a memory cell is divided into the left and right half cells as shown in Fig.4 (a). In the case of 6T SRAM, each half cell consists of 3 transistors. Therefore, we first laid out 16 half cells (L1-8, R1-8) as shown in Fig.4 (b). The area occupied when all transistors have the maximum (30μm) is ensured as the size of footprint for all half cells. Then we develop the array reference of half cells to obtain the MOSAIC array as shown in Fig.4 (c). This method can reduce the layout cost to the order of square root of memory size.

978-1-4673-4845-4/13 $31.00 © 2013 IEEE

4. Developed MOSAIC SRM Cell TEGs

We developed 3 MOSAIC SRAM TEGs. Figure 5 (a) shows the 64-bit MOSAIC SRAM TEG which includes conventional 6 transistor SRAM cells. Figure 5(b) Shows 256-bit MOSAIC SRAM TEG which includes 8T SRAM cells. Figure 6 (a) shows 1024-bit MOSAIC SRAM TEG which includes 1024 ratio-less 10T cells. The cell layout examples in 1024-bit ratio-less MOSAIC SRAM TEG are shown in Fig. 6(b)-(d). Cell_0 in Fig. 6(b) is the case in which all transistors have the minimum size of 0.3μm, Cell_1023 in Fig. 6(d) is the case in which all transistors have the maximum size of 30μm, and Cell_374 in Fig. 6(c) is an example of a mix of 0.3 and 30μm transistors.

5. Experimental Results

Figure 7(a)-(e) are measured fail bit maps (FBMs) obtained by applying marching pattern tests to our MOSAIC SRAM TEGs. In the MOSAIC SRAM design, the position of memory cell corresponds to the combination of transistor sizes as indicated to the right and bottom of the FBMs. Figures 7(a), and (b) respectively show the results obtained in measurements of the MOSAIC SRAM TEGs with 6T SRAM cells and 8T SRAM cells. The hatched parts in the figures are failed cells. These results show that there are many failed cells in spite of the low-operating frequency of 1MHz. Since these circuits require ratio design, many combinations are inoperative. Figure 7(c) shows the result obtained in measurement with 8T MOSAIC SRAM TEG with the write after read technique [4]. Because this operation can relax the operation margin in write-half-select cells, failed cells are slightly decreased when compared with the result shown in Fig. 7(b). In contrast, only a few cells are failed in our ratio-less SRAM with the high-operating frequency of 15MHz as shown in Fig. 7(d). The operation of whole cells in our ratio-less SRAM is experimentally confirmed at an operating frequency less than 10MHz as shown in Fig. 7(e). The result of MOSAIC SRAM with our ratio-less SRAM cell shows that operation of the SRAM is independent of the size of transistors in the memory cell.

6. Conclusions

This paper proposed a new SRAM TEG to confirm the ratio-less SRAM operation. The MOSAIC SRAM Cell TEGs consisting of memory cells having all combinations of gate sizes of transistors differing by two orders of magnitude were developed with 0.18um CMOS process. The measured result of MOSAIC SRAM Cell TEGs with the ratio-less SRAM cell shows that operation of the SRAM is independent of the size of transistors in the memory cell.

Acknowledgments

This work was supported by VLSI Design and Education Center (VDEC), the University of Tokyo, in collaboration with Cadence Design Systems, Inc., Mentor Graphics, Inc., the Rohm Corporation and the Toppan Printing Corporation. This research was partially supported by funds from the Japanese Ministry of Education, Culture, Sports, Science and Technology, MEXT, via the Kyushu knowledge-based cluster project.

References

[1] S. Inaba et al., *IEDM Tech. Digest*, pp. 487 - 490, Dec. 2007.
[2] E. Seevinck et al., *IEEE J. Solid-State Circuits*, vol. SC-22, no. 5, pp. 748-754, Oct. 1987.
[3] H. Yamauchi, *Journal of Semiconductor Technology and Science*, Vol.9, No.1, Mar. 2009.
[4] Y. Morita et al., *Symposium on VLSI Circuits*, pp. 256-257, June 2007.
[5] T. Saito et al., *IEEE International Memory Workshop*, pp. 167-170, May 2012.
[6] M. Suzuki et al., *Symposium on VLSI Technology*, pp. 191-192, June 2010.

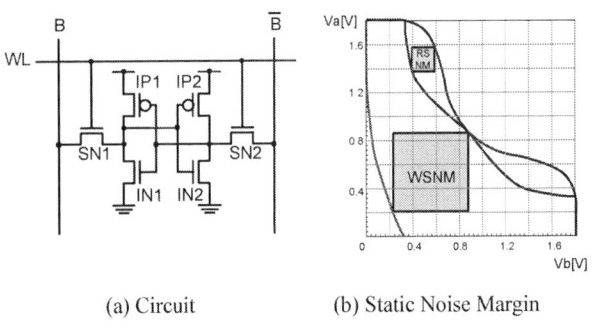

(a) Circuit   (b) Static Noise Margin

Fig. 1 Conventional 6 Transistor SRAM Cell

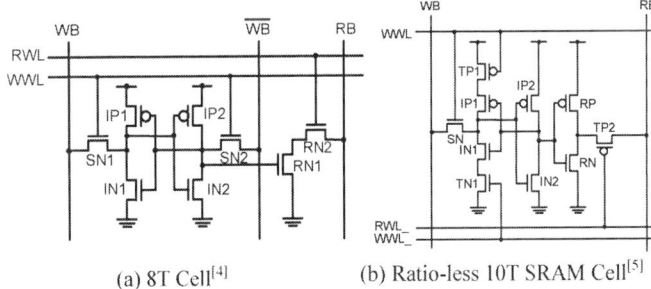

(a) 8T Cell[4]   (b) Ratio-less 10T SRAM Cell[5]

Fig.2 Improved SRAM Cells

978-1-4673-4845-4/13 $31.00 © 2013 IEEE

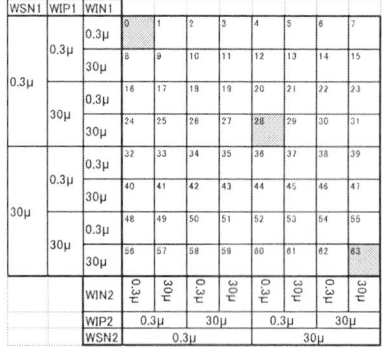

(a) Design Allocation Array

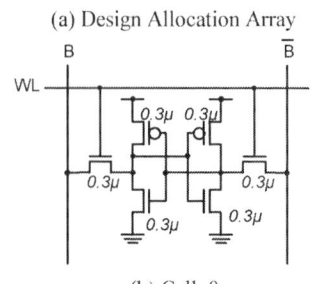

(b) Cell_0

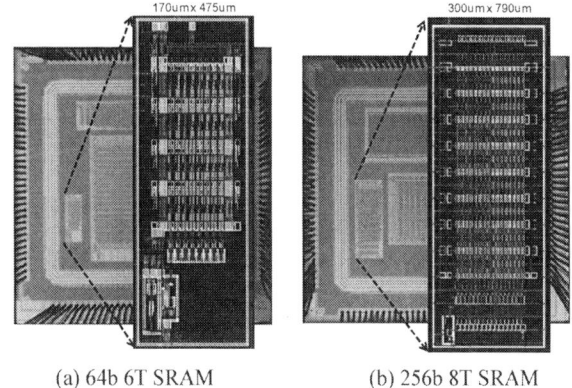

(c) Cell_63　　　　(d) Cell_28

Fig.3 Design of MOSAIC SRAM Cell Array

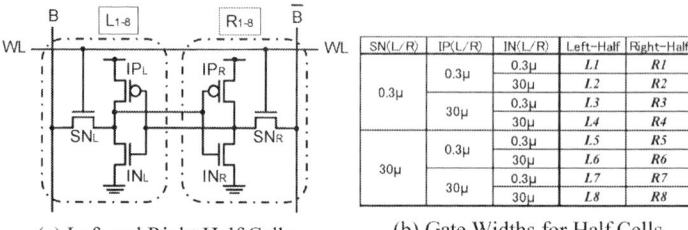

(a) Left and Right Half Cells　　　(b) Gate Widths for Half Cells

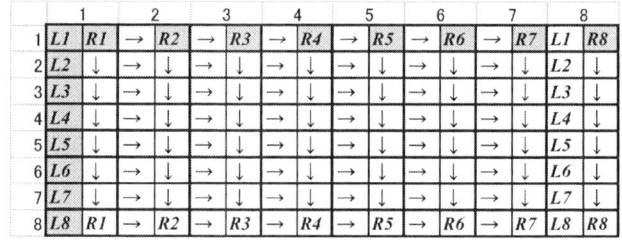

(c) Array Layout of Half Cells

Fig.4 Layout Method of MOSAIC Cell Array

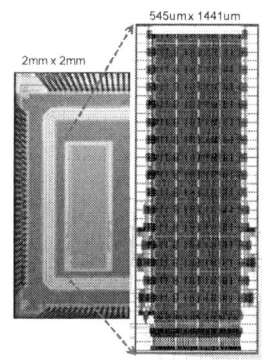

(a) 1024b 10T Ratio-less SRAM

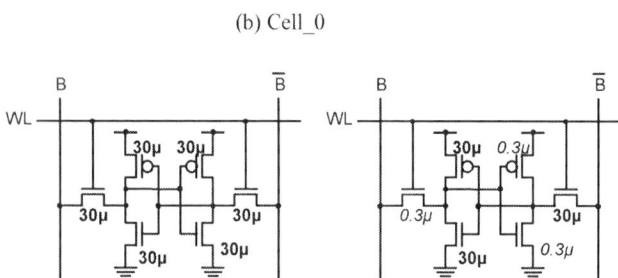

(a) 64b 6T SRAM　　　　(b) 256b 8T SRAM

Fig.5 Chip Photograph of MOSAIC 6T/8T SRAMs

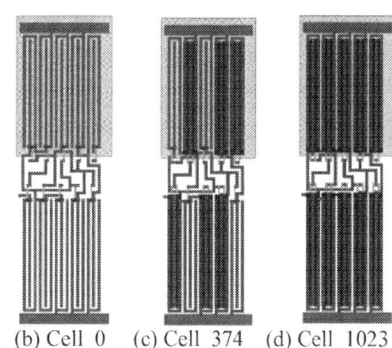

(b) Cell_0　(c) Cell_374　(d) Cell_1023

Fig.6 Chip Photograph of 10T Ratio-less MOSAIC SRAM

978-1-4673-4845-4/13 $31.00 © 2013 IEEE　　　214

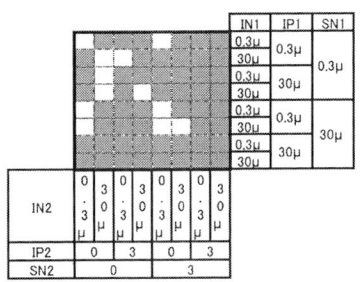

(a) 6T SRAM (1MHz, 1.8V)

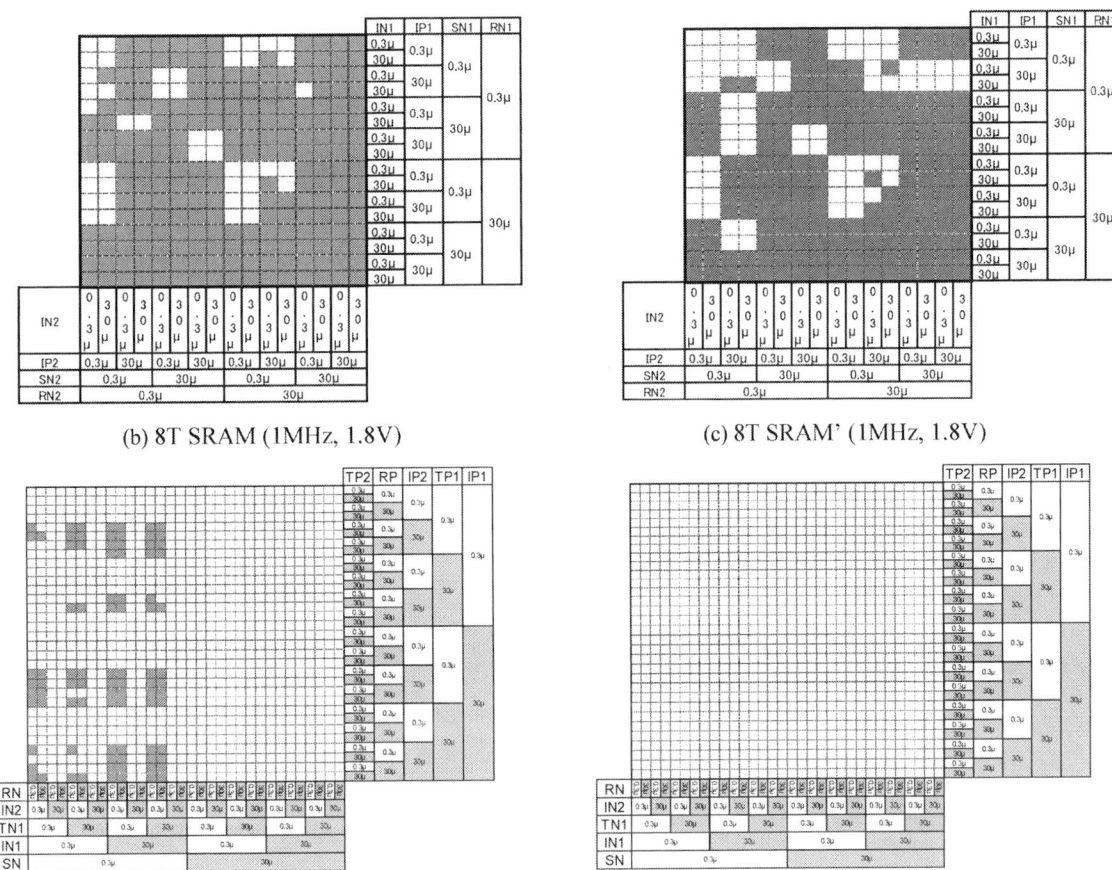

(b) 8T SRAM (1MHz, 1.8V)

(c) 8T SRAM' (1MHz, 1.8V)

(d) 10T Ratio-less SRAM (15MHz, 1.8V)

(e) 10T Ratio-less SRAM (10MHz, 1.8V)

Fig.7 Measurement Results of MOSAIC TEGs

Characterization and Simulation of NMOS Pass Transistor Reliability for FPGA Routing Circuits

Christopher S. Chen and Jeffrey T. Watt

Process Technology Development
Altera Corporation
San Jose, CA 95134, USA
chrchen@altera.com

Abstract— **In this work, the impact of bias temperature instability is evaluated for routing pass gate circuits. A simple test structure is proposed and measured data is compared to aging models to demonstrate the importance of modeling circuit level aging effects. Aging models which are shown to be accurate at the transistor level are inadequate at the circuit level unless frequency dependent aging effects are taken into account.**

Keywords— MOSFETs; reliability; bias temperature instability; field-programmable gate arrays

I. INTRODUCTION

Wafer level reliability characterization has become increasingly important for designing robust semiconductor devices and circuits. The growing interest in reliability characterization can be attested by performing a basic search through recent issues of electron device journals [1,2]. As process technology continues to advance, successful product development will rely increasingly on strong collaborations between process technologists, reliability physicists, and integrated circuit designers [3]. These industry trends underscore the importance of developing test structures and characterization methods to evaluate transistor and circuit reliability as a part of process qualification.

The reliability characteristics of a metal oxide semiconductor field effect transistor (MOSFET) are closely linked to the material and processing conditions used to manufacture the gate dielectric. While hot carrier injection (HCI) and time dependent dielectric breakdown (TDDB) are common to most modern MOS technologies, the bias temperature instability effect (BTI) varies greatly from one technology to another.

Until the 65nm technology node, MOSFETs were manufactured using a silicon oxynitride (SiON) gate dielectric. For SiON gate dielectrics, negative bias temperature instability (NBTI) is a reliability concern for PMOS devices. Starting from the 45nm technology node, the use of high-k gate dielectrics and metal gate electrodes (HKMG) introduces a positive BTI (PBTI) degradation mechanism for NMOS devices [4]. Thus, for HKMG processes, both NMOS and PMOS devices can degrade due to BTI effects.

In a field programmable gate array (FPGA), programmability is achieved using configuration memory cells which control both logic functions and routing circuitry [5]. Since routing speed is dominated by the pass transistor, degradation in this transistor could lead to timing failures or functional failures [6]. For NMOS pass transistors fabricated in HKMG processes, PBTI is a significant reliability concern. The expected aging of a transistor subjected to PBTI can be expressed using a metric called PBTI lifetime. The PBTI lifetime of a transistor is typically predicted by measuring the shift in drain current or threshold voltage of a single transistor subjected to DC voltage stress. A large number of single transistor test structures are stressed at voltages much higher than the nominal operating voltage to accelerate degradation until a given failure criteria is met. Using this approach, reliability engineers can predict the transistor lifetimes under ordinary use conditions.

In this work, the impact of PBTI is evaluated for routing pass gate circuits fabricated in a 28nm HKMG process. The paper is organized as follows. In Section II, an overview of our test structures will be given, with attention to specific features of our ring oscillator which afford unique advantages for reliability evaluations. In Section III, we describe an approach to modeling transistor aging and compare simulated results to measurement data for both circuits and single devices. A discussion of results is given in Section IV, with an explanation for why frequency dependent aging effects must be accounted for. Finally, a conclusion is presented in Section V.

II. TEST STRUCTURES

As discussed in Section I, single transistor test structures are typically employed for characterizing the reliability and expected lifetime of a transistor. However, while characterization of single transistors is necessary, it is also important to consider circuit level effects for determining aging of transistors in real circuits. To accurately predict reliability of product circuits, it is imperative to characterize aging at the circuit level [7,8].

Aging on circuits is typically characterized using either ring oscillators or delay chains [7,9]. Of these two approaches, ring oscillators are generally preferred for their ease of measurement [7]. Moreover, ring oscillator test structures which are commonly used during process development can also be used for basic reliability analysis. For reliability studies, the inverter supply voltage is typically increased to accelerate aging and circuit frequency is monitored without interrupting the stress. In this configuration, frequency measurements can be employed to monitor the total circuit degradation caused by NBTI, PBTI, HCI, and TDDB. Although the simplicity of the inverter ring oscillator is attractive, it is of limited usefulness for isolating one aging mechanism (e.g., PBTI) from the others (e.g., NBTI, HCI).

To address this issue, various ring oscillator test structures have been proposed to isolate aging effects on one device type at a time [10,11]. However, the structures described in [10] and [11] require additional device connections and control voltages which may not be readily available. Moreover, to date, we are not aware of any publications which evaluate the reliability of routing pass transistor circuits typically used in FPGAs.

In this work, we propose a simple ring oscillator test structure which can be used to isolate NMOS PBTI aging at the circuit level. In addition, single transistor test structures are also used for device level characterization to correlate to circuit level results.

A. Ring oscillator test structures

A simple ring oscillator test structure is employed to characterize PBTI at the circuit level (Fig. 1). The ring oscillators in this work were designed specifically to evaluate the speed and reliability of routing circuitry in an FPGA. The oscillator consists of a ring of inverters with two NMOS pass transistors inserted before each inverter to mimic routing multiplexers in an FPGA routing circuit [5].

The presence of NMOS pass transistors in this circuit has significant implications for both speed and reliability. During a switching event, the channel resistance of the pass transistor is much higher than the channel resistance of the on-state transistor in the inverter. Therefore, circuit speed is dominated by the NMOS pass transistors.

Moreover, the reliability of the circuit is dominated by PBTI on the NMOS pass transistor. When a traditional inverter ring oscillator is stressed with a high supply voltage, both the NMOS and PMOS devices contribute to the circuit reliability due to HCI and BTI [7]. However, since the stress voltage in our circuit is applied only to the NMOS pass transistor, it is the pass transistor reliability which dominates overall circuit aging. Moreover, due to our circuit topology, HCI effects are minimized because the pass transistors are not stressed by a high electric field across the channel. Therefore, our ring oscillator test structure is useful for isolating PBTI on the pass transistor.

In summary, since circuit speed and reliability are dominated by the NMOS pass transistors, this test structure is a convenient vehicle for evaluating circuit aging as a function of gate voltage.

Ring oscillator characterization was performed using an Agilent 4072A parametric tester and an automated wafer prober. A stress voltage is applied to the gate of the pass transistors and circuit speed is measured via an on-chip frequency divider and an off-chip spectrum analyzer. To minimize unwanted recovery effects, the circuit frequency is monitored without interrupting stress.

B. Single transistor test structures

Single transistor test structures are characterized to validate aging models at the device level (Fig. 2). To minimize changes in device characteristics caused by layout style, the transistor test structures were laid out to match the ring oscillator layout environment.

Transistors were stressed at a high gate voltage using an Agilent B1530A fast measurement unit on a semi-automatic wafer prober. To minimize recovery effects, stress was interrupted for 1 micro-second to measure drain current at the nominal gate voltage. The transistor current-voltage (IV) characteristics were measured before and after stress to monitor the shift in threshold voltage caused by PBTI. Although the post-stress IV curve includes some inevitable recovery, the pre- and post-stress curves are nevertheless helpful indicators of device degradation.

III. MEASUREMENT AND SIMULATION DATA

Measurements on both single transistors and ring oscillators were performed at room temperature. Measured results were compared to SPICE simulations where transistor aging was accounted for using transistor aging models.

A. Aging models

The Synopsys MOSRA aging model was chosen for ease of integration with commercial SPICE simulators [12]. The aging model accounts for NBTI, PBTI, and HCI using model parameters which were fitted to experimental data measured on single transistors. The aging model was scaled to the mean time to failure (MTTF) lifetime condition for comparison to silicon data.

Transistor BTI effects are modeled as a shift in threshold voltage caused by charge trapping. The aging model accounts for sensitivities to voltage, temperature, and stress time. In AC operation, the simulation accounts for duty cycle effects by reducing the total stress time. Recovery effects are modeled based on recovery data measured on single transistors.

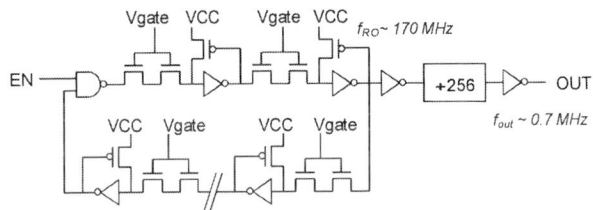

Fig. 1. Ring Oscillator Test Structure. Number of stages is 127.

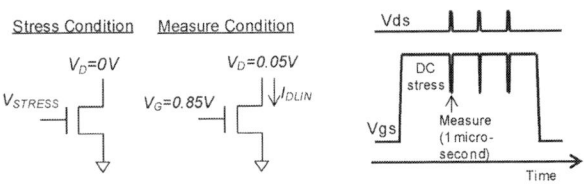

Fig. 2. Transistor Stress Conditions. Waveform shows interruption of DC stress for a fast IV measurement.

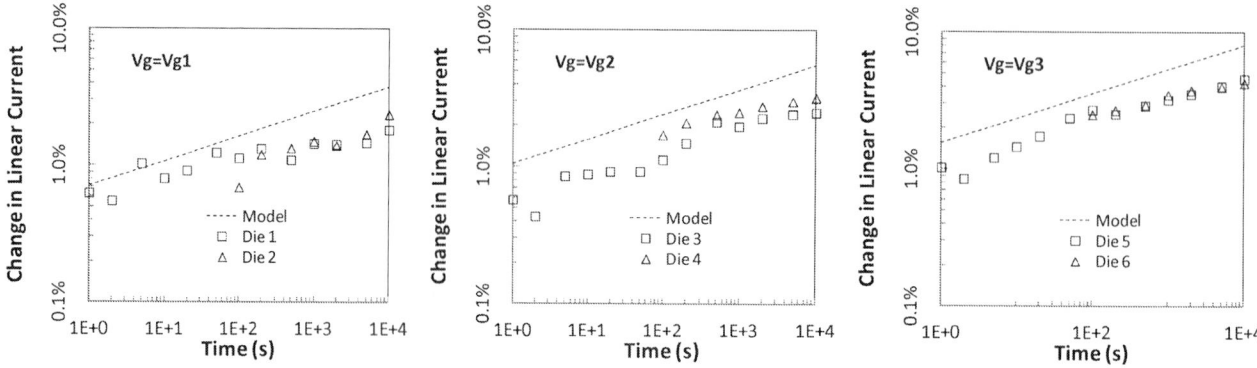

Fig. 3. Transistor Stress Data and Model. Each measurement done on a fresh die.

B. Transistor data

Transistor data measured using the fast IV method is shown in Fig. 3. Measured and simulated data are plotted for three different stress voltages (where Vg1<Vg2<Vg3). At each stress voltage, measured data is shown for two fresh dies, where each die was stressed for 10,000 seconds. Although the data is noisy, the silicon data is consistent with the model for stress times up to 100 seconds. For some die, a change in slope is observed for stress times beyond 100 sec. It is unclear whether this effect is real, or a measurement artifact. Overall, the transistor level fast IV data demonstrates that the aging model overestimates the degradation in linear current by up to 2X for single transistors under DC stress conditions.

Some of the discrepancy between simulated and measured results may be attributed to unavoidable recovery effects during measurement which are not included in our transistor level simulations. To minimize recovery effects, measurements were taken in logarithmic time steps with a 1 micro-second measurement window. However, any charge trap activity which occurs on a timescale smaller than 1 micro-second could induce unwanted recovery to our measurement data [13].

To evaluate the impact of recovery effects on our transistor measurements, half of the dies were sampled three times per decade starting at 0.001 seconds, while the other half of the dies were sampled three times per decade starting at 100 seconds. If the measurement were to induce recovery, one would expect the dies which were sampled at earlier times would show reduced degradation. Due to variations in our data, it is difficult to quantify the amount of recovery induced by our measurement. Further study may be required to quantify the recovery effects in our transistor measurements.

C. Ring oscillator data

Ring oscillators with NMOS pass transistors were stressed at the same three stress voltages used for single transistors. The degradation in frequency was monitored in-situ while a stress voltage was applied to the gate of the pass transistors. To minimize stress on the inverter stages, the inverter supply voltage was kept at the nominal operating voltage.

Simulations were adjusted using a model parameter to account for AC recovery effects. However, the model does not account for frequency dependent aging effects.

In Fig. 5, the degradation in stage delay is plotted versus stress time for both the model and silicon. Since the degradation measured on silicon is significantly lower than the model prediction, PBTI lifetimes based on measured data are over 100X longer than lifetimes predicted by the model. Both silicon and model show weak sensitivity to pass transistor gate voltage. The speed degradation measured on silicon decreases with increasing gate voltage, which suggests that degradation in threshold voltage may be muted at higher gate voltages.

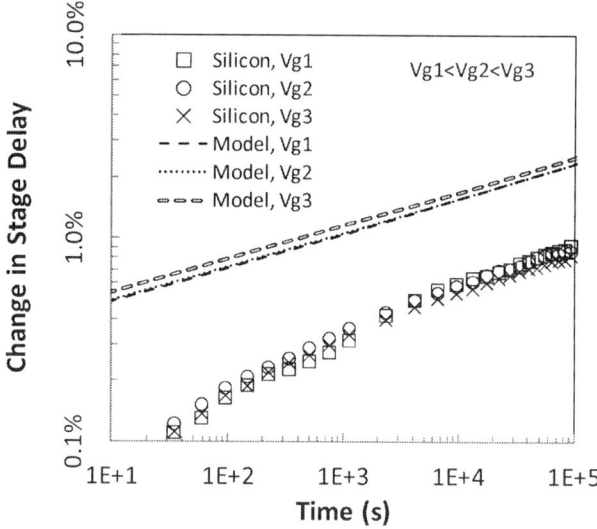

Fig. 5. Ring Oscillator Stress Data versus Model. Plot of delay degradation versus stress time for various stress voltages applied to NMOS pass transistor gate. Silicon lifetime is over 100 times longer than model.

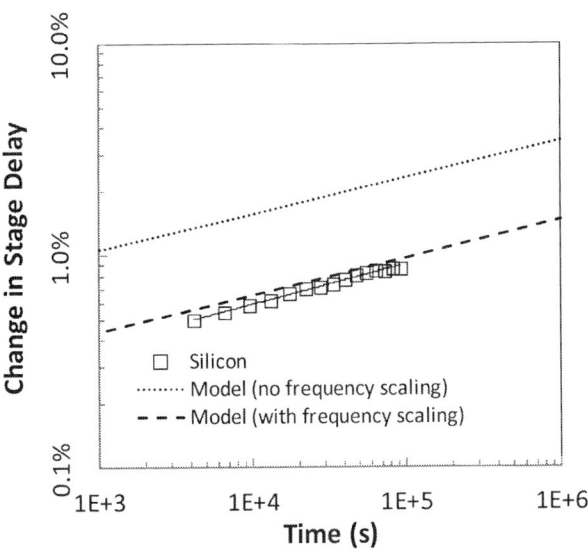

Fig. 6. **Model Adjusted for Frequency Effects.** Plot of delay degradation versus stress time for Vg2 stress voltage applied to NMOS pass transistor gate. Aging model matches data after accounting for recovery at 170 MHz.

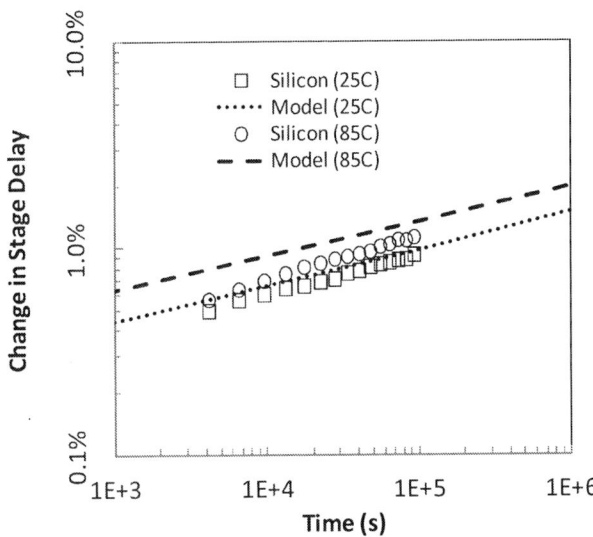

Fig. 7. **Ring Oscillator Stress at High Temperature.** Model slightly overestimates degradation at high temperature.

IV. DISCUSSION

The underlying physics of the BTI effect remains a hotly debated topic. Most authors agree that BTI is caused by charge trapping, although several different trapping models have been proposed [13,14,15]. Moreover, recent publications on bias temperature instability have reported significant frequency dependence of the BTI aging mechanism [16]. For example, Tsai reported NBTI lifetime improved over 1000X for circuits running at frequencies above 1 MHz [17].

In Tsai's work, frequency dependent data was extrapolated to the actual circuit frequency to predict the lifetime improvement at the circuit operating frequency. Using a similar approach, we predict that PBTI lifetime would improve about two orders of magnitude at our operating frequency of 170 MHz. Fig. 6 shows the measured and simulated data after adjusting the aging model to account for frequency dependent recovery at 170 MHz. Table I shows that the simulated lifetime matches the measured lifetime within 2% after properly modeling frequency dependent aging effects. Moreover, the data matches the model very closely across all three stress voltages (Table II). The analysis is also verified at high temperature, where the model slightly overestimates the degradation increase from 25°C to 85°C (Fig. 7, Table III). These results demonstrate the importance of modeling circuit level effects in aging simulations.

TABLE I.   RING OSCILLATOR LIFETIME AT 1% DELAY DEGRADATION.

Frequency Dependent Effects	Lifetime (sec)		Lifetime Difference
	Model	*Silicon*	
Not modeled	6.3E+2	1.4E+5	218X
Modeled	1.4E+5	1.4E+5	1.02X

Lifetime extrapolated from ~1.5 decades of data for Vstress=Vg2.

TABLE II.   RING OSCILLATOR DELAY DEGRADATION AFTER 25 HOUR STRESS, MODEL ADJUSTED TO FREQUENCY OF 170 MHz, T=25°C.

Delay Degradation	Pass Gate Voltage		
	Vg1	*Vg2*	*Vg3*
Silicon	0.93%	0.87%	0.83%
Model	0.95%	0.94%	1.02%
Error (X)	1.0	1.1	1.2

Model: MTTF, with NBTI and PBTI recovery at 170 MHz.

TABLE III.   RING OSCILLATOR DEGRADATION ACROSS TEMPERATURES, AFTER 25 HOURS, MODEL ADJUSTED TO 170 MHz.

Delay Degradation	Temperature		
	25°C	*85°C*	*Ratio (85°C/25°C)*
Silicon	0.9%	1.1%	1.21
Model	0.9%	1.4%	1.43
Error (X)	1.0	1.2	1.18

Model: MTTF, with NBTI and PBTI recovery at 170 MHz.

V. CONCLUSION

In this work, measured data was compared to aging models to demonstrate the necessity of modeling circuit level aging effects. A simple test structure was proposed as a vehicle for evaluating reliability of FPGA routing circuits. Aging models which were validated at the transistor level were over 100X off in predicting lifetime at the circuit level unless frequency dependent recovery effects were taken into account. After properly accounting for frequency effects, the simulated lifetime matched silicon lifetime within 2%.

ACKNOWLEDGMENT

The authors would like to express thanks to Sue Chen and Queennie Lim for layout of the test structure used in this work.

REFERENCES

[1] S.K. Lee, M. Jo, C.-W. Sohn, C.Y. Kang, J.C. Lee, Y.-H. Jeong, and B.H. Lee, "New insight into PBTI evaluation method for nMOSFETs with stacked high-k/IL gate dielectric," *Electron Device Letters*, vol. 33, no. 11, pp. 1517-1519, November 2012.

[2] U. Monga, S. Khandelwal, J. Aghassi, J. Sedlmeir, and T.A. Fjeldly, "Assessment of NBTI in presence of self-heating in high-k SOI FinFETs," *Electron Device Letters*, vol. 33, no. 11, pp. 1532-1534, November 2012.

[3] A.S. Oates, "Reliability challenges for the continued scaling of IC technologies,' in *Proc. Custom Integrated Circuits Conf.*, 2012.

[4] M. Cho, M. Aoulaiche, R. Degraeve, B. Kaczer, J. Franco, T. Kauerauf, P. Roussel, L.A. Ragnarsson, J. Tseng, T.Y. Hoffmann, and G. Groeseneken, "Positive and negative bias temperature instability on sub-nanometer EOT high-k MOSFETs," in *Proc. Int. Reliability Physics Symp.*, 2010, pp. 1095-1098.

[5] D. Lewis and J. Chromczak, "Process technology implications for FPGAs," in *IEDM Tech. Dig.*, 2012, pp. 565-568.

[6] W. Wolf, "FPGA Fabrics," in *FPGA-Based System Design*, Upper Saddle River, NJ: Prentice Hall, 2004, pp. 117-119.

[7] T. Nigam, "Pulse-stress dependence of NBTI degradation and its impact on circuits," *Trans. Device Materials Reliability*, vol. 8, no. 1, pp. 72-78, March 2008.

[8] J. Keane, W. Zhang, and C.H. Kim, "An array-based odometer system for statistically significant circuit aging characterization," *J. Solid State Circuits*, vol. 46, no. 10, pp. 2374-2385, October 2011.

[9] M. Chen, V. Reddy, J. Carulli, S. Krishnan, V. Rentala, V. Srinivasan, and Y. Cao, "A TDC-based test platform for dynamic circuit aging characterization," in *Proc. Int. Reliability Physics Symp.*, 2011, pp. 36-40.

[10] J. Keane, X. Wang, D. Persaud, and C.H. Kim, "An all-in-one silicon odometer for separately monitoring HCI, BTI, and TDDB," *J. Solid State Circuits*, vol. 45, no. 4, pp.817-829, April 2010.

[11] F. Ahmed and L. Milor, "Ring oscillator based embedded structure for decoupling PMOS/NMOS degradation with switching activity replication," in *Proc. Int. Conf. Microelectronic Test Structures*, 2010, pp. 118-121.

[12] L. Li, J. Watt, J. Wang, B. Tudor, J. Peng, and W. Liu, "28nm MOSFET aging modeling and simulation using HSPICE and HSIM MOSRA," presented at Synopsys User Group San Jose, 2011, unpublished.

[13] T. Grasser, B. Kaczer, W. Goes, H. Reisinger, T. Aichinger, P. Hehenberger, P.-J. Wagner, F. Schanovsky, J. Franco, M. Toledano Luque, and M. Nelhiebel, "The paradigm shift in understanding the bias temperature instability: from reaction-diffusion to switching oxide traps," *Trans. Electron Devices*, vol. 58, no. 11, pp. 3652-3666, November 2011.

[14] V. Huard, C. Parthasarathy, C. Guerin, T. Valentin, M. Mammasse, N. Planes, and L. Camus, "NBTI degradation: from transistor to SRAM arrays," in *Proc. Int. Reliability Physics Symp.*, 2008.

[15] M.A. Alam, H. Kufluoglu, D. Varghese, and S. Mahapatra, "A comprehensive model for PMOS NBTI degradation: recent progress," *Microelectronics Reliability*, vol. 47, pp. 853-862, 2007.

[16] T. Grasser, B. Kaczer, H. Reisinger, P.-J. Wagner, and M. Toledano-Luque, "On the frequency dependence of the bias temperature instability," in *Proc. Int. Reliability Physics Symp.*, 2012.

[17] Y.S. Tsai, N.K. Jha, Y.-H. Lee, R. Ranjan, W. Wang, J.R. Shih, M.J. Chen, J.H. Lee, and K. Wu, "Prediction of NBTI degradation for circuit under AC operation, in *Proc. Int. Reliability Physics Symp.*, 2010, pp. 665-669.

9781467348454